Handbook of Statistics
Volume 35

Cognitive Computing: Theory and Applications

Handbook of Statistics

Series Editor

C.R. Rao
C.R. Rao AIMSCS, University of Hyderabad Campus,
Hyderabad, India

Handbook of Statistics

Volume 35

Cognitive Computing: Theory and Applications

Edited by

Venkat N. Gudivada
East Carolina University, Greenville, NC, United States

Vijay V. Raghavan
University of Louisiana at Lafayette, Lafayette, LA, United States

Venu Govindaraju
University at Buffalo Amherst, NY, United States

C.R. Rao
C.R. Rao AIMSCS, University of Hyderabad Campus, Hyderabad, India

ELSEVIER

Amsterdam • Boston • Heidelberg • London • New York • Oxford
Paris • San Diego • San Francisco • Singapore • Sydney • Tokyo
North-Holland is an imprint of Elsevier

N·H

North-Holland is an imprint of Elsevier
Radarweg 29, PO Box 211, 1000 AE Amsterdam, Netherlands
The Boulevard, Langford Lane, Kidlington, Oxford OX5 1GB, United Kingdom

ISBN: 978-0-444-63744-4
ISSN: 0169-7161

For information on all North-Holland publications visit our website at https://www.elsevier.com/

Publisher: Zoe Kruze
Acquisition Editor: Kirsten Shankland
Editorial Project Manager: Hannah Colford
Production Project Manager: Radhakrishnan Lakshmanan
Cover Designer: Victoria Pearson

Typeset by SPi Global, India

Contents

Contributors

Numbers in Parentheses indicate the pages on which the author's contributions begin.

S. Chakraborty (229), University of Missouri, Columbia, MO, United States

E. Fathi (169), East Carolina University, Greenville, NC, United States

J.-B. Fiot (255), IBM Research, Dublin, Ireland

R.N. Gottumukkala (81), University of Louisiana at Lafayette, Lafayette, LA, United States

V.N. Gudivada (3, 169, 283, 301), East Carolina University, Greenville, NC, United States

E. Harley (117), The MITRE Corporation, McLean, VA, United States

F.D. Hudson (341), University Corporation for Advanced Internet Development, White Plains, NY, United States

M.T. Irfan (169, 283), Bowdoin College, Brunswick, ME, United States

S. Jothilakshmi (301), Annamalai University, Chidambaram, Tamil Nadu, India

M. Limiero (117), The MITRE Corporation, McLean, VA, United States

V.P.A. Lonij (255), IBM Research, Dublin, Ireland

A.S. Maida (39), The Center for Advanced Computer Studies, The University of Louisiana at Lafayette, Lafayette, LA, United States

E.W. Nichols (341), University Corporation for Advanced Internet Development, White Plains, NY, United States

S. Noel (117), The MITRE Corporation, McLean, VA, United States

V.V. Raghavan (81), University of Louisiana at Lafayette, Lafayette, LA, United States

D.L. Rao (169), East Carolina University, Greenville, NC, United States

M. Share (117), The MITRE Corporation, McLean, VA, United States

S. Suthaharan (207), The University of North Carolina at Greensboro, Greensboro, NC, United States

K.H. Tam (117), The MITRE Corporation, McLean, VA, United States

S.R. Venna (81), University of Louisiana at Lafayette, Lafayette, LA, United States

Preface

Cognitive computing is a nascent interdisciplinary domain. It is a confluence of cognitive science, neuroscience, data science, and cloud computing. Cognitive science is the study of mind and offers theories, and mathematical and computational models of human cognition. Cognitive science itself is an interdisciplinary domain and draws upon philosophy, linguistics, psychology, and anthropology, among others.

Neuroscience is the study of the nervous system including its development, structure, and function. More specifically, neuroscientists study the structure of the brain, and how behavior and cognitive functions are regulated by the brain. Brain imaging techniques such as positron emission tomography, functional magnetic resonance imaging, diffusion tensor imaging, and arterial spin labeling enable probing brain functions both qualitatively and quantitatively.

Data science is also an interdisciplinary domain. It deals with processes and systems to extract information and knowledge from structured and unstructured data using machine learning algorithms and statistical methods. The end goal is to discover patterns, generate actionable insights, and answer predictive questions.

Cloud computing provides turnkey solutions such as platform-as-a-service, infrastructure-as-a-service, and software-as-a-service. It uses high-performance CPUs, GPUs, neuromorphic processors, virtually unlimited memory and storage, and high-speed networks to provide computing resources on demand. A fixed pool of these resources is dynamically provisioned among various applications and continually adjusted so that the applications can guarantee performance amid fluctuating workloads. Cloud computing achieves economies of scale and helps cognitive computing applications to perform at scale without upfront computing investments. Applications are billed for only the resources they actually use.

Broadly, there are two lines of research in the cognitive computing discipline. The first one is centered on cognitive science as the foundation and encompasses neuroscience, philosophy, psychology, anthropology, and linguistics research. The second one is more recent and is based on computer science as the foundation. It encompasses data science, statistics, and subdisciplines of computer science such as high-performance computing, cloud computing, natural language processing, computer vision, machine learning, information retrieval, and data management. These two lines of research are not only complementary but mutually helping to accelerate discoveries and innovation.

It is this synergistic confluence that makes cognitive computing powerful and has the potential for groundbreaking discoveries and advances. Especially the advances in the computing discipline are poised to bring about transformational changes to the way research is conducted in the discipline. For example, IBM's TrueNorth cognitive computing system is a case in point. Its design is inspired by the function and efficiency of the human brain. The TrueNorth architecture provides a spiking neuron model as a building block. Its programming paradigm is based on an abstraction called *corelet*, which represents a network of neurosynaptic cores. The corelet encapsulates all details except the external inputs and outputs. An object-oriented language is available for programming corelets. A library of reusable corelets as well as an integrated development environment help accelerate the development of cognitive computing applications. Using this environment, IBM has already implemented several algorithms including hidden Markov models, convolution networks, and restricted Boltzmann machines. These algorithms have been incorporated into applications such as speaker recognition, sequence prediction, and collision avoidance.

As of this writing, Nvidia released Tesla P100 GPU, which specifically targets machine learning algorithms that employ deep learning. P100 features 150 billion transistors on a single chip. Also, Google released Natural Language API, a cloud service that provides application developers access to pretrained algorithms for sentiment analysis, entity recognition, and syntax analysis. Likewise, Speech API, Translate API, and Vision API are public cloud services for speech to text conversion, translation between natural languages, and image analysis, respectively. The Speech API enables converting audio to text for over 80 languages and the Translate API provides machine translation between these languages. Applications can use the Vision API to perform image analysis tasks including object detection and classification. These computing advances will propel further research in cognitive and neurosciences.

The goal of this handbook is to bring together a coherent body of knowledge and recent research in cognitive computing. It promotes a unified view of the domain and lays the foundation for cognitive computing as an academic discipline and a research enterprise. To the best of the editors' knowledge, this handbook is the first in formally defining cognitive computing and providing an academic exposition of the field. The handbook aims to serve as a catalyst for advancing research in cognitive computing.

AUDIENCE

The handbook aims to meet the needs of both students and industry practitioners. Especially it is suited for students in advanced undergraduate and beginning graduate courses on cognitive computing, neuroscience, and cognitive science. It is also a good source for graduate students who plan to pursue research in cognitive computing. The handbook is also a good reference for industry practitioners who desire to learn about cognitive computing.

ORGANIZATION

The handbook is comprised of 11 chapters, which are organized into 3 sections. *Section A: Fundamentals and Principles* consists of two chapters, provides an introduction to cognitive computing, and sets the backdrop for reading rest of the handbook. *Section B: Complex Analytics and Machine Learning* is comprised of five chapters. Complex analytics and machine learning areas are discussed in this section. Lastly, *Section C: Applications* discusses applications of cognitive computing and four chapters are devoted to these topics.

Section A: Fundamentals and Principles

Chapter 1: Cognitive Computing: Concepts, Architectures, Systems, and Applications

This chapter provides an interdisciplinary introduction to cognitive computing. The aim of the chapter is to provide a unified view of the discipline. It begins with an overview of cognitive science, data science, and cognitive computing. Principal technology enablers of cognitive computing, an overview of three major categories of cognitive architectures, cognitive computing systems, and their applications are discussed. Current trends and future research directions in cognitive computing are indicated. The chapter concludes by listing various cognitive computing resources.

Chapter 2: Cognitive Computing and Neural Networks: Reverse Engineering the Brain

IBM, Nvidia, and Qualcomm have developed microprocessors which mimic neurons and synapses of the human brain. These microprocessors are called neuromorphic chips, and IBM's TrueNorth and the HumanBrain Project's SpiNNaker are examples. This chapter presents principles and theory needed as a backdrop to understanding these advances from a cognitive science perspective. Neural networks found within the mammalian neocortex, and associated formal and computational models that appear to form the basis of human cognition are described.

Section B: Complex Analytics and Machine Learning

Chapter 3: Visual Analytic Decision-Making Environments for Large-Scale Time-Evolving Graphs

Data scientists are faced with the challenge of analyzing large-scale graphs that change dynamically. Existing tools and metaphors for data collection, processing, storage, and analysis are not suitable for handling large-scale evolutionary graphs. This chapter describes visual analytics as a cognitive computing approach to improving decision making with large-scale dynamic graphs. It provides a conceptual introduction to time-varying graphs, describes functional components of systems for visual analytics including performance considerations, and presents a visual graph analytics sandbox architecture and sample applications implemented within it.

Chapter 4: CyGraph: Graph-Based Analytics and Visualization for Cybersecurity

The adversarial nature and complex interdependencies of networked machines demand a cognitive systems approach to cybersecurity. This chapter describes CyGraph, a graph-based cognitive system for protecting mission-critical computing assets and applications. CyGraph brings together isolated data and events into a comprehensive property-graph model, providing an overall picture for decision support and situational awareness. CyGraph features CyQL (CyGraph Query Language), a domain-specific query language for expressing graph patterns of interest, with interactive visualization of query results. CyGraph integrates with third-party tools for visualizing graph state changes. CyGraph can also synthesize graph models with particular statistical properties.

Chapter 5: Cognitive Analytics: Going Beyond Big Data Analytics and Machine Learning

Traditional data analytics evolved from the database domain and exclusively focused on structured data stored in relational databases. It was propelled to the next stage in its evolution with the advent of data warehouses and data mining. Cognitive analytics is the third stage in this evolutionary path and goes beyond structured data. It integrates semistructured and unstructured data into the analytic process. This chapter provides an introduction to cognitive analytics. It describes types of learning and classes of machine learning algorithms in the context of cognitive analytics. It proposes a reference architecture for cognitive analytics and indicates ways to implementing it. It also describes a few cognitive analytics applications.

Chapter 6: A Cognitive Random Forest: An Intra- and Intercognitive Computing for Big Data Classification Under Cune Condition

This chapter address the classification problem in the big data context. The data is often noisy, inconsistent, and incomplete. To solve the classification problem, a cognitive model (called STE-M) is proposed in this chapter. Also, a cognitive computing architecture, called Cognitive Random Forest, is proposed to implement STE-M. The architecture amalgamates the STE-M model and a set of random forest classifiers to enhance continuous learning. The architecture is implemented and validated.

Chapter 7: Bayesian Additive Regression Tree for Seemingly Unrelated Regression with Automatic Tree Selection

This chapter introduces a flexible Bayesian additive regression tree (seemingly unrelated regression) model, called BART-SUR, which is suitable for situations where the response variable is a vector and the components of the vector are highly correlated. BART-SUR can jointly model the correlation structure among the related response variables and provide a highly flexible and nonlinear regression structure for each of the individual regression functions. The number of trees in BART-SUR is selected adaptively by treating it as a model parameter and assigning a prior distribution on it. The adaptive tree selection makes BART-SUR extremely fast. The author demonstrates the superiority of BART-SUR over several out of the shelve popular methods like the random forest, neural network, wavelet regression, and support vector machine through two simulation studies and three real data applications.

Section C: Applications

Chapter 8: Cognitive Systems for the Food–Water–Energy Nexus

Meeting the food, water, and energy needs of a growing world population is a grand challenge. These resources are often not produced in places where they are consumed, which entails transportation and storage costs. One can avoid storing a resource, if good forecast models for supply and demand exist. Developing such models requires handling large-scale datasets efficiently, building forecasting models using machine learning methods, and leveraging optimization techniques to help incorporate forecasting results into a decision-making process. Toward these goals, this chapter discusses methods to make the most of the sensor data, including data from satellites and ground-based cameras. Forecasting methods ranging from minutes-ahead to days-ahead are described. Finally, ways to use the outputs of these analytics tools to help decision-making processes are discussed in the context of energy.

Chapter 9: Cognitive Computing Applications in Education and Learning

Education and learning applications stand out among many uses of cognitive computing due to their practical appeal and research challenges. This chapter discusses the role of cognitive computing in teaching and learning environments. More specifically, the chapter examines the important roles played by the Educational Data Mining (EDM) and Learning Analytics (LA) researchers in improving student learning. It describes an architecture for personalized eLearning and summarizes relevant research.

Chapter 10: Large-Scale Data Enabled Evolution of Spoken Language Research and Applications

Human languages are used in two forms: written and spoken. Text and speech are the mediums for written and spoken languages, respectively. Human languages are the most natural means of communication between cognitive computing systems and their users. The emergence of big data and data science is accelerating research and applications in the analysis and understanding of human/natural languages. This chapter provides an introductory tutorial on the core tasks in speech processing, reviews recent large-scale data-driven approaches to solving problems in spoken languages, describes current trends in speech research, and indicates future research directions.

Chapter 11: The Internet of Things and Cognitive Computing

Internet of Things (IoT) technologies are now more widely deployed. The confluence of IoT and cognitive computing provides unprecedented opportunities to develop deeper insights from the data generated by IoT devices. These actionable insights have the potential for transformational changes that affect people, cities, and industry. This chapter explores the state of the art and future opportunities to bring IoT and cognitive computing together to solve a range of problems including smart cities and connected health care.

Venkat N. Gudivada
Vijay V. Raghavan
Venu Govindaraju
C.R. Rao

Section A

Fundamentals and Principles

Chapter 1

Cognitive Computing: Concepts, Architectures, Systems, and Applications

V.N. Gudivada[1]
East Carolina University, Greenville, NC, United States
[1]*Corresponding author: e-mail: gudivadav15@ecu.edu*

ABSTRACT

Cognitive computing is an emerging field ushered in by the synergistic confluence of cognitive science, data science, and an array of computing technologies. Cognitive science theories provide frameworks to describe various models of human cognition including how information is represented and processed by the brain. Data science provides processes and systems to extract knowledge from both structured and unstructured data. Cognitive computing employs the computing discipline's theories, methods, and tools to model human cognition. The recent advances in data science and computing disciplines—neuromorphic processors, big data, predictive modeling, machine learning, natural language understanding, and cloud computing—are accelerating advances in cognitive science and cognitive computing.

The overarching goal of this chapter is to provide an interdisciplinary introduction to cognitive computing. The focus is on breadth to provide a unified view of the discipline. The chapter begins with an overview of cognitive science, data science, and cognitive computing. The principal technology enablers of cognitive computing are presented next. An overview of three major categories of cognitive architectures is presented, which is followed by a description of cognitive computing systems and their applications. Trends and future research directions in cognitive computing are discussed. The chapter concludes by listing various cognitive computing resources.

Keywords: Cognitive computing, Cognitive architectures, Cognitive models, Cognitive systems, Cognitive applications, Cognitive computing systems, Data science

1 INTRODUCTION

An *autonomous system* is a self-contained and self-regulated entity. The system continually reconstitutes itself in real time in response to changes in its

Handbook of Statistics, Vol. 35. http://dx.doi.org/10.1016/bs.host.2016.07.004

environment (Vernon, 2014). The self-reorganization aspect embodies learning, development, and evolution. *Cognition* is the process by which an autonomous system acquires its knowledge and improves its behavior through senses, thoughts, and experiences (Anderson, 1983). *Cognitive processes* are critical to autonomous systems for their realization and existence (Franklin et al., 2014; Newell, 1994).

Human cognition refers to the cognitive processes which enable humans to perform various tasks, both mundane and highly specialized (Chipman, 2015). A collection of processes that enable computers to accomplish tasks that target performance at human cognition levels is referred to as *machine cognition*. Human cognition employs biological and natural means—brain and mind—for its realization. On the other hand, machine cognition views cognition as a type of computation and uses *cognitive computing* techniques for its realization.

Cognitive science is an interdisciplinary approach to the study of human and animal cognition (Frankish and Ramsey, 2012; Friedenberg and Silverman, 2015). Abrahamsen and Bechtel (2012) provide an exposition and core themes of cognitive science. *Cognitive computing* is an emerging field ushered in by the synergistic confluence of cognitive science, data science, and an array of computing technologies (Hurwitz et al., 2015). The recent advances in the computing discipline—high-performance computers, neuromorphic chips, neurocams, big data, machine learning, predictive modeling, natural language processing (NLP), and cloud computing—are accelerating advances in cognitive science and cognitive computing disciplines.

Given the interdisciplinary origin of cognitive science and data science, there are multiple perspectives on cognitive computing. These perspectives are shaped by diverse domain-specific applications and fast evolution of enabling technologies. There is no consensus on what exactly comprises the field of cognitive computing. Our exposition of cognitive computing in this chapter is driven by big data, information retrieval, machine learning, and natural language understanding and applications.

1.1 Chapter Organization

The overarching goal for this chapter is to provide a unified introduction to cognitive computing by drawing on multiple perspectives. Section 2 provides an overview of cognitive science as an interdisciplinary domain. The primary characteristics of cognitive computing systems and a preview of cognitive applications are provided in Section 3. Principal technology enablers of cognitive computing are discussed in Section 5. Concepts of knowledge representation are presented in Section 4.

Cognitive architectures model human performance on multiple cognitive tasks. They are computational frameworks which specify structure and functions of cognitive systems as well as how structure and functions interact. Section 6 discusses cognitive architectures and approaches to cognitive tasks.

Cognitive computing systems and their applications are presented in Section 7. Trends and future research directions in cognitive computing are discussed in Section 8. Finally, Section 9 ends the chapter by listing various cognitive computing resources.

2 INTERDISCIPLINARY NATURE OF COGNITIVE SCIENCE

Cognitive science theories provide frameworks to describe various models of human cognition including how information is represented and processed by the brain. The human brain is perhaps the most complex system in terms of its structure and function. The *mental processes* of the brain span a broad spectrum ranging from visual and auditory perception, attention, memory, imagery, problem solving, and natural language understanding. We use the terms mental processes and *cognitive tasks* synonymously.

Cognitive science encompasses academic disciplines including philosophy, psychology, neuroscience, linguistics, artificial intelligence (AI), and robotics. Philosophers pose broad questions about the nature of the mind and the relationship between the mind and thought processes (Thagard, 2009). They also offer hypotheses to explain the mind and its mental processes. Philosophers' primary method of inquiry is through deductive and inductive reasoning.

Cognitive psychologists design experiments and execute them under controlled conditions to validate hypotheses and develop cognitive theories (Neisser, 2014). Cognitive psychology studies aim to discover how thinking works. Such studies encompass, for example, how experts solve problems compared with novices, how short is short-term memory, and why people who are the most incompetent are the least aware of their own incompetence. The discipline of evolutionary psychology explains human mental processes using the selection theory. More specifically, it uses evolutionary principles to explain psychological adaptations such as changes in our thinking to improve our survival.

Neuroscientists employ engineering instruments and scientific methods to measure brain activity in response to external stimuli (McClelland and Ralph, 2015). For example, functional magnetic resonance imaging (fMRI), positron emission tomography (PET), and computerized axial tomography (CAT) techniques are used to identify specific brain regions associated with various cognitive tasks. A neurocam is a head-mounted camera which monitors brain waves (Neurowear, n.d). When the person wearing this device looks at something that causes the brain activity to spike, the activity is automatically recorded. This camera is not yet available in the market.

Linguists study various aspects of natural languages including language acquisition and understanding (Evans, 2012; Isac and Reiss, 2013). Cognitive linguists investigate the interaction between language and cognition. How can we explain the fact that a 5-year old in one culture can do with ease a simple task such as pointing in the direction of the north that eminent scientists in other cultures struggle with? The notion that different languages may impart different cognitive skills dates back centuries and there is empirical evidence for this causal relation (Boroditsky, 2011). This notion is formally stated as Sapir–Whorf hypothesis, which states that the structure of a language affects its speakers' cognition or world view (Kay and Kempton, 1984). However, it appears that language is only one factor that influences cognition and behavior.

AI (Russell and Norvig, 2009) and robotics (Samani, 2015) researchers investigate how robots can be endowed with human-like intelligent behavior to perform various cognitive tasks. In recent years, the practice of developing intelligent systems by implicitly or explicitly embedding knowledge through advanced programming techniques is not a dominant AI practice. Though some of the AI systems incorporate learning into their design, the primary effort has been on codifying domain knowledge, specifying integrity constraints, and designing inference rules. Also, such systems are strongly coupled with the domain, and the effort required for *domain adaption* is as much as developing the system from scratch for the new domain.

Another AI approach to developing intelligent systems is *data driven* and eases the domain knowledge encoding and rule specification effort (Abu-Mostafa et al., 2012). Though this approach existed for quite some time, the recent emergence of big data created renewed interest (Hastie et al., 2003). It emphasizes primarily semi-supervised and unsupervised *machine learning* algorithms. This approach entails relatively less effort for domain adaptation.

The terms brain and mind are often used interchangeably. Cognitive scientists from philosophy, psychology, and linguistics backgrounds typically use the term *mind*. These domains investigate cognition at a more abstract and logical level and are less concerned about the underlying apparatus that enables cognition. On the other hand, cognitive scientists from the neuroscience and computing disciplines use the term *brain*. The apparatus that enables cognition is central to their investigations.

3 COGNITIVE COMPUTING SYSTEMS

Cognitive computing employs the computing discipline's theories, methods, and tools to model cognitive tasks. It views the mind as a highly parallel information processor, uses various models for representing information, and employs algorithms for transforming and reasoning with the information. The means to represent and store information in a computer bears little or no resemblance to its counterparts in the human brain.

Technologies that enable cognitive computing systems include AI, machine learning, computer vision, robotics, written and spoken language recognition and understanding, information retrieval, big data, Internet of Things (IoT), and cloud computing. Some of these are enabling technologies and others are technologies in their own right.

Cognitive computing systems are fundamentally different from traditional computing systems. Cognitive systems are adaptive, learn and evolve over time, and incorporate context into the computation. They sense their environment, think and act autonomously, and deal with uncertain, ambiguous, and incomplete information.

Cognitive computing systems do not use brute force approaches. For example, the IBM's Deep Blue system which defeated the world Chess champion Garry Kasparov in 1997 is not considered a cognitive computing system. Deep Blue used exhaustive search in planning its moves. In contrast, the IBM Watson of 2011 is a cognitive computing system. It uses deep natural language understanding, incorporates contextual information into its decision making, and reasons with incomplete and uncertain data. It performs spatial and temporal reasoning and can recognize statistical paraphrasing of natural language text.

Cognitive computing systems span a broad spectrum in terms of their capabilities. With rapid advances in cognitive science and data science, current computing applications embody varying degrees of cognitive capabilities. For example, an assortment of cognitive capabilities is essential for self-driving cars. Cognitive capabilities enable self-driving cars to learn from past experiences and use contextual information in making decisions in real time.

Cognitive assistants such as the *Google Now* predict and suggest next context-dependent actions. Emerging information extraction and search technologies provide evidence-based answers and explain their answers. Cognitive technologies for translating webpage content to different languages have achieved unprecedented levels of accuracy. Transformative advances in speech recognition and language understanding are used for real-time speech understanding and translation tasks. Cognitive IoT (Wu et al., 2014) and big data technologies are used in smart cities for improving public safety and efficiency of infrastructure operations. Finally, humanoid robots are learning difficult tasks such as archery (Kormushev et al., 2010).

4 REPRESENTATIONS FOR INFORMATION AND KNOWLEDGE

Cognitive computing views the brain as an information processor. Therefore, suitable representations are needed to represent and transform information (Davis et al., 1993). In fact, how to represent information/knowledge is one of the challenges in developing autonomous systems. According to Friedenberg and Silverman (2015), there are four categories of representations: concept, proposition, rule, and analogy. *Concepts* denote objects of interest in the domain such as people, places, and events. Words in a natural language are good examples of concepts. Propositions are statements about the domain. They are always associated with a truth value (*true* or *false*). For example, the sentence "Cognitive computing is an emerging area of computer science," is a proposition. Concepts are the building blocks of propositions. Propositions can be combined using logical connectives to generate compound propositions.

Rules specify relationships between propositions. Rules enable *inferring* new information from existing information. Rules help to lessen the need to exhaustively and explicitly store factual information about the domain. These type of rules are called *inference* or *reasoning rules*. There is another type of rules referred to as *integrity constraints*. Their purpose is to verify the consistency of the information and to identify incompatibilities. The third type of rules is called *procedural knowledge*, represent more complex and abstract rules which describe the sequences of steps involved in performing cognitive tasks.

An *analogy* is a comparison between two things, typically based on their structure. *Analogical representations* store information about analogies. Such representations are used to solve problems through *analogical reasoning*. If two problem situations are similar and a solution is known to the first problem, analogical reasoning proposes the solution to the first problem as a solution to the second problem. Analogical reasoning solutions are typically specified using a certainty factor. The knowledge represented using analogies is termed *heuristic knowledge*.

An *ontology* is another knowledge representation scheme (Stephan et al., 2007). They are explicit formal specifications of the terms and concepts in the domain and relations among them (Gruber, 1993; Guarino et al., 2009). They provide a consistent and formal vocabulary to describe the domain and facilitate reasoning. They promote domain knowledge reuse as well as enable analyzing the domain knowledge. Ontologies help in making domain assumptions explicit. Ontologies are not suitable for representing certain types of knowledge such as diagrammatic and procedural knowledge (Brewster and O'Hara, 2004). WordNet (Miller, 1995) is a widely used lexical ontology in cognitive computing. DBpedia is a knowledge base of information extracted from Wikipedia through crowd-sourcing methods (DBpedia, n.d). In addition to enabling posing of sophisticated queries against Wikipedia, DBpedia is an excellent knowledge base for developing certain types of cognitive computing applications.

Using declarative methods to represent facts and relationships about entities in the domain have their limitations. Not everything in the domain is amenable for facts and relationships representation. Relationships are too many for explicit representation and exceptions are common. The above approaches to knowledge representation are called *symbolic representations*.

There is another class of representations termed *distributed representations*, which is used in the neural network-based cognitive computing architectures. A neural network is a weighted directed graph comprised of nodes and edges in a predefined configuration. Typically, a neural network consists of a layer of *input nodes* and another layer of *output nodes*. Input layer nodes may be directly connected to the nodes in the output layer or there can be a number of hidden layers between them. A neural network represents knowledge as the weights associated with the edges. A network needs to be trained on inputs to learn edge weights. In a multilayer network, the layers are learned simultaneously in a nonlinear fashion. In essence, knowledge representation emerges as a result of training the network. For example, distributed word representations are used in Bowman et al. (2015) to support the rich, diverse logical reasoning captured by natural logic.

An advantage of distributed representations is that they are more resilient to noisy input and performance degradation is more graceful. However, it is difficult to explain the behavior of the system using the internal structure of the network. In applications such as personalized medicine, an explanation about how a decision has been made is critical. Deep learning, which is a type of machine learning, heavily relies on multiscale distributed representations. Input data is characterized using multiple features, and each feature is represented at multiple levels of scale. In passing, it should be noted that there is a strong coupling between the cognitive computing architectures and the knowledge representations used.

5 PRINCIPAL TECHNOLOGY ENABLERS OF COGNITIVE COMPUTING

Cognitive science has been in existence for long as an interdisciplinary discipline whose research focus has been understanding cognition and functioning of the human brain. In contrast, computing is a young discipline. However, during the last few years, there were transformational advances in the computing discipline. These advances, in turn, are providing unprecedented and unique opportunities for advancing research in cognitive science and data science. This section provides an overview of computing technologies which are central to realizing cognitive computing systems.

5.1 Big Data and Data Science

Recent advances in storage technologies, high performance computing, gigabit networks, and pervasive sensing are driving the production of unprecedented volumes of data (Gudivada et al., 2015a). Some of this is streaming data which is produced at high velocities. Furthermore, most of this data is *unstructured* and *heterogeneous* in the form of written and spoken documents, images, and videos (Berman, 2013). This data is referred to as *big data* and numerous systems have been developed for its storage and retrieval (Gudivada et al., 2016a,b). Big data has enabled several new and innovative applications (McCreary and Kelly, 2013).

Data Science refers to big data enabled approaches to research and applications development (Grus, 2015). Data Science provides innovative algorithms and workflows for analysis, visualization, and interpretation of big data to enable scientific breakthroughs (Hey et al., 2009). Dhar (2013) defines data science as *the systematic study of the extraction of generalizable knowledge from data*.

Big data provides new ways to solve problems using data-driven approaches (Gudivada et al., 2015b). In Halevy et al. (2009), it is argued that the accurate selection of a mathematical model ceases its importance when compensated by *big enough* data. This insight is particularly significant for solving various problems in AI, machine learning, and autonomous systems. These problems are typically ill-posed for mathematically precise algorithmic solutions. For example, in NLP, such problems include parsing, part-of-speech (POS) tagging, named entity recognition, information extraction, topic modeling, machine translation, and language modeling.

To illustrate how big data and data science are changing the course of research in NLP, consider the POS tagging problem. This involves assigning a correct POS tag for each word in a document. For example, given the input sentence—Big Data is changing the course of NLP research and enabling new applications—a POS tagger may produce the following: Big/NNP Data/NNP is/VBZ changing/VBG the/DT course/NN of/IN natural/JJ language/NN processing/NN (/(NLP/IN)/) research/NN and/CC enabling/VBG new/JJ applications/NNS ./. The notation Big/NNP means that the POS tag of Big is NNP (proper noun, singular). The meaning of other POS tags is: CC = conjunction, coordinating; DT = determiner/pronoun, singular; IN = preposition; JJ = adjective; NN = noun, singular, common; NNS = noun, plural, common; VBG = verb, present participle or gerund; VBZ = verb, present tense, third person singular. There is no standard for POS tag sets.

The POS tagging problem is difficult because the same word can be assigned different tags depending on the context. Therefore, assigning a tag to a word must consider the definition of the word as well as the *context* in which the word is used. Furthermore, many nouns can also be used as verbs. Also, POS tagging rules vary from one language to another.

There are two broad categories of algorithms for POS tagging: *rule based* and *stochastic*. Algorithms in the first category employ rules. The Brill tagger uses a form of supervised learning which aims to minimize error (Brill, 1995). Initial POS tags assigned to words are iteratively changed using a set of predefined rules that take context into consideration. This approach requires hundreds of rules which are developed by linguists or synthesized using machine learning algorithms and training data. This is an error-prone and labor-intensive process. Furthermore, the rules are bound to the language and domain adaptation is difficult.

Stochastic POS algorithms, on the other hand, are data driven. They are based on supervised learning models such as Hidden Markov Model, Log-linear Model, and Conditional Random Field (CRF). More recent stochastic approaches strive to transition from supervised to semi-supervised and unsupervised algorithms. For example, the approach to POS tagging in Ling et al. (2015) obviate the need for manually engineering lexical features in words. The work presented in Andor et al. (2016) is another data-driven, globally normalized transition-based neural network model that achieves state-of-the-art performance on POS tagging.

TensorFlow is an open source software library for developing machine learning-centric applications (TensorFlow, n.d). SyntaxNet (n.d) is an open-source neural network framework for TensorFlow. SyntaxNet provides a library of neural models for developing Natural Language Understanding (NLU) systems. Parsey McParseface is a component of SyntaxNet, which is a pretrained parser for analyzing English text including POS tagging.

5.2 Performance, Scalability, and Elasticity

Cognitive computing applications typically deal with large, unstructured data. Moreover, the data may be fraught with quality problems. The data may be incomplete, inconsistent, conflicting, uncertain, and ambiguous. Furthermore, the data may also contain duplicates and determining them is not trivial. Processing unstructured data to extract information is computationally expensive. Therefore, substantial computing resources are needed to clean the data and to extract meaning.

Performance, *scalability*, and *elasticity* are three attributes that are used to characterize the computing needs of cognitive applications. *Performance* refers to stringent requirements placed on the time to process the data. For example, consider IBM Watson and its capabilities as of 2011. Watson is a question-answering cognitive application whose capabilities include natural language understanding. To enable IBM Watson to play the Jeopardy! game, Watson must answer all questions in less than 3 s. The *scalability* parameter refers to a computing systems ability to perform under increased workload without requiring any changes to the software. For example, how does a system's performance gets affected when the size of the input is doubled? Finally, the term *elasticity* refers to how a computing system dynamically and automatically provisions and deprovisions resources such as processor cycles, primary memory, and secondary storage to meet fluctuations in the system workload. The dynamic aspect is key to provisioning just enough resources to operate a cognitive system at a specified performance level despite unpredictable fluctuations in the system workload. Elasticity is critical to minimize the operating costs of a cognitive system.

5.3 Distributed Computing Architectures

One way cognitive computing systems meet performance and scalability requirements is through distributed computing architectures. Such architectures consist of several processing nodes, where a node is a self-contained computer comprised of compute cores, primary memory, and secondary

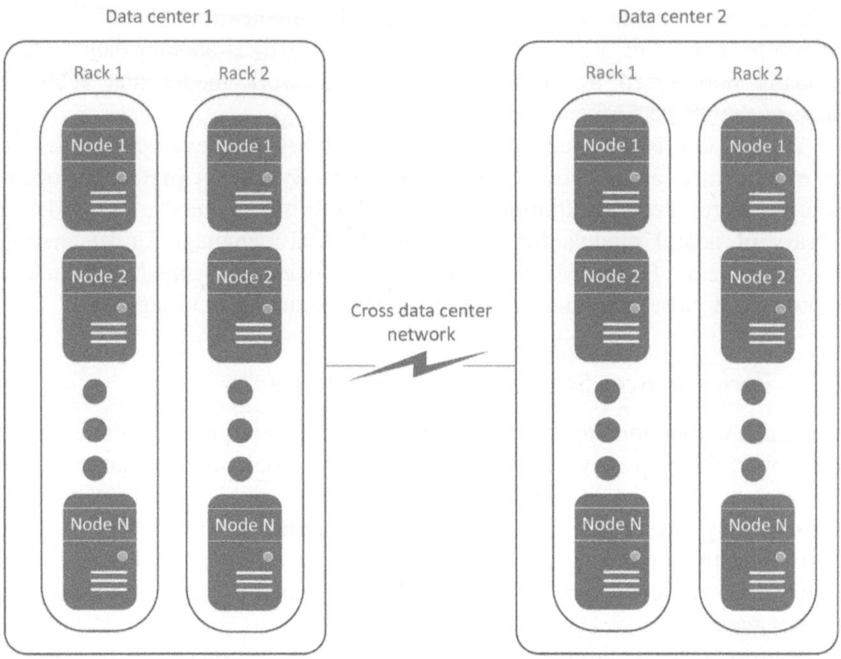

FIG. 1 A compute cluster whose nodes span across geographically separated data centers.

storage as shown in Fig. 1. The nodes communicate and coordinate their actions to achieve a common goal through mechanisms such as *shared memory* and *message passing*. The nodes are interconnected through a high-speed computer network. A logical collection of nodes is called a *cluster*. Several nodes are physically mounted on a *rack*. Some cognitive computing systems run on clusters whose nodes reside in geographically separated data centers.

Client-server architecture is a widely used computing model for cognitive computing applications. A *server* provides a service which is made available to the *clients* through an Application Programming Interface (API) or a protocol such as REST (Fielding, 2000). Typically, the server and the clients reside on physically different computers and communicate over a network. However, the server and the clients may also reside on the same physical computer. The workload is divided between the server and the clients.

In production environments, a cognitive computing application server typically runs on a cluster. The responsibility for processing client requests as well as distributing and coordinating workload among various nodes can be centralized or distributed. Fig. 2 shows both these models. Shown on the left is the *master-worker* architecture. A specific node is designated as the master and is responsible for intercepting client requests and delegating them to worker nodes. In this sense, the master node acts as a load balancer. It is also responsible for coordinating the activities of the entire cluster.

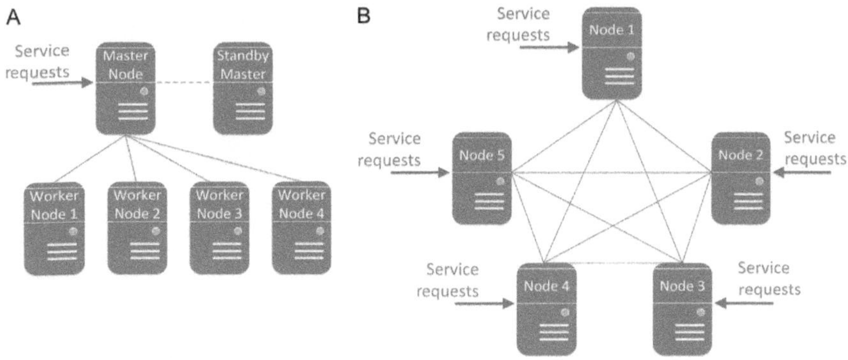

FIG. 2 A shared-nothing architecture. (A) Master–worker shared-nothing architecture. (B) Master–master shared-nothing architecture.

This architecture simplifies cluster management, but the master node becomes the single point of failure. If the master node fails, a *standby master* takes over the responsibility.

Shown in Fig. 2B is an alternative to the master-worker architecture. This is called *master-master* or *peer-to-peer* architecture. All nodes in the cluster are treated as equals. At any given point of time, a specific node is accorded the role of a master. If the master node fails, one of the remaining nodes is elected as the new master. Another architecture called *multi-master* employs a hierarchy of masters and master–worker style is used at the lowest level.

Master–worker and master–master configurations are called *shared-nothing architectures* since the nodes are self-contained and do not share resources. Both architectures distribute data and processing across the nodes in the cluster to achieve *performance at scale*. Data is also replicated to a subset of the nodes to ensure *high availability*. Some systems allow adding new nodes or removing existing ones (intentionally or due to a node failure) without service interruption. Computing systems based on shared-nothing architecture accommodate increased workloads by adding new nodes. Testing of cognitive computing systems that use master–master architecture is easier than the ones that use master–worker architecture.

5.4 Massive Parallel Processing Through MapReduce

Given the massive volumes of unstructured data that cognitive systems process in near real time, a high degree of parallel processing is required. *MapReduce* is a distributed programming model designed for processing massive amounts of data using cluster computers (Ryza et al., Year; White, 2015). This model is inspired by the *map* and *reduce* functions commonly used in *functional programming* languages.

Fig. 3 shows processing steps involved in a MapReduce computation. They are best illustrated through an example. Consider the problem of

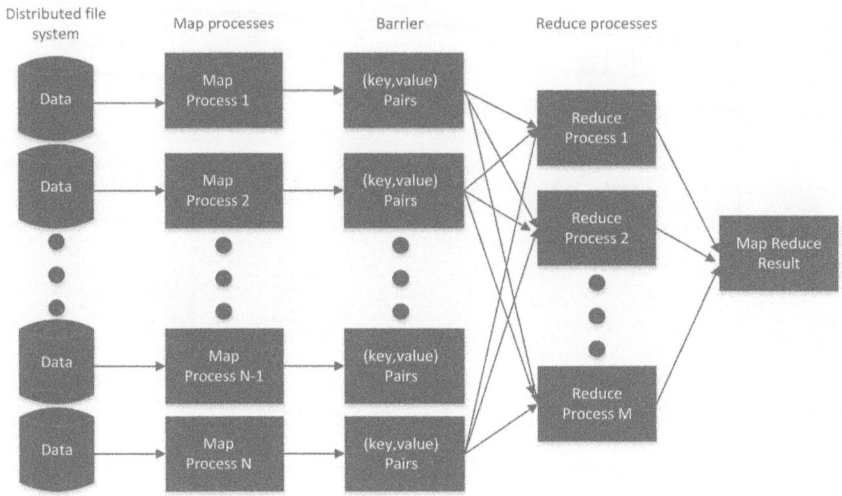

FIG. 3 MapReduce architecture.

computing an *inverted index* for a very large textual document collection. Each document in the collection is identified with a unique identifier. For each significant word in the collection, the inverted index lists all the documents in which the word occurs.

As a first step, the documents in the collection are partitioned into nonoverlapping sets and each set is assigned to a *map process* (see Fig. 3). Optimal partitioning of the input files into sets and assigning each set to a map process depends on the problem characteristics. Map processes execute in parallel on different nodes (step 1). They read documents assigned to them and extract ordered pairs of the form (word, doc_id). In other words, for each instance of a significant word that appears in any document, an ordered pair is generated.

In the *second step*, the Barrier column in Fig. 3 acts as a barricade by ensuring that all the mapper processes have completed their work before moving to the third step. The second step also collects the generated key-value pairs from each mapper process, sorts them on the key value, and partitions the sorted key-value pairs. In the last step, each partition is assigned to a different *reduce* process. A function which performs this assignment is called *shard*.

Each reduce process essentially receives *all* the key-value pairs corresponding to one or more words (e.g., (authentic, doc_id_14), (authentic, doc_id_3), (eclectic, doc_id_111), (eclectic, doc_id_21), (eclectic, doc_id_45)). The output of each reduce process is a subset of the inverted index (e.g., (authentic, (4, 13)), (eclectic, (21, 45, 111))). No synchronization is required for the reduce processes.

5.5 Cloud Computing

For individuals and even for many organizations, building and maintaining cluster-based distributed computing infrastructure is not desirable for many reasons. First, a significant upfront investment is needed to build the cluster. Second, highly specialized, technical personnel are required to build and operate the cluster. Third, it is difficult to achieve scalability without service interruption. Fourth, given the unpredictable fluctuations in workload, it is extremely difficult to achieve scalability in a cost-effective manner. Often, scalability requires overprovisioning of resources to compensate for unpredictable fluctuations in the workload. Cognitive computing applications need *elastic scalability*—provisioning resources optimally to guarantee performance requirements at all times regardless of fluctuations in the workload. This is where *cloud computing* comes into play.

The primary goal of cloud computing is to provide computing infrastructure as well as turnkey software applications as utilities (Armbrust et al., 2010). Terms such as *infrastructure as a service* and *software as a service* are used to describe these utility services. Cloud computing is a model for providing on-demand access to a shared pool of computing resources. It leverages the fact that a fixed pool of resources can be dynamically provisioned and deprovisioned across multiple applications to meet their fluctuating workloads. In other words, an economy of scale is achieved through dynamically sharing resources. Each application is provided a virtual machine (VM) whose local resources are configured and dynamically adjusted using the global shared pool.

Organizations pay for only the resources that their applications use, irrespective of the size of the resource pool. Several vendors including Microsoft, IBM, Nvidia, and Amazon provide cloud services for hosting applications. Features of these services vary along the facets such as service level agreements, authorization and authentication services, data encryption, data transfer bottlenecks, application monitoring, and appearance of infinite computing resources on demand. Cloud computing is especially important for cognitive computing applications as vendors such as IBM and Numenta make their cognitive APIs available via cloud hosting.

5.6 AI and Machine Learning

There is an analogy between the state of AI and machine learning today to that of the Information Retrieval (IR) (Gudivada et al., 1997) research between 1990 and 1995. Before 1990, IR research was mostly confined to academia. There were few commercial products such as those offered by LexisNexis, which provided search functionality to a collection of legal documents and journal publications. These documents were stored together on a disk. The advent of the World-wide Web and the need for searching documents distributed across the Web provided the necessary impetus for advancing IR research and distributed search. The Text REtrieval Conference

(TREC) is an annual conference and competition, whose goal is to promote IR research by providing test data sets. TREC-1 started in 1992 (Harman, 1993) and continues to attract researchers globally even today. The dramatic advances in the Web search are in part credited to the TREC competitions.

A similar situation exists today with cognitive computing. For example, consider the very first autonomous ground vehicle Grand Challenge sponsored by the Defense Advanced Research Projects Agency (DARPA) in 2004. The goal was to promote research and development of autonomous, self-driving ground vehicles capable of reaching a destination within specified time constraints using off-road terrain in the Mojave Desert region of the United States. Twenty teams participated in this event and none of the vehicles reached the destination. The farthest distance traveled by any vehicle was 7.32 miles. Contrast this with the DARPA Urban Challenge competition held in 2007. The course involved a 60-mile stretch of an urban area and the journey needed to be completed in less than 6 h. Six vehicles successfully completed the entire course. This is a remarkable improvement in just 3 years. DARPA continues to conduct these challenges, and the most recent one being the 2013 Fast Adaptable Next-Generation Ground Vehicle (FANG) Challenge.

This unprecedented interest in autonomous and intelligent systems resulted in the creation of several frameworks and libraries for developing such systems. Notable is the availability of various data-driven machine learning libraries. They include TensorFlow, a high-level neural network library; DeepLearning4J, a framework for deep learning; NuPIC, a library for a theory of the neocortex called Hierarchical Temporal Memory (HTM); Caffe, a machine learning library for computer vision applications; and scikit-learn, a low-level library for data mining and data analysis research.

Intended use cases and functional features of machine learning algorithms vary greatly. The facets that differentiate them include types of preprocessing required on input data to achieve data quality; amount of data required for training, testing, and model refinement; approaches to cross-validating models to prevent overfitting; types of learning algorithms—supervised vs unsupervised, and ensemble learning; quality and pervasiveness of learning; and ease of domain adaptation. Based on these facets, machine learning algorithms are categorized into the following classes: decision trees, associative rule learning, genetic algorithms, refinement learning, random forest, support vector machines, Bayesian networks, and deep learning.

5.7 Neuromorphic Computing

Cognitive computing applications typically require sophisticated processing of noisy and unstructured real-world data under stringent time constraints. Neuromorphic computing and neural network accelerators are solutions to meet these processing challenges. The goal of neuromorphic computing is to use very large scale integration (VLSI) systems that are driven by electronic analog circuits to simulate neurobiological architectures present in the nervous system (Williams, 2016). These VLSI chips are characterized by ultra-low power consumption and high performance. They are referred to as neuromorphic chips or *brain chips*.

IBM's funding through the DARPA Synapse program resulted in the creation of a neuromorphic chip called *True North*. It is a 4096-core chip consisting of 256 *programmable neurons* which function much like synapses in the brain. True North uses digital spikes to perform neuromorphic computing.

Kim et al. (2015) describe a reconfigurable digital neuromorphic processor (DNP) architecture for large-scale spiking neural networks. In another study, Du et al. (2015) investigate relative merits of a hardware implementation of two neural network accelerators. The first accelerator's design is inspired by the machine learning domain, whereas that of the second by the neuroscience discipline. They analyze these two classes of accelerators in terms of energy consumption, speed gains, area cost, accuracy, and functionality. One more study reported in Chen et al. (2015) discusses implementing machine learning algorithms on a chip. Another study (Liu et al., 2013) describes how an ultra-high power efficiency beyond One-TeraFlops-Per-Watt was achieved in a bioinspired neuromorphic embedded computing engine named Centaur.

In April 2016, Nvidia released a state-of-the-art chip, Tesla P100 GPU, which specifically targets machine learning algorithms that employ deep learning. The GPU features 150 billion transistors on a single chip. DGX-1 is Nvidia's newest supercomputer which is powered by eight Tesla P100 GPUs. DX-1 comes with deep-learning software preinstalled and costs less than $130,000.

Neuromorphic computing opens up new possibilities for advancing cognitive computing. The field is new and the methods lack maturity. For example, the low quantization resolution of the synaptic weights and spikes significantly limits the inference in the True North chip. To alleviate this problem, Wen et al. (2016) propose a new learning method which constrains a random variance in each computation copy. This results in a 68.8% reduction in the number of neurosynaptic cores or equivalently a 6.5× speedup.

The recent European Union funded Human Brain Project aims to develop scientific research infrastructure to accelerate research in neuroscience, computing, and brain-related medicine. A counterpart in the US is the White House Brain Research through Advancing Innovative Neurotechnologies (BRAIN) initiate, whose focus is dynamic understanding of brain functions.

6 COGNITIVE COMPUTING ARCHITECTURES AND APPROACHES

A *cognitive architecture* is a blueprint for developing cognitive systems. It specifies fixed structures and interactions among them with the goal of achieving functions of the mind. The knowledge embodied in the architecture drives the interactions among the structures to achieve intelligent behavior.

A *cognitive model*, in contrast with a cognitive architecture, focuses on a single cognitive process such as language acquisition. Cognitive models are also used to study the interaction between cognitive processes such as language understanding and problem-solving. Furthermore, they are used for behavioral predictions for tasks. For example, how does an increase in training time affect air traffic controllers' performance?

Cognitive architectures tend to focus on structural aspects of cognitive systems. They constrain the types of cognitive models that can be developed. Likewise, cognitive models help to reveal the limitations of cognitive architectures. Thus, there is a strong interplay between cognitive architectures and models. In the literature, the terms cognitive architecture and cognitive model are not used consistently, and are often used synonymously. The context should help to reveal the intended meaning.

Cognitive architectures are an area of intensive research. In Langley et al. (2009), motivations for research on cognitive architectures are discussed. The study enumerates capabilities that cognitive architecture should provide related to representation, organization, performance, and learning. It also specifies criteria for evaluating cognitive architectures at the systems level and points out open research problems in cognitive architectures. A critical survey of the state-of-the-art in cognitive architectures is presented in Duch et al. (2008). This study provides useful insights into the usefulness of existing architectures for creating artificial general intelligence. The focus of cognitive architectures research is moving away from the functional capabilities of architectures to their ability to model details of human behavior and brain activity (Taatgen and Anderson, 2010).

There are three major classes of cognitive architectures: cognitivist, connectionist, and hybrid. They are discussed in the following sections. At the core of any cognitive system lies a cognitive architecture. A cognitive system is realized by creating a cognitive model using a cognitive computing architecture.

6.1 Cognitivist Architectures and Approaches

Cognitivist architectures represent information using explicit symbolic representations (Anderson, 1983). These representations use an absolute ontology to symbolize external objects. Representations are synthesized by human designers and are directly placed into artificial cognitive systems. Cognitivist architectures are also called *symbolic architectures* and *Artificial Intelligence* (AI) approaches. Cognitive systems based on this architecture are quite successful in solving specific problems. However, they lack generality to be useful across domains.

6.1.1 ACT-R

Adaptive Control of Thought-Rational (ACT-R) (Anderson, 1996; Anderson et al., 2004) is a theory about the mind and also a basis for several cognitive architectures of the cognitivist type. According to this theory, complex cognition is enabled by interaction among a set of integrated knowledge modules in the mind. Each module is associated with a distinct cortical region in the brain. Some modules represent procedural knowledge whereas others represent declarative knowledge.

Procedural knowledge is represented using *production rules* and modules called *chunks* represent *declarative knowledge*. Production rules capture transformations that occur in the cognitive system's environment. Declarative knowledge about objects in the environment is captured by chunks.

A large collection of modules in the brain is the underlying infrastructure for functions of the mind. Depending on the context and cognitive task, appropriate modules are selected and activated. Anderson et al. (2004) illustrates how these modules function individually and in unison in realizing simple and complex cognitive tasks.

6.1.2 Soar

Soar is another general architecture for developing cognitive systems (Lehman et al., 1996, 2006). It has been in use since 1983 and has evolved over the years. Soar uses a single architecture for all its tasks and subtasks. In the initial versions, Soar used a single representation for production rules and temporary knowledge, another single mechanism for generating goals and subgoals, and *chunking* for learning. In the recent releases, Soar uses multiple learning mechanisms—chunking, reinforcement learning, episodic learning, and semantic learning (Laird, 2008). Also, multiple representations of long-term knowledge are used—productions for procedural knowledge, semantic memory, and episodic memory.

Soar's goal is to support cognitive functions required to implement a general intelligent agent. Soar makes its decisions dynamically based on relevant knowledge. Decisions are based on current interpretation of sensory data, contents of working memory accumulated through previous experiences, and relevant knowledge retrieved from the long-term memory.

6.1.3 Mala

Mala is a multientity cognitive architecture for developing robots that must work in dynamic, unstructured environments (Haber and Sammut, 2013). Mala bridges the gap between a robot's sensorimotor and cognitive components. Mala supports modular and asynchronous processing, specialized representations, translations between representations, relational reasoning, and multiple types of integrated learning.

6.1.4 GOMS

GOMS is a family of predictive models of human performance (Card et al., 1983). GOMS stands for Goals (which can be accomplished with the system),

Operators (basic actions that can be performed on the system), Methods (sequences of operators that can be used to accomplish a goal), and Selection rules (selecting right methods to accomplish a goal). The models in the GOMS family are Keystroke-Level Model (KLM), Critical-Path Method GOMS (CPM-GOMS), Natural GOMS Language (NGOMSL)/Cognitive Complexity Theory, and Executable GOMS Language (GOMSL)/GLEAN. Each model in the family provides different sets of operators.

GOMS models are used to evaluate and improve human–computer interaction (HCI) (Sears and Jacko, 2009). They are used to describe the knowledge of procedures that a user must possess to operate a system such as a software application. GOMS models help to identify and eliminate unnecessary user actions. In other words, GOMS is a technique for task analysis. GOMS models are limited to describing procedural knowledge only.

6.1.5 *Limitations*

Symbolic representations reflect the designers' understanding of the domain and may bias the system. Also, it is difficult for the designers to come up with all relevant representations which are adequate to realize the desired cognitive behaviors of the system. Another issue with representations is the *symbol grounding problem* (Harnard, 2003), which is related to the problem of how meanings are assigned to words and what these meanings really are. The thing that a word refers to (its referent) is not its meaning.

6.2 Connectionist Architectures and Approaches

Connectionist architectures are inspired by the information processing that occurs in biological neural systems (Flusberg and McClelland, 2014). In the latter, information is processed by simple, networked computational units called neurons, which communicate in parallel with each other using electrochemical signals (Laughlin and Sejnowski, 2003). A synapse is a junction between two neurons with a minute gap across which signals pass by diffusion of a neurotransmitter.

The brain is made up of neurons, which are estimated to be in the 10–100 billion range. Each neuron is predicted to have over 10,000 connections to other neurons. A neuron receives stimuli from other neurons through its *incoming connections*. Neurons perform nonlinear computations using the received stimuli. The effect of this computation activates other neurons through its *outgoing connections*. Strengths of connection activations are quantified on a numeric scale and adjusted to reflect the state of network learning. Architectures based on this approach are called *connectionist* or *emergent architectures*.

6.2.1 *ANN*

Artificial neural networks (ANN) are a family of computational models based on connectionist architectures. In recent years, there is a renaissance of neural networks as powerful machine learning models (Goldberg, 2015). Though neural models have been used for tasks such as speech processing and image recognition for many decades, their widespread and intense use in NLP is relatively new. Goldberg (2015) provides a tutorial survey of neural network models for NLP including input encoding; feed-forward, convolutional, recurrent, and recursive networks; and computation graph abstraction for automatic gradient computation.

6.2.2 Lexical and Compositional Semantics

The mathematical representation of semantics is an important open issue in NLP. Semantics are of two types: lexical and compositional. *Lexical semantics* focuses on the meaning of individual words, whereas *compositional semantics* represent the meaning at the level of larger units such as phrases, sentences, and paragraphs.

Vector space is one way to represent lexical semantics. Lexical semantics are useful, but compositional semantics are critical for many natural language understanding (NLU) tasks such as text summarization, statistical paraphrasing, and textual entailment. Representations that draw upon cooccurrence statistics of large corpora are called *distributed representations*. Hermann (2014) describes several approaches to distributed representations and learning compositional semantics. He also discusses neural models that use distributed representations for various NLP tasks. A related study (Bowman et al., 2015) discusses how distributed representations-driven neural models learn the basic algebra of natural logic relations.

6.2.3 NEF

Neural Engineering Framework (NEF) is a general methodology for developing large-scale, biologically plausible, neural models of cognition (Stewart, 2012). NEF functions like a neural compiler. Given suitable inputs such as neuron properties and functions to be computed, NEF solves for the connection weights between neurons that will perform the desired functions. Nengo is a Python software library for developing and simulating large-scale brain models based on the NEF (Bekolay et al., 2014).

Development of a 2.5-million-neuron model of the brain, called Spaun, is described in Eliasmith et al. (2012). Spaun is built using the NEF and models many aspects of neuroanatomy, neurophysiology, and psychological behavior. These behaviors are illustrated via eight tasks—copying a drawing; image recognition; performing a three-armed bandit task; reproducing a given arbitrary length list; summing two values; question-answering on a list of numbers; given a number of syntactic input/output patterns, determining output pattern for a novel input pattern; and performing a syntactic or semantic reasoning task that is similar to the induction problems from the Raven's Progressive Matrices test for fluid intelligence.

An overview and comparison of several recent large-scale brain models are described in Eliasmith and Trujillo (2014). In his recent book, Eliasmith (2015) provides a guided exploration of a new cognitive architecture termed *Semantic Pointer Architecture*. The book also provides tools for constructing a wide range of perceptual, cognitive, and motor models.

6.2.4 Deep Learning

The recent advent of computing processors specially designed for neural network computations (see Section 5.7) ushered in *deep learning networks*. These networks employ multiple layers of neural processing units. The core of almost all deep learning algorithms is *backpropagation*, which is used for training the network. Backpropagation is a gradient descent computation distributed over the neural network.

Fundamental concepts such as *representations* are not formally defined for deep learning (Balduzzi, 2015). Also, there is no common language for describing and analyzing deep learning algorithms. Balduzzi (2015) proposes an abstract framework which enables formalizing current deep learning algorithms and approaches. Deep learning networks have been quite successful in solving diverse classification problems.

It is argued in Hawkins et al. (2016) that neural networks deviate from known brain principles and thus do not truly reflect the way the brain functions. Therefore, from a neuroscience perspective, it can be stated that systems based on connectionist architectures also lack generality in solving problems the way the human brain does.

6.3 Hybrid Architectures and Approaches

The last class of cognitive computing architectures encompasses those that employ a combination of symbolic and connectionist architectures. This class also includes neocortex inspired approaches (Mountcastle, 1998). The neocortex is a part of the cerebral cortex (brain) concerned with sight and hearing in mammals. Neocortex employs an extremely high degree of parallelism.

Neocortex neurons perform simple functions. These characteristics bode well for highly efficient hardware implementations of the neocortex (Rice et al., 2009). Also, every region of the neocortex has both sensory and motor functions. This observation suggests that cognitive computing systems be built by integrating sensory-motor senses. Furthermore, neocortex of the brain uses common principles and algorithms for diverse cognitive tasks such as vision, hearing, language, touch, and behavior. This suits well for developing neocortex inspired cognitive systems that are not tied to specific domains. We use the umbrella term *hybrid architectures* to refer to these approaches.

6.3.1 LIDA

Learning Intelligent Distribution Agent (LIDA) is a hybrid cognitive architecture as well as a model of cognition. LIDA is grounded in cognitive science and cognitive neuroscience and is intended to model a significant portion of human cognition (Franklin et al., 2014). Human cognition functions through cascading *cycles* of recurring brain events. Each cognitive cycle assesses the current state, interprets the state with reference to current goals of a cognitive task, and selects an internal or external response.

LIDA implements a three-phase cycle: understanding phase, attention (consciousness) phase, and action selection and learning phase. To achieve a human-level performance, a cognitive system must be capable of a theory of mind. How LIDA accomplished its version of a theory of mind is discussed in Friedlander and Franklin (2008). The suitability of LIDA's characteristics for developing cognitive architectures for artificial general intelligence (AGI) is discussed in Snaider et al. (2011).

6.3.2 Sigma

Sigma (Σ) is a recent cognitive system based on a novel cognitive architecture (Rosenbloom, 2013). It is comprised of three layers: a cognitive architecture, which is a fixed structure; knowledge and skills component which is positioned on top of the architecture; and an equivalent of firmware architecture. Predicates and conditionals provided by the cognitive architecture integrate the functionality of rule-based systems and probabilistic networks. The firmware architecture connects its implementation language (Lisp) and the cognitive architecture using a language of factor graphs and piecewise continuous functions. Sigma's capability has been demonstrated for the following cognitive functions: perception, mental imagery, decision making and problem solving, and NLP tasks such as such as word sense disambiguation and part of speech tagging.

6.3.3 HTM

Hierarchical Temporal Memory (HTM) is a theoretical framework for both biological and machine intelligence (Hawkins et al., 2016). The HTM models structural and algorithmic properties of the neocortex. It combines existing ideas to simulate the neocortex with a simple design, and yet provides a large range of cognitive capabilities. HTM integrates and extends approaches used in sparse distributed memory, Bayesian networks, spatial and temporal clustering algorithms. HTM structure resembles a tree-shaped hierarchy of nodes as in some neural networks.

6.3.4 IBM Watson

IBM Watson is perhaps the best known cognitive computing system in terms of its broader impact on society at large (IBM, n.d; Ferrucci et al., 2010). It is also the first cognitive computing system that leveraged the synergy between cognitive science and an array of computing technologies. Watson performed at a level to win Jeopardy! game against two all-time Jeopardy! human champions in 2011.

Watson cognitive computing capabilities are available to organizations and businesses through BlueMix API. There are more than two dozen Watson APIs which encapsulate over 50 cognitive technologies. IBM is also marketing Watson technologies under themes such as cognitive analytics, cognitive businesses, cognitive homes, and cognitive cars. Underlying these themes is a transparent cognitive computing infrastructure that provides services to build cognitive agents.

6.3.5 *Hierarchical Bayesian Model*

George and Hawkins (2005) describe a hierarchical Bayesian model for invariant pattern recognition in the visual cortex. Rice et al. (2009) investigate parallel software- and hardware-accelerated implementations of George and Hawkins model and also perform scaling analysis using Cray XD1 supercomputer. They report that hardware-acceleration provides an average throughput gain of 75× over software-only implementations of the networks they studied.

6.3.6 *Textual Narrative to 3D Geometry Models*

Constructing 3D geometric representations from textual narratives has applications in arts, robotics, and education. A hybrid cognitive architecture is used to accomplish this task in Chang et al. (2015). A dataset of 3D scenes annotated with natural language descriptions is used as training data for a neural network classifier and extract significant features that ground lexical terms to 3D models. These features are then integrated into a rule-based scene generation system.

7 COGNITIVE COMPUTING SYSTEMS AND APPLICATIONS

Cognitive computing systems have been used to solve a range of problems in diverse disciplines.

7.1 Intelligent Tutoring Systems

The earliest cognitive systems incarnated as intelligent tutoring systems (Anderson et al., 1995). Using the ACT cognitive architecture (Anderson, 1996) and advanced computer tutoring theory (Anderson, 1983), three ACT-based production models were developed (Anderson et al., 1995). The models reflect how students solve problems in algebra, geometry, and Lisp. Best case evaluations showed that students were able to achieve the same level of performance with these cognitive tutors compared to conventional instruction in one-third of the time.

PAT (Koedinger et al., 1997) is a cognitive tutor for high school algebra. It is also based on the ACT cognitive architecture. PAT's evaluation was performed on 470 students in experimental classes. Students in these classes outperformed students in comparison classes (who did not use PAT) by

15% on standardized tests. Supporting guided learning using a cognitive tutor is discussed in Aleven and Koedinger (2002). The study reports that students who explained their steps during problem-solving practice learned with greater understanding compared to those who did not explain their steps.

7.2 Problem Solving Systems

GEOS is a cognitive system for solving unaltered SAT geometry questions (Seo et al., 2015). It employs both text understanding and diagram interpretation in solving geometry problems. The problem is modeled as one of submodular optimization. It identifies a formal problem description which is likely to be compatible with question text and diagram. GEOS scored 49% on official SAT questions and 61% on practice questions.

SimStudent is a teachable agent which learns skills such as solving linear equations (Matsuda et al., 2013). It learns from examples as well as from feedback on its performance. SimStudent is used to investigate how and when students learn or not learn when they teach. Analysis of results indicates that several cognitive and social factors are correlated with SimStudent learning (equivalently, student learning). Such factors include accuracy of students' feedback and hints, quality of students' explanations during tutoring, and appropriateness of problem selection.

7.3 Question Answering

The IBM DeepQA project is IBM Watson's genesis (Ferrucci et al., 2010; Zadrozny et al., 2015). Knowledge representation and reasoning, machine learning, natural language understanding, and information retrieval are Watson's foundational technologies. DeepQA technology is at the core of Watson, whose functions include hypothesis generation, evidence gathering across heterogeneous data sources, analyzing evidence, and assigning a score for each hypothesis. Highly parallel underlying hardware and network provided the compute power.

7.4 Health Care

IBM Watson Health is a version of Watson for the health care and wellness domains. In 2011 Watson made forays into personalized cancer treatments. Medtronic is a medical technology, services, and solutions company. Using IBM Watson Health, Medtronic developed a glucose monitor and an insulin pump. The glucose monitor tracks sugar levels in people with diabetes and the insulin pump automatically supplies insulin. 125 million patient days of anonymous data from insulin pumps and glucose monitors has been collected by Medtronic to date. Watson Health will analyze this data to discover insights which can lead to new treatments for diabetes.

The Massachusetts General Hospital plans to use Nvidia GPUs to spot anomalies in CT scans and other medical images (Don, n.d). This project will draw upon 10 billion existing images to train a cognitive computing system that will use deep learning algorithms. The system is intended to help doctors more accurately detect diseases such as cancer and Alzheimers when the diseases are in their early stages.

7.5 Cognitive Businesses

Cognitive businesses are characterized by learning systems explicitly designed to collaborate with people in conversational style and understand natural language. Cognitive businesses also benefit from cognitive analytics. Instead of using predefined rules and structured queries to identify data to enable decision making, cognitive analytics utilizes cognitive computing technologies to generate multiple hypotheses, gather and weigh evidence from multiple data sources in support of each hypothesis, and rank hypotheses. Hypotheses that score above a certain threshold are presented as recommendations along with a numeric value to indicate the system's confidence in each hypothesis. The quality of insights generated by cognitive analytics increases with more data, which is used to train machine learning algorithms. Versions of IBM Watson are being marketed as cognitive business solutions.

7.6 Human–Robot Interaction

An important aspect of a human–robot interaction is *perspective-taking*, which encompasses an understanding of a concept or assessing a situation from a different point-of-view. Trafton et al. (2005) illustrate how perspective-taking enables astronauts to work in a collaborative project environment. They describe a cognitive architecture termed Polyscheme for perspective taking. They also develop a cognitive system based on Polyscheme and integrate it with a working robot system. The cognitive system is successful in solving a series of perspective-taking problems.

7.7 Cognitive Robots

IBM Watson is used at SoftBank Robotics, which is a provider of robotics platforms. IBM Watson-powered robot, called Pepper, is under development at SoftBank Robotics. Watson will help Pepper to extract meaning from unstructured data including text, images, video, and social media.

7.8 Deep Learning and Image Search

Machine learning algorithms play a prominent role in cognitive computing. However, many of them require hand crafting of *features*, which requires substantial effort. A new class of neural algorithms, called deep learning algorithms, is becoming popular since they can automatically learn feature representations. This automation obviates the need for labeled data required for training the neural networks.

Deep learning algorithms require massive neural networks similar to that of biological neural networks. The Google Brain project, Google DeepMind, and GPU-accelerated Torch scientific computing framework (with extensive support for machine learning algorithms) attest to the enormous interest in deep learning research and applications. Deep learning algorithms have been successfully used for solving a range of problems.

The MSR-Bing grand challenge (Hua et al., 2014) is a competition to advance content-based image retrieval research (Gudivada and Raghavan, 1995). MSR-Bing grand challenge goals are similar to those of TREC conferences (Harman, 1993). The image dataset for training is generated by sampling 1-year click logs of the Microsoft Bing image search engine. The dataset consists of 23.1 million triads of the form (I, Q, C). A triad specifies that for the image query Q, image I was clicked (as the relevant image) C times.

A deep neural network (DNN) approach presented in Bai et al. (2014) for computing similarity between images, maps image raw pixels to a bag-of-words vector space. Images are represented as vectors in the bag-of-words space and relevance between two images is computed using a cosine similarity measure. The DNN model also extracts high-level visual features, which are also used for computing image similarity. The DNN model is trained using the MSR-Bing grand challenge dataset. In another study, Wan et al. (2014) developed a framework for deep learning and used it to investigate approaches for learning feature representations and similarity measures in the context of content-based image retrieval applications.

7.9 Cross-media Retrieval

Cross-media retrieval is a challenging problem. It allows users to search for documents across various media by specifying a query in any media. The term *document* is used in a generic sense and refers to any media content such as text, image, audio, or video. Searching for images using a query expressed in text is an example of cross-media query. To support such queries, generating text that best describes the content of an image is required. Dong et al. (2016) propose a deep neural network-based approach, called *Word2VisualVec*, for cross-media retrieval. Evaluation is performed using three classes of queries: text-to-image, image-to-text, and text-to-text.

7.10 Brain–Computer Interfaces

Brain–computer interface (BCI) enables interaction between a brain and an external device. For example, brain signals can be used to control external devices such as cars and prosthetic limbs. Several analytic platforms are available for studying brain–computer interface (BCI). These platforms use electroencephalogram (EEG), magnetoencephalography (MEG), and functional near-infrared spectroscopy (fNIRS) to record brain signals. The signals are used to estimate a person's cognitive state, response, or intent for various purposes such as opening a door. Martinez et al. (2007) propose a new multistage procedure for a real-time BCI. They develop a system based on the procedure, which allows a BCI user to navigate a small car on a computer screen in real time. In a related study, Negueruela et al. (2011) investigate the use of BCI in the space environments, where an astronaut can use mental commands to control semi-automatic manipulators.

7.11 Autonomous Vehicle Navigation

Cognitive systems are central to autonomous vehicle navigation. They use several technologies such as radar, lidar, GPS, computer vision, and odometry to detect their surroundings. Advanced control systems analyze sensory data to distinguish different objects on the road. Google self-driving car, Tesla Motors Autopilot, and Mercedes-Benz Drive Pilot are active projects towards realizing the dream of self-driving cars.

8 TRENDS AND FUTURE RESEARCH DIRECTIONS

The synergistic confluence of cognitive science, data science, and computing will accelerate cognitive computing research and applications. Cognitive science will continue to provide theoretical underpinnings, the data science will provide cognitive analytics, and the computing discipline will bring advances in hardware technologies, big data, and machine learning to enable developing extremely large scale cognitive systems. The latter is critical to advancing the field by validating cognitive theories through empirical studies.

Biological cognitive systems have an advantage in that we do not understand their structure and function at a level to replicate them artificially. On the other hand, cognitive computing systems do not have the limitations of biological cognitive systems in terms of memory size and performance degradation due to aging and overwork.

The distinction between the symbolic AI approaches and neural network approaches will continue to blur. They complement each other rather than compete. Symbolic approaches have an edge over neural approaches in explaining the reasoning, whereas the neural network approaches contribute to building better models by leveraging big data and machine learning algorithms. Neural approaches can use symbolic inference rules to extract reasoning steps. On the other hand, neural approaches can be used to synthesize symbolic inference rules using semi-supervised and unsupervised learning.

Schatsky et al. (2015) conducted a study to assess how cognitive technologies are used in over 100 organizations that span 17 industry sectors. Their study revealed that cognitive technologies use falls into three categories. The first case involves embedding the technology into a product or service such as self-driving cars and other autonomous vehicles. Second, process applications embed cognitive technologies into workflows to improve operational efficiency. The third category encompasses using cognitive analytics to gain insights that will drive operational and strategic decisions across an organization. The current use of cognitive technologies by businesses is the proverbial tip of the iceberg.

According to EMC Digital Universe, based on research & analysis performed by International Data Corporation (IDC) (Turner, n.d), data is growing at an annual rate of 40% into the next decade. For example, the impending Airbus A380-1000 airliner will be capable of carrying about 1000 passengers. It will be equipped with 10,000 sensors in each wing, which will generate about 8 Tb of data daily. There will be many more sensors in the engines and other parts of the aircraft. Cognitive analytics will play a prominent role in utilizing this data for a range of tasks including preventive maintenance, improving passenger comfort and safety, and enhancing the operational efficiency of airports. More and more companies, ranging from Netflix, Amazon, and IBM to Google are branding themselves machine learning companies. Machine learning algorithms are at the core of their products and services.

Garrett (2014) discusses how big data analytics and cognitive computing will propel research and applications in astronomy. He argues that factors such as performance at scale, machine learning, and cognitive science will contribute to the rapid progress of the field. Along the same lines, Noor (2014) states that cognitive computing will be the game changer for engineering systems. Patel et al. (2001) explain how cognitive science can help gain insights into the nature of human–computer interaction processes in medical informatics, which can be used to improve the design of medical information systems.

Though the state-of-the-art cognitive systems are not anywhere close to performing at the level of humans, the gap is rapidly diminishing. For example, consider the *Go* game, which has long been viewed as the most challenging problem for AI and machine learning researchers. Compared to the Chess game, *Go* has enormous search space and this makes brute force evaluation of board positions and moves difficult. Silver et al. (2016) discuss how they have applied DeepMind cognitive technologies to develop *AlphaGo*, which defeated the human European *Go* champion by 5 games to 0. Furthermore, AlphaGo achieved a 99.8% winning rate against other *Go* computer programs. This event is extremely significant given that this is the first time ever a computer program defeated a human professional player in a full-sized *Go* game.

In closing, cognitive computing is the confluence of cognitive science, data science, and computing. Dramatic advances in natural language understanding, image, and video analysis hold keys for propelling cognitive computing to the next level. Cognitive computing is a double-edged sword with the power to transform human lives or become an instrument for misuse and destruction.

9 COGNITIVE COMPUTING RESOURCES

9.1 Open Source Frameworks, Tools, and Digital Libraries

1. 1-billion-word-language-modeling-benchmark, a standard training and test setup for natural language modeling experiments. Google Corporation. https://github.com/ciprian-chelba/1-billion-word-language-modeling-benchmark.
2. Apache Lucene Core, a full-featured text search engine Java library. http://lucene.apache.org/core/.
3. Apache UIMA project. Open source frameworks, tools, and annotators for facilitating the analysis of unstructured content such as text, audio and video. http://uima.apache.org/.
4. ACL Anthology: A Digital Archive of Research Papers in Computational Linguistics. Association for Computational Linguistics, http://www.aclweb.org/anthology/.
5. Biologically Inspired Cognitive Architectures (BICA) Society. Promotes and facilitates the transdisciplinary study of biologically inspired cognitive architectures. http://bicasociety.org/.
6. British National Corpus. A 100 million word collection of samples of written and spoken language designed to represent a wide cross-section of British English. http://www.natcorp.ox.ac.uk/.
7. The Cognitive Computing Consortium, a forum for researchers, developers and practitioners of cognitive computing and its allied technologies. https://cognitivecomputingconsortium.com/.
8. Cognitive Linguistics Journal, a publication of the International Cognitive Linguistics Association. Publishes linguistic research which addresses the interaction between language and cognition. http://www.cognitivelinguistics.org/en/journal.
9. Cognitive Science Society. A professional organization for researchers whose goal is to understanding the nature of the human mind. http://www.cognitivesciencesociety.org/.
10. Database Management Systems (DBMS). This Web application provides a ranking of over 300 software systems for data management. http://db-engines.com/en/.
11. DBpedia, Towards a Public Data Infrastructure for a Large, Multilingual, Semantic Knowledge Graph. http://wiki.dbpedia.org/.
12. Europarl, a Parallel Corpus for Statistical Machine Translation. Includes versions for 21 European languages. http://www.statmt.org/europarl/.
13. eWAVE, an interactive database on morphosyntactic variations in 50 varieties of spontaneous spoken English. Kortmann, Bernd & Lunkenheimer, Kerstin (eds.), The Electronic World Atlas of Varieties of English, Leipzig: Max Planck Institute for Evolutionary Anthropology. http://ewave-atlas.org/.
14. GATE, an open source software for solving natural language problems, The University of Sheffield. http://gate.ac.uk/.

15. IBM Watson Academy. Cognitive computing resources including a public access Cognitive Computing MOOC. http://www-304.ibm.com/services/weblectures/watsonacademy/#intro.
16. IBM Research: Cognitive computing. http://www.research.ibm.com/cognitive-computing/#fbid=-yy66f-kYIh.
17. International Cognitive Linguistics Association. Promotes approaches to linguistics research that are based on the perspective that language is an integral part of cognition. http://www.cognitivelinguistics.org/en.
18. ImageJ, an open source platform for scientific image analysis. http://imagej.net/Welcome.
19. ImmuNet, data science tools for predicting the role of genes in immunology. http://immunet.princeton.edu/.
20. MITCogNet, a research tool for scholars in the Brain & Cognitive Sciences. http://cognet.mit.edu/.
21. NoSQL. A list of systems for nonrelational data management. http://nosql-database.org/.
22. NuPIC, a platform and community for machine intelligence based on HTM theory. http://numenta.org/.
23. OpenCV, an open source computer vision and machine learning software library. http://opencv.org/.
24. openNLP, a machine learning based toolkit for NLP. he Apache Software Foundation. http://opennlp.apache.org/.
25. openSMILE, a tool for extracting audio feature spaces in real time. http://audeering.com/research/opensmile/.
26. Open Source Speech Software, Carnegie Mellon University. http://www.speech.cs.cmu.edu/.
27. Open Text Summarizer, a library and a command line tool for multilingual text summarization. http://libots.sourceforge.net/.
28. Praat, a tool for speech manipulation, analysis, and synthesis. http://www.fon.hum.uva.nl/praat/.
29. Parsey McParseface, a pretrained SyntaxNet model for parsing the standard English language. https://github.com/tensorflow/models/tree/master/syntaxnet.
30. Project Gutenberg, offers over 50,000 free ebooks. https://www.gutenberg.org/.
31. The R Project for Statistical Computing. Provides open source tools for statistical computing and data visualization. https://www.r-project.org/.
32. The Stanford Encyclopedia of Philosophy, Edward N. Zalta (ed.), http://plato.stanford.edu/.
33. The Stanford Natural Language Processing Group's statistical NLP, deep learning NLP, and rule-based NLP software tools for solving major computational linguistics problems. http://nlp.stanford.edu/software/index.shtml.

34. Syntactic N-grams, which are dependency tree fragments extracted from the English portion of the Google Books corpus. http://storage.googleapis.com/books/syntactic-ngrams/index.html.
35. SyntaxNet, an open-source neural network framework for developing Natural Language Understanding (NLU) systems. SyntaxNet is implemented in TensorFlow. https://research.googleblog.com/2016/05/announcing-syntaxnet-worlds-most.html.
36. TensorFlow, an open source software library for machine intelligence. https://www.tensorflow.org/.
37. Weka 3: Data Mining Software in Java, a collection of machine learning algorithms for data mining tasks. The University of Waikato. http://www.cs.waikato.ac.nz/ml/weka/.
38. World Atlas of Language Structures (WALS), a large database of phonological, grammatical, lexical properties of languages gathered from descriptive materials such as reference grammars. Max Planck Institute for Evolutionary Anthropology. http://wals.info/.
39. WordNet, a lexical database for English. https://wordnet.princeton.edu/wordnet/.

9.2 Nonopen Source Infrastructures and Cognitive Computing Systems

1. Arimo, data intelligence for all. https://arimo.com/.
2. BSBS Axis, improving healthcare quality through data science. http://www.bcbsaxis.com/.
3. D2L, a learning management system for delivering personalized learning experiences. http://www.d2l.com/.
4. databricks, a hosted end-to-end data platform for data analytics applications. https://databricks.com.
5. IBM Bluemix API Catalog including Watson APIs. https://console.ng.bluemix.net/catalog/.
6. Linguistic Data Consortium. Data collections, corpora, software, research papers and specifications for language research and education. https://www.ldc.upenn.edu/language-resources.
7. SAS Analytics Software & Solutions. http://www.sas.com/en_us/home.html.
8. Synthesys, a cognitive computing system for stooping rule-breaking before it happens. http://www.digitalreasoning.com/.
9. Tableau software for data analysis and visualization. http://www.tableau.com/.
10. VoiceBox Technologies, provides conversational natural language solutions for applications such as *connected car*. http://www.voicebox.com/.

REFERENCES

Abrahamsen, A., Bechtel, W., 2012. History and core themes. In: Frankish, K., Ramsey, W.M. (Eds.), The Cambridge Handbook of Cognitive Science. Cambridge University Press, New York, NY.

Abu-Mostafa, Y.S., Magdon-Ismail, M., Lin, H.T., 2012. Learning from Data. AMLBook, Pasadena, CA.

Aleven, V., Koedinger, K.R., 2002. An effective metacognitive strategy: learning by doing and explaining with a computer-based cognitive tutor. Cogn. Sci. 26, 147–179.

Anderson, J.R., 1983. The Architecture of Cognition. Psychology Press, Hove, UK.

Anderson, J.R., 1996. Act: a simple theory of complex cognition. Am. Psychol. 51 (4), 355–365.

Anderson, J.R., Corbett, A.T., Koedinger, K.R., Pelletier, R., 1995. Cognitive tutors: lessons learned. J. Learn. Sci. 4 (2), 167–207.

Anderson, J.R., Bothell, D., Byrne, M.D., Douglass, S., Lebiere, C., Qin, Y., 2004. An integrated theory of the mind. Psychol. Rev. 111 (4), 1036.

Andor, D., Alberti, C., Weiss, D., Severyn, A., Presta, A., Ganchev, K., Petrov, S., Collins, M., 2016. Globally normalized transition-based neural networks. CoRR, abs/1603.06042, http://arxiv.org/abs/1603.06042.

Armbrust, M., Fox, A., Griffith, R., Joseph, A.D., Katz, R., Konwinski, A., Lee, G., Patterson, D., Rabkin, A., Stoica, I., Zaharia, M., 2010. A view of cloud computing. Commun. ACM 53 (4), 50–58. http://dx.doi.org/10.1145/1721654.1721672.

Bai, Y., Yu, W., Xiao, T., Xu, C., Yang, K., Ma, W.Y., Zhao, T., 2014. Bag-of-words based deep neural network for image retrieval. In: Proceedings of the 22nd ACM International Conference on Multimedia. ACM, Orlando, FL, pp. 229–232.

Balduzzi, D., 2015. Semantics, representations and grammars for deep learning. CoRR, abs/1509.08627.

Bekolay, T., Bergstra, J., Hunsberger, E., DeWolf, T., Stewart, T.C., Rasmussen, D., Choo, X., Voelker, A., Eliasmith, C., 2014. Nengo: a python tool for building large-scale functional brain models. Front. Neuroinform. 7(48). http://dx.doi.org/10.3389/fninf.2013.00048.

Berman, J.J., 2013. Principles of Big Data: Preparing, Sharing, and Analyzing Complex Information. Morgan Kaufmann, Burlington, MA.

Boroditsky, L., 2011. How language shapes thought. Sci. Am. 304 (2), 62–65.

Bowman, S.R., Potts, C., Manning, C.D., 2015. Learning distributed word representations for natural logic reasoning. In: Knowledge Representation and Reasoning: Integrating Symbolic and Neural Approaches: Papers from the 2015 AAAI Spring Symposium. AAAI Publications, Stanford, CA, pp. 10–13.

Brewster, C., O'Hara, K., 2004. Knowledge representation with ontologies: the present and future. IEEE Intell. Syst. 19 (1), 72–81. http://dx.doi.org/10.1109/MIS.2004.1265889.

Brill, E., 1995. Transformation-based error-driven learning and natural language processing: a case study in part-of-speech tagging. Comput. Linguist. 21 (4), 543–565.

Card, S.K., Newell, A., Moran, T.P., 1983. The Psychology of Human-Computer Interaction. L. Erlbaum Associates Inc., Hillsdale, NJ

Chang, A., Monroe, W., Savva, M., Potts, C., Manning, C.D., 2015. Text to 3D scene generation with rich lexical grounding. In: Proceedings of the 53rd Annual Meeting of the Association for Computational Linguistics and the 7th International Joint Conference on Natural Language Processing. Association for Computational Linguistics, Stroudsburg, PA, pp. 53–62.

Chen, P.Y., Kadetotad, D., Xu, Z., Mohanty, A., Lin, B., Ye, J., Vrudhula, S., Seo, J.s., Cao, Y., Yu, S., 2015. Technology-design co-optimization of resistive cross-point array for accelerating learning algorithms on chip. In: Proceedings of the 2015 Design, Automation & Test in Europe Conference & Exhibition. EDA Consortium, San Jose, CA, pp. 854–859.

Chipman, S., 2015. The Oxford Handbook of Cognitive Science, Volume 1. Oxford University Press, Oxford.

Davis, A., Shrobe, H., Szolovits, P., 1993. What is a knowledge representation. AI Mag. 14, 17–33.

DBpedia, n.d. Towards a public data infrastructure for a large, multilingual, semantic knowledge graph. http://wiki.dbpedia.org/.

Dhar, V., 2013. Data science and prediction. Commun. ACM 56 (12), 64–73.

Don, C., n.d. New chips propel machine learning: Nvidia microchips are helping in detection of anomalies on CT scans. http://www.wsj.com/articles/new-chips-propel-machine-learning-1463957238.

Dong, J., Li, X., Snoek, C.G.M., 2016. Word2visualvec: cross-media retrieval by visual feature prediction. http://arxiv.org/abs/1604.06838.

Du, Z., Ben-Dayan Rubin, D.D., Chen, Y., He, L., Chen, T., Zhang, L., Wu, C., Temam, O., 2015. Neuromorphic accelerators: a comparison between neuroscience and machine-learning approaches. In: Proceedings of the 48th International Symposium on Microarchitecture, MICRO-48. ACM, New York, NY, pp. 494–507.

Duch, W., Oentaryo, R.J., Pasquier, M., 2008. Cognitive architectures: where do we go from here? In: Proceedings of the 2008 Conference on Artificial General Intelligence 2008: Proceedings of the First AGI Conference. IOS Press, Amsterdam, The Netherlands, pp. 122–136.

Eliasmith, C., 2015. How to Build a Brain: A Neural Architecture for Biological Cognition. Oxford University Press, Oxford.

Eliasmith, C., Trujillo, O., 2014. The use and abuse of large-scale brain models. Curr. Opin. Neurobiol. 25, 1–6. (Theoretical and computational neuroscience). ISSN 0959-4388. http://dx.doi.org/10.1016/j.conb.2013.09.009..

Eliasmith, C., Stewart, T.C., Choo, X., Bekolay, T., DeWolf, T., Tang, Y., Rasmussen, D., 2012. A large-scale model of the functioning brain. Science 338 (6111), 1202–1205. http://dx.doi.org/10.1126/science.1225266.

Evans, V., 2012. Wiley interdisciplinary reviews: cognitive science. Cogn. Linguist. 3. http://dx.doi.org/10.1002/wcs.1163.

Ferrucci, D., Brown, E., Chu-Carroll, J., Fan, J., Gondek, D., Kalyanpur, A.A., Lally, A., Murdock, J.W., Nyberg, E., Prager, J., Schlaefer, N., Welty, C., 2010. Building watson: an overview of the deepqa project. AI Mag. 31 (3), 59–79. http://dx.doi.org/10.1609/aimag.v31i3.2303.

Fielding, R.T., 2000. Architectural Styles and the Design of Network-Based Software Architectures, Chapter 5: Representational State Transfer (REST). Ph.D. thesis, University of California, Irvine.

Flusberg, S.J., McClelland, J.L., 2014. Connectionism and the emergence of mind. In: Chipman, S. (Ed.), The Oxford Handbook of Cognitive Science, vol. 1. Oxford University Press.

Frankish, K., Ramsey, W.M., (Eds.) 2012. The Cambridge Handbook of Cognitive Science. Cambridge University Press, Cambridge.

Franklin, S., Madl, T., D'Mello, S., Snaider, J., 2014. Lida: a systems-level architecture for cognition, emotion, and learning. IEEE Trans. Auton. Ment. Dev. 6 (1), 19–41. ISSN 1943-0604.

Friedenberg, J.D., Silverman, G.W., 2015. Cognitive Science: An Introduction to the Study of Mind, third ed. SAGE Publications, Thousand Oaks, CA.

Friedlander, D., Franklin, S., 2008. Lida and a theory of mind. In: Proceedings of the 2008 Conference on Artificial General Intelligence 2008: Proceedings of the First AGI Conference. IOS Press, Amsterdam, The Netherlands, pp. 137–148.

Garrett, M., 2014. Big data analytics and cognitive computing: future opportunities for astronomical research. Mater. Sci. Eng. Conf. Ser. 67(1). http://dx.doi.org/10.1088/1757-899X/67/1/012017.

George, D., Hawkins, J., 2005. A hierarchical bayesian model of invariant pattern recognition in the visual cortex. In: Proceedings. 2005 IEEE International Joint Conference on Neural Networks, 2005, vol. 3. pp. 1812–1817.

Goldberg, Y., 2015. A primer on neural network models for natural language processing. CoRR, abs/1510.00726.

Gruber, T.R., 1993. A translation approach to portable ontology specifications. Knowl. Acquis. 5 (2), 199–220. http://dx.doi.org/10.1006/knac.1993.1008.

Grus, J., 2015. Data Science from Scratch: First Principles with Python. O'Reilly Media, Sebastopol, CA.

Guarino, N., Oberle, D., Staab, S., 2009. What is an ontology? In: Staab, S., Studer, R. (Eds.), Handbook on Ontologies. Springer.

Gudivada, V., Raghavan, V., 1995. Content-based image retrieval systems. IEEE Comput. 28 (9), 18–22.

Gudivada, V., Raghavan, V., Grosky, W., Kasanagottu, R., 1997. Information retrieval on the world-wide web. IEEE Internet Comput. 58–68.

Gudivada, V., Baeza-Yates, R., Raghavan, V., 2015a. Big data: promises and problems. IEEE Comput. 48 (3), 20–23.

Gudivada, V., Rao, D., Raghavan, V., 2015b. Big data driven natural language processing research and applications. In: Govindaraju, V., Raghavan, V., Rao, C.R. (Eds.), Big Data Analytics. Elsevier, New York, NY, pp. 203–238.

Gudivada, V., Apon, A., Rao, D., 2016a. Database systems for big data storage and retrieval. In: Segall, R., Cook, J., Gupta, N. (Eds.), Big Data Storage and Visualization Techniques. IDG Global, Boston, MA (in press).

Gudivada, V., Rao, D., Raghavan, V., 2016b. Renaissance in database management: navigating the landscape of candidate systems. IEEE Comput. 49 (4), 31–42.

Haber, A., Sammut, C., 2013. A cognitive architecture for autonomous robots. Adv. Cogn. Syst. 257–275.

Halevy, A., Norvig, P., Pereira, F., 2009. The unreasonable effectiveness of data. IEEE Intell. Syst. 24 (2), 8–12.

Harman, D., 1993. Overview of the first trec conference. In: Proceedings of the 16th Annual International ACM SIGIR Conference on Research and Development in Information Retrieval, SIGIR '93. ACM, New York, NY, pp. 36–47.

Harnard, S., 2003. The symbol grounding problem. In: Nadel, L. (Ed.), Encyclopedia of Cognitive Science. Nature Publishing Group/Macmillan, New York, NY, pp. 335–346.

Hastie, T., Tibshirani, R., Friedman, J., 2003. The Elements of Statistical Learning: Data Mining, Inference, and Prediction, second ed. Springer, New York, NY.

Hawkins, J., Ahmad, S., Purdy, S., Lavin, A., 2016. Biological and machine intelligence (BAMI). Initial online release 0.4, http://numenta.com/biological-and-machine-intelligence/.

Hermann, K.M., 2014. Distributed representations for compositional semantics. http://arxiv.org/abs/1411.3146.

Hey, T., Tansley, S., Tolle, K., (Eds.) 2009. The Fourth Paradigm: Data-Intensive Scientific Discovery. Microsoft, Redmond, WA.

Hua, X.S., Ye, M., Li, J., 2014. Mining knowledge from clicks: MSR-Bing image retrieval challenge. In: IEEE International Conference on Multimedia and Expo Workshops, pp. 1–4.

Hurwitz, J., Kaufman, M., Bowles, A., 2015. Cognitive Computing and Big Data Analytics. Wiley, Hoboken, NJ.

IBM, n.d. IBM watson™developer cloud (WDC). http://www.ibm.com/smarterplanet/us/en/ibmwatson/developercloud/doc/getting_started/.

Isac, D., Reiss, C., 2013. I-Language: An Introduction to Linguistics as Cognitive Science, second ed. Oxford University Press, Oxford.

Kay, P., Kempton, W., 1984. What is the sapir-whorf hypothesis? Am. Anthropol. 65–79.

Kim, Y., Zhang, Y., Li, P., 2015. A reconfigurable digital neuromorphic processor with memristive synaptic crossbar for cognitive computing. ACM J. Emerg. Technol. Comput. Syst. 11 (4), 38:1–38:25. http://dx.doi.org/10.1145/2700234.

Koedinger, K.R., Anderson, J.R., Hadley, W.H., Mark, M.A., 1997. Intelligent tutoring goes to school in the big city. Int. J. Artif. Intell. Educ. 8, 30–43.

Kormushev, P., Calinon, S., Saegusa, R., Metta, G., 2010. Learning the skill of archery by a humanoid robot iCub. In: Proc. IEEE Intl Conf. on Humanoid Robots (Humanoids), Nashville, USA, pp. 417–423.

Laird, J.E., 2008. Extending the soar cognitive architecture. In: Proceedings of the 2008 Conference on Artificial General Intelligence 2008: Proceedings of the First AGI Conference. IOS Press, Amsterdam, The Netherlands, pp. 224–235.

Langley, P., Laird, J.E., Rogers, S., 2009. Cognitive architectures: research issues and challenges. Cogn. Syst. Res. 10 (2), 141–160. http://dx.doi.org/10.1016/j.cogsys.2006.07.004.

Laughlin, S.B., Sejnowski, T.J., 2003. Communication in neuronal networks. Science 301, 1870–1874.

Lehman, J.F., Laird, J., Rosenbloom, P., 1996. A gentle introduction to soar, an architecture for human cognition. In: Sternberg, S., Scarborough, D. (Eds.), Invitation to Cognitive Science, vol. 4. MIT Press.

Lehman, J.F., Laird, J., Rosenbloom, P., 2006. A gentle introduction to soar, an architecture for human cognition, 2006 update. University of Michigan.

Ling, W., Luís, T., Marujo, L., Astudillo, R.F., Amir, S., Dyer, C., Black, A.W., Trancoso, I., 2015. Finding function in form: compositional character models for open vocabulary word representation. CoRR abs/1508.02096. http://arxiv.org/abs/1508.02096.

Liu, B., Hu, M., Li, H., Chen, Y., Xue, C.J., 2013. Bio-inspired ultra lower-power neuromorphic computing engine for embedded systems. In: Proceedings of the Ninth IEEE/ACM/IFIP International Conference on Hardware/Software Codesign and System Synthesis, CODES+ISSS '13. IEEE Press, Piscataway, NJ, pp. 23:1.

Martinez, P., Bakardjian, H., Cichocki, A., 2007. Fully online multicommand brain-computer interface with visual neurofeedback using ssvep paradigm. Intell. Neurosci. 2007, 1–9. http://dx.doi.org/10.1155/2007/94561.

Matsuda, N., Yarzebinski, E., Keiser, V., Raizada, R., William, W.C., Stylianides, G.J., Koedinge, K.R., 2013. Cognitive anatomy of tutor learning: lessons learned with SimStudent. J. Educ. Psychol. 105 (4), 1152–1163.

McClelland, J.L., Ralph, M.A., 2015. Cognitive neuroscience. In: Wright, J. (Ed.), International Encyclopedia of the Social & Behavioral Sciences, vol. 4. pp. 95–102.

McCreary, D., Kelly, A., 2013. Making Sense of NoSQL: A Guide for Managers and the Rest of Us. Manning Publications, Greenwich, CT.

Miller, G.A., 1995. Wordnet: a lexical database for English. Commun. ACM 38 (11), 39–41. http://dx.doi.org/10.1145/ 219717.219748.

Mountcastle, V.B., 1998. Perceptual Neuroscience: The Cerebral Cortex. Harvard University Press, Cambridge, MA.

Negueruela, C., Broschart, M., Menon, C., Millán, J.R., 2011. Brain-computer interfaces for space applications. Pers. Ubiquit. Comput. 15 (5), 527–537. http://dx.doi.org/10.1007/s00779-010-0322-8.

Neisser, U., 2014. Cognitive Psychology, Classic Edition. Psychology Press, Hove, UK.

Neurowear, n.d. Neurocam: a head mounted camera that detects human emotions. http://neurowear.com/projects_detail/neurocam.html.

Newell, A., 1994. Unified Theories of Cognition. Harvard University Press, Cambridge, MA.

Noor, A.K., 2014. Potential of cognitive computing and cognitive systems. Open Eng. 5 (1), 75–88. http://dx.doi.org/10.1515/ eng-2015-0008.

Patel, V.L., Arocha, J.F., Kaufman, D.R., 2001. A primer on aspects of cognition for medical informatics. J. Am. Med. Inform. Assoc. 8 (4), 324–343.

Rice, K.L., Taha, T.M., Vutsinas, C.N., 2009. Scaling analysis of a neocortex inspired cognitive model on the cray XD1. J. Supercomput. 47 (1), 21–43. http://dx.doi.org/10.1007/ s11227-008-0195-z.

Rosenbloom, P.S., 2013. The Sigma cognitive architecture and system. AISB Q. 136, 4–13.

Russell, S.J., Norvig, P., 2009. Artificial Intelligence: A Modern Approach, third ed. Pearson Education, Upper Saddle River, NJ.

Ryza, S., Laserson, U., Owen, S., Wills, J., Year. Advanced Analytics with Spark: Patterns for Learning from Data at Scale. O'Reilly Media, Sebastopol, CA.

Samani, H., 2015. Cognitive Robotics. CRC Press, Boca Raton, FL.

Schatsky, D., Muraskin, C., Gurumurthy, R., 2015. Cognitive technologies: the real opportunities for business. Deloitte Rev 113–129.

Sears, A., Jacko, J.A., (Eds.) 2009. Human-Computer Interaction: Fundamentals. CRC Press, Boca Raton, FL.

Seo, M.J., Hajishirzi, H., Farhadi, A., Etzioni, O., Malcolm, C., 2015. Solving geometry problems: combining text and diagram interpretation. In: MÃărquez, L., Callison-Burch, C., Su, J., Pighin, D., Marton, Y. (Eds.), EMNLP. The Association for Computational Linguistics, pp. 1466–1476.

Silver, D., Huang, A., Maddison, C.J., Guez, A., Sifre, L., van den Driessche, G., Schrittwieser, J., Antonoglou, I., Panneershelvam, V., Lanctot, M., Dieleman, S., Grewe, D., Nham, J., Kalchbrenner, N., Sutskever, I., Lillicrap, T., Leach, M., Kavukcuoglu, K., Graepel, T., Hassabis, D., 2016. Mastering the game of Go with deep neural networks and tree search. Nature 529 (7587), 484–489. http://dx.doi.org/10.1038/nature16961.

Snaider, J., McCall, R., Franklin, S., 2011. The lida framework as a general tool for agi. In: Schmidhuber, J., Thórisson, K.R., Looks, M. (Eds.), Artificial General Intelligence: 4th International Conference, AGI 2011. Springer, Berlin, Heidelberg, pp. 133–142.

Stephan, âst.G., Pascal, âst.H., Andreas, âst.A., 2007. Knowledge representation and ontologies. In: Studer, R., Grimm, S., Abecker, A. (Eds.), Semantic Web Services: Concepts, Technologies, and Applications. Springer, Berlin, Heidelberg, pp. 51–105.

Stewart, T.C., 2012. The neural engineering framework. AISB Q. 135, 2–7.

SyntaxNet, n.d. An open source neural network framework for tensorflow for developing natural language understanding (NLU) systems. https://github.com/tensorflow/models/tree/master/syntaxnet.

Taatgen, N., Anderson, J.R., 2010. The past, present, and future of cognitive architectures. In: Topics in Cognitive Science, vol. 2. Blackwell Publishing Ltd, pp. 693–704.

TensorFlow, n.d. An open source software library for numerical computation using data flow graphs. https://www.tensorflow.org.

Thagard, P., 2009. Why cognitive science needs philosophy and vice versa. In: Topics in Cognitive Science, vol. 1. Blackwell Publishing Ltd, pp. 237–254.

Trafton, J.G., Cassimatis, N.L., Bugajska, M.D., Brock, D.P., Mintz, F.E., Schultz, A.C., 2005. Enabling effective human-robot interaction using perspective-taking in robots. IEEE Trans. Syst. Man Cybern. Syst. Hum. 35 (4), 460–470.

Turner Vernon, n.d. The digital universe of opportunities: rich data and the increasing value of the internet of things [retrieved: July, 2016], http://www.emc.com/leadership/digital-universe/2014iview/digital-universe-of-opportunities-vernon-turner.htm.

Vernon, D., 2014. Artificial Cognitive Systems: A Primer. The MIT Press, Cambridge, MA.

Wan, J., Wang, D., Hoi, S.C.H., Wu, P., Zhu, J., Zhang, Y., Li, J., 2014. Deep learning for content-based image retrieval: a comprehensive study. In: Proceedings of the 22nd ACM International Conference on Multimedia. ACM, New York, NY, pp. 157–166.

Wen, W., Wu, C., Wang, Y., Nixon, K., Wu, Q., Barnell, M., Li, H., Chen, Y., 2016. A new learning method for inference accuracy, core occupation, and performance co-optimization on truenorth chip. In: Proceedings of the 53rd Annual Design Automation Conference. ACM, New York, NY, pp. 18:1–18:6.

White, T., 2015. Hadoop: The Definitive Guide, fourth ed. O'Reilly Media, Sebastopol, CA.

Williams, R.S., 2016. Brain inspired computing. ACM SIGOPS Oper. Syst. Rev. 50 (2), 295. http://dx.doi.org/10.1145/2954680.2872417.

Wu, Q., Ding, G., Xu, Y., Feng, S., Du, Z., Wang, J., Long, K., 2014. Cognitive internet of things: a new paradigm beyond connection. CoRR, abs/1403.2498. http://arxiv.org/abs/1403.2498.

Zadrozny, W., Paiva, V.D., Moss, L.S., 2015. Explaining watson: polymath style. In: Proceedings of the Twenty-Ninth AAAI Conference on Artificial Intelligence. Association for the Advancement of Artificial Intelligence, pp. 4078–4082.

Chapter 2

Cognitive Computing and Neural Networks: Reverse Engineering the Brain

A.S. Maida
The Center for Advanced Computer Studies, The University of Louisiana at Lafayette, Lafayette, LA, United States

ABSTRACT

Cognitive computing seeks to build applications which model and mimic human thinking. One approach toward achieving this goal is to develop brain-inspired computational models. A prime example of such a model is the class of deep convolutional networks which is currently used in pattern recognition, machine vision, and machine learning.

We offer a brief review of the mammalian neocortex, the minicolumn, and the ventral pathway. We provide descriptions of abstract neural circuits that have been used to model these areas of the brain. This include Poisson spiking networks, liquid computing networks, spiking models of feature discovery in the ventral pathway, spike-timing-dependent plasticity learning, restricted Boltzmann machines, deep belief networks, and deep convolutional networks.

In summary, this chapter explores abstractions of neural networks found within the mammalian neocortex that support cognition and the beginnings of cognitive computation.

Keywords: Brain simulation, Deep belief networks, Convolutional networks, Liquid computing, Biological neural networks, Neocortex

1 INTRODUCTION

One goal of cognitive computing is to build applications which model and mimic the characteristics of human thinking. A natural approach toward achieving this goal is to develop brain-inspired computational models of cognitive function. In an endeavor to reverse engineer the brain, this chapter explores neural networks found within the mammalian neocortex, and associated formal and computational models, that appear to form the basis of such cognition.

Handbook of Statistics, Vol. 35. http://dx.doi.org/10.1016/bs.host.2016.07.011

1.1 Chapter Outline

This chapter will discuss neural computations in the neocortex that appear to play a role in pattern classification. We shall begin by noting the scalability of brain architecture and then focus on the evolutionary invention that enables this scalability, namely, the neocortical microcircuit. We will briefly look at some aspects of global brain organization and then look at the elementary brain computing elements—neurons and synapses—and state-of-the-art ways to model them. We will look at modeling techniques for abstract neurons and synapses and look at direct attempts to create formal models of cortical microcircuits. After this, we move on to look at how neocortical microcircuits are grouped to form areas which perform computations over feature maps. This will set the stage for comparing brain models with deep convolutional networks that have recently become the state of the art in pattern classification.

2 BRAIN SCALABILITY

To start, we examine the part of the brain that is most scalable and also responsible for the higher-level cognition which is characteristic of mammals, primates, and humans. This is the six-layered, or laminar, neocortex. Structurally, the neocortex is a sheet of neurons whose thickness is approximately uniform. It varies in thickness only slightly, from 2 to 4 mm for nearly most mammals, including the mouse, the primate, and human. Despite the uniform thickness across mammalian species, the surface area varies drastically across species. In the mouse, the surface area of the neocortex is about 1 cm^2. In humans, the surface area is about 2600 cm^2 (Mountcastle, 1998). Qualitatively, the differences in local circuitry within the neocortical sheet are too small to account for the large qualitative differences in intelligence across species. It appears that the quantity of circuitry (surface area), combined with long-distance connectivity between different parts of the sheet, governs the difference in intelligence across species. Although six layered, these lamina cannot be modeled by layers in an artificial neural network (ANN) because the neural circuitry interconnecting the layers is too dense. In the following we will refer to the cortical layers as lamina to avoid confusion with layers in ANNs.

We have explained that the neocortex is a highly scalable computer. During evolution, the neocortex scaled its computational power primarily by increasing surface area and, secondarily, by adjusting its intrinsic circuitry. This scalability is unique to mammals. The hypothesized neocortical building block that enables this scalability is the minicolumn (Buxhoeveden and Casanova, 2002; Mountcastle, 1997). In the literature, the words "minicolumn" and "microcolumn" are used interchangeably. The neocortex is tiled with minicolumns. The tiling implements repeating canonical cortical microcircuits. Minicolumns are oriented vertically within the neocortex and are packed in a honeycomb-like pattern to tile the neocortical surface. They consist of about 100 neurons and their width varies from 20 to 60 μm, depending on species. The 2–4 mm thickness of the neocortical surface is determined by the height of the minicolumn. The minicolumns have adjacent connections with other minicolumns as well as very long-distance connections. The number of minicolumns within the neocortex scales linearly with the neocortical surface area, and the minicolumns become specialized for the types of computations that they perform, either by learning or by evolution.

A core hypothesis about brain organization is that the minicolumn, which emerged at the advent of mammalian evolution, may encode a quasi-universal algorithm for adaptive computation and learning (Allman, 1991). This hypothesis has found its way into recent books for the general audience (Domingos, 2015; Kurzweil, 2013). The hypothesis states that the learning algorithm(s) implemented in the minicolumn can adapt to open-ended computations. More specifically, they can perform sensory functions such as auditory processing, as well as all aspects of visual processing (shape, motion, color), as well as the more abstract forms of computation related to language, logic, and higher-level thinking that is unique to humans.

When referring to a quasi-universal algorithm, we must proceed cautiously. The machine learning literature contains many theorems about the limitations of what is learnable. Particularly famous are the "curse of dimensionality" and the "no free lunch theorem." The curse of dimensionality states that the amount of data required to train a learning machine grows exponentially with the dimensionality of the input feature space. Since training data are finite, there will never be enough training data for sufficiently large problems. The "no free lunch theorem" informally states that for any machine learning algorithm, there is a data distribution that it will perform poorly on (Wolpert, 1996). This result does not rule out quasi-universal algorithms that operate in real-world environments that organisms must survive in. This includes algorithms for visual and auditory processing, reasoning, planning, motor control, navigation, and language understanding. This has been called the AI set of functions (Bengio and LeCun, 2007).

Another thing that is unique about the neocortex is that no sensory inputs reach the neocortex without passing through the relevant nucleus in the thalamus (Sherman and Guillery, 2004). For instance, visual input passes through the lateral geniculate nucleus (LGN) within the thalamus and auditory input passes through the medial geniculate nucleus (MGN). Apparently, the respective nuclei recode visual and auditory input into a more abstract representation that allows the neocortex to understand it. Support for this idea appears in the study described below.

A classic demonstration that supports the "universality" of neocortical circuitry was demonstrated in a series of studies from the lab of Mriganka Sur during the 1990s (Sharma et al., 2000). He and colleagues induced the brains of newborn ferrets to be rewired so that the retina projected to the auditory thalamus or MGN (instead of the visual thalamus which is the LGN) which in turn projected to the primary (input area) auditory cortex. The auditory cortex is built from honeycomb-packed minicolumns, as are all neocortical areas. The neurons in the auditory cortex acquired response and organizational characteristics resembling those of a normally functioning primary visual cortex. It appears that, somehow, the genetically determined auditory minicolumns analyzed the statistical structure of their input during development to discover features (i.e., unsupervised learning) useful in analyzing visual input.

From an algorithmic viewpoint, the above suggests that the minicolumn has one or more general learning algorithms embedded within its circuitry to cope with high-dimensional complex data. It would be of great relevance to AI to know what these are.

3 NEOCORTICAL BRAIN ORGANIZATION

The reader needs some basic information about brain organization at the systems level to understand this chapter. We quickly survey the multiple levels of anatomical organization of the neocortex starting from the highest description level and moving down to the minicolumn and neuronal circuit levels. Later, we look at machine learning algorithms that have been used to model the minicolumn as well as more global brain pathways, such as the ventral visual pathway for visual object classification and recognition. The intent is to get some idea, in a machine learning sense, what kind of computation the minicolumn may perform and the role it plays in larger scale, more global brain computations.

The 2600 cm^2 neocortical surface area can be subdivided into anatomically and functionally distinct regions called cortical areas. When discussing the ferret in the previous section, we made reference to the term cortical area but did not define it. There are anatomical and functional criteria to identify a cortical area (Krüger et al., 2012) that we need not worry about. We only need to know that, functionally, the many minicolumns within a given cortical area perform a uniform parallel computation. They are modeling candidates to serve as layers in an ANN.

In 1909, Brodmann used cytoarchitectonic studies (anatomical studies of cell architecture) to identify the 6 lamina in neocortex and to catalog 52 separate areas in the human neocortex. We now know that the separate cortical areas perform highly specialized computations, whether it is edge detection in the visual modality or phonetic analysis in the auditory modality. At later processing stages, other cortical areas perform unimodal information processing and cross-modal information fusion to form coherent representations of objects and events. We now know that there are over 30 areas committed to visual processing. They form a network with both feedforward and separate feedback connections.

Modern studies of cortical areas and their interaction have much higher resolution than those of Brodmann. In fact, for the brain of the rhesus macaque monkey, modern studies (Modha and Singh, 2010) catalog as many as 383 hierarchically organized areas with 6600 connection pathways between areas that form a small world graph with six degrees of separation. Humans have more areas and more interconnections, but let us use these numbers to perform some thumbnail calculations. Cortical areas can vary drastically in size. For the sake of discussion, let us also assume they are all of the same size. On average, then, the surface area of each cortical area is 6–7 cm^2. Recent studies have counted 16 billion neurons in the human adult male neocortex (Azevedo et al., 2009). Thus, on average there are 41 million neurons per cortical area. There are roughly 100 neurons in a minicolumn, suggesting roughly 400,000 minicolumns per cortical area. All of these cortical areas are built from minicolumns. Despite residing in different cortical areas, these minicolumns have very similar structure.

The brain contains a large diversity of neuron types. Within a minicolumn, there are two umbrella classes of neurons. These are excitatory and inhibitory neurons. From a biological perspective, these classes can be divided into many different subtypes of neurons using continually evolving taxonomies based on morphological features, electrophysiology, and expression of genes and proteins (Harris and Shepard, 2015). The abstract models in this chapter will use generic excitatory and inhibitory neurons. However, we will describe in a later section one detailed model of a cortical volume that uses over 200 highly detailed neural subtypes.

To understand the nature of human cognition, it is imperative to understand how the 300 plus cortical areas interoperate to form cohesive computations that might contribute to the AI set of functions. However, more fundamentally, although each cortical area computes a different function, it uses essentially the same quasi-universal building block, namely, the minicolumn. One goal of this chapter is to characterize the state-of-the-art concerning theories of how the minicolumn might function. We will emphasize theories that have connections to machine learning and statistical pattern recognition (Bishop, 2007). In the short run, these theories inform neuroscience. On a longer time scale, insights from the brain may inform theories of machine learning and pattern recognition, as has already happened in the field of deep learning (LeCun et al., 2015; Schmidhuber, 2015).

3.1 Ventral Visual Pathway

An important example of a network of cortical areas is the ventral visual pathway in the primate brain. Experiments have shown that this pathway performs visual object recognition. It consists of at least eight feedforward connected brain areas and is at least as many synapses downstream from the retina. Its function is specialized for visual object recognition and is distinct from visual areas for visual-motor coordination or navigating in space. This pathway begins with the retina and ends in area IT of the inferior temporal cortex, where the object recognition output is located. It is of interest to us both because of its function and because we will make direct comparisons to the state-of-the-art deep convolutional networks for object classification later in this chapter.

This pathway consists of several cortical areas connected in both a feedforward and feedback fashion. Successive layers in the feedforward pathway extract increasing complex features that are position and scale invariant. Neuroscientists have implanted electrodes in macaques and recorded the cell activity from neurons in area IT at the top of the feedforward path finding neurons whose responses are selective to complex objects such as cars, animals, and faces (Tanaka, 1996).

An impressive experiment showed that these biological neuron populations have enough information content to serve as training input to an artificial classifier. Specifically, the activity of small populations of neurons in area IT of macaque monkeys, recorded over small time intervals as low as 100 ms after stimulus onset, contained information about object identity and category (Hung et al., 2005). The feature portfolio was sufficiently complete that it could successfully be used as input to an artificial classifier. Specifically, the macaques were presented with 77 stimulus examples from five possible categories: toys, foods, monkey faces, cats and dogs, and vehicles. These authors showed that this information could be read out and used as feature input to an artificial classifier that was trained to either classify the object or recognize the object's identity. Categorization accuracy was as high as 80% and recognition accuracy was as high as 60% while showing invariance to a range of position and sizes. This makes it, in principle, possible for downstream brain areas which receive the outputs of IT to also act as classifiers.

4 THE CONCEPT OF A BASIC CIRCUIT

We have painted a picture that neocortical scalability is enabled by simply adding minicolumns which then group themselves into cortical areas that perform specialized computations. We now examine the circuitry within a minicolumn. We will use the term *minicolumn* to refer to the anatomical entity found in the neocortex and the term *basic circuit* or *neocortical microcircuit* to refer to neural subnetworks that may implement one or more information processing algorithms to support learning within the minicolumn. This section will review some models of basic circuits.

Let us start with one of the main elements of a basic circuit, the neuron. To a first approximation, neurons are nonlinear threshold devices adapted for long-distance signal transmission by emitting spikes (voltage pulses with duration about 1 ms) upon reaching threshold. By threshold, we mean a specific value of the membrane potential at a specific location in the neuron.

The membrane potential is the voltage difference (millivolts) between the inside and outside of the neuron, with the outside of the neuron taken as ground. A typical resting potential is about −65 mV and a typical threshold is about −55 mV. In abstract neural models, these values are rescaled to something more convenient. The spike generator is located at the beginning of the axon, that is, where the axon meets the soma of the cell. When the neuron generates a spike, the membrane potential suddenly and briefly jumps to a positive value (e.g., 30 mV) and then returns to the resting potential.

To a first approximation, the spike shape and amplitude do not convey information. The presence and timing of the spike is what matters. This insight leads to the question of how a neuron encodes information in its spike train. Does it depend on the firing rate of the neuron over some small time interval, or is it some kind of temporal code involving the timing of the spikes (e.g., like some form of Morse code). There are many nonexclusive possibilities and the question is not currently resolved (Gerstner et al., 1997; Shadlen, 2006).

After emitting a spike, the neuron enters an absolute refractory period of about 2 ms, where it is unable to generate another spike. This is because the sodium channels embedded in the cell membrane cause the spike generation event by opening. Once open, they close again after about 1 ms and cannot reopen for another 2 or 3 ms because of kinetic constraints. The actual spike generation dynamics are in fact very complicated (Izhikevich, 2004), but they have not made it into the abstract network models discussed in this chapter.

Neurons form networks by projecting to other neurons, which receive their output spikes. From a morphological viewpoint, a neuron's main components are soma, dendrites, and axon, as shown in Fig. 1. The soma generates the spike, the dendrites receive inputs from other neurons, and the axon is the output fiber that acts as a wire. That is, the axon broadcasts its output by

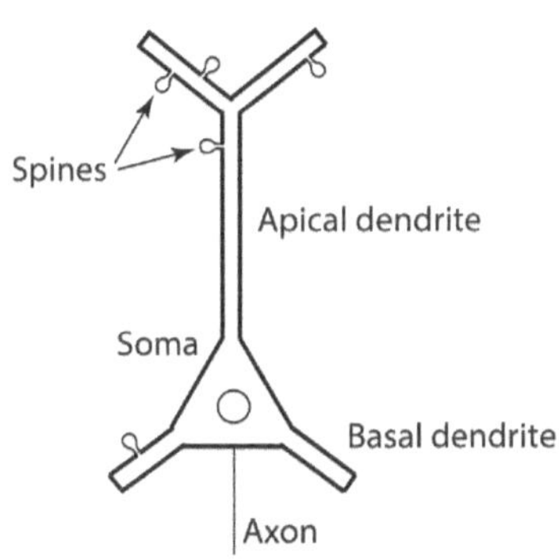

FIG. 1 A template for a pyramidal neuron. The spike is generated at the point where the soma meets the axon. The neuron is oriented vertically in the neocortex with the start of the axon pointing away from the skull. The inputs occur along the dendrites. The spines receive excitatory inputs and are the sites where synapses undergo learning.

branching and projecting to other neurons. From a network viewpoint, a neuron can be viewed as a node in a directed graph where the fan-out can be more than 100,000 in some parts of the brain (the cerebellum). For the principal neuron in the neocortex, the pyramidal neuron, the fan-out is typically 10,000. Theoretically, a deterministic spiking network can mimic a universal Turing machine (Maass, 1996).

Just as important as the neuron are the contact points between neurons. These are known as synapses. A synapse transfers an electrochemical signal from a presynaptic neuron to the postsynaptic neuron, influencing the postsynaptic neuron's probability of reaching threshold and emitting a spike. Using the directed graph model, the synapses are represented by the edges in the graph. The weights in an ANN coarsely model the strength of biological synapses (Haykin, 2008). Adaptive synapses between pyramidal neurons are the sites where learning occurs in the brain and where experienced-based knowledge is stored. In large-scale brain simulations, it is common to include 1000 or 10,000 times as many synapses as neurons (Modha et al., 2011).

The brain's computing elements are slow by computing standards. Everything at the neuronal level and above occurs at the millisecond time scale or longer. The baseline firing rate of a neuron is almost always much lower than 100 Hz. It is more common for the activity rate to be between 20 and 40 Hz. Another rule of thumb is that it takes about 10 ms for a signal to travel across a synapse to go from one cortical area to another. With these constraints, it may be shocking to learn that neurons in IT can make a binary classification such as *animal* vs *nonanimal* within 100 ms after stimulus onset. Since IT is about 10 synapses downstream from the retina, the neurons within each cortical area in the feedforward pathway are constrained to make their decision within about 10 ms after receiving their local input. With this constraint, there is only time for each area to receive at most one spike from each of its inputs. Despite the existence of feedback circuitry in the ventral pathway, there is no time to use it to help with the immediate classification decision. These considerations of slow computing elements being used as building blocks for networks that have fast response times force us to use highly constrained models. One such model is the time-to-first-spike classification model (Masquelier and Thorpe, 2007, 2010).

Presumably, within the minicolumn are one or more basic circuits that underlie the highly general learning which supports the scalability of the neocortex. We now consider how the minicolumn and basic circuit(s) may be embedded within the neocortex. Recall that the neocortical sheet is about 3 mm in thickness. From the vantage point of histology and cytoarchitectonics, it is classically considered to have six layers. Some computer scientists count four layers using the viewpoint of functional circuitry (Modha et al., 2011).

The principal neurons in the neocortex—pyramidal neurons—are oriented vertically and span some or all of these layers. A schematic depiction of a pyramidal neuron is shown in Fig. 1. In the brain, pyramidal neurons are often found grouped into small clusters adjacently interlinked by inhibitory neurons. This is the source of the term minicolumn and it is tempting to hypothesize that locally encoded within the minicolumn are the basic circuits that implement key learning computations.

5 ABSTRACTIONS OF CORTICAL BASIC CIRCUITS

We will now discuss methods for modeling cortical basic circuits at a high level of abstraction.

5.1 Abstract Neurons

All network models discussed in this section are spiking neural networks in contrast to traditional ANNs that use a sigmoidal or rectified linear output function. In abstract models, there are two common model neurons in use. The simplest is the leaky integrate-and-fire (LIF) neuron. The LIF neuron is modeled as a parallel RC circuit that charges in response to its input.

It has a voltage threshold and when this is reached, the circuit emits a spike and then resets to its resting level, which is usually zero. The LIF neuron has a time constant that is associated with its RC circuit. If the resistance in the circuit is high, so that leakage is negligible, then it is said to be an IF neuron instead of an LIF neuron. The capacitor model is motivated by the fact the cell membrane is a good insulator and is very thin, so has properties of a parallel plate capacitor. For increased biological realism, an absolute refractory period can be implemented by simply constraining the LIF neuron not to emit another spike for 2 or 3 ms whenever it emits a spike. The LIF model neuron, in its simplest form, is deterministic and it gives the same output on repeated trials with the same input. Under some conditions, the temporal precision of the spike times of an isolated neuron can be extremely high (Jolivet et al., 2006). However, biological neurons in vivo generally show trial-to-trial variability. A more sophisticated model that captures this variability is the Poisson spiking neuron, explained below.

In a Poisson spiking model neuron, the neuron stochastically emits discrete spikes as its output. A spiking neuron simulation usually advances in 1-ms time steps. On a particular time step, the neuron either emits a spike (1) or not (0). The spiking activity of a neuron exhibits trial-to-trial variability and baseline spiking activity. For modeling purposes, spike events can be generated by a Poisson process. The firing rate of the neuron is controlled by the rate parameter of an inhomogeneous Poisson process (Ermentrout and Terman, 2010). These models are classified "as simplified conceptual models amenable to mathematical analysis" (Gurstner et al., 2012).

5.2 Excitatory and Inhibitory Neurons

The roughly 16 billion neurons in the neocortex can be divided into two broad categories. Excitatory neurons enhance the firing probabilities of neurons to which they project and inhibitory neurons do the opposite. Excitatory neurons compose about 80–85% of the neurons in the neocortex and inhibitory neurons compose about 15–20%. Excitatory neurons correspond to artificial neurons that have only positive output weights (synapses) and, similarly, inhibitory neurons only negative (inhibitory) output weights. This constraint is known as Dale's law. This "law" posits that individual neurons use only one type of output neurotransmitter, which is either excitatory or inhibitory. The weights found in ANNs that can switch between excitatory and inhibitory are not found in biology. Of the excitatory neurons in the neocortex, most are pyramidal neurons. The results of learning are encoded in the excitatory synapses of the pyramidal neurons (Mountcastle, 1997). The inhibitory synapses of the nonexcitatory neurons do not appear to undergo learning. In the case of feedforward spiking networks, no computational power is lost if learning only occurs within the synapses of excitatory neurons (Maass, 2000).

5.3 Receptive Fields and Visual Feature Maps

Neurons in early sensory pathways have a property known as their local *receptive field* (RF), a feature which is mimicked in the architecture of convolutional neural networks. We will limit our discussion to the visual pathway although neurons for other sensory modalities can also have receptive fields. A classical visual receptive field of a neuron is the region of visual space that plays a role in the firing of the neuron, along with a description of the structural properties of the visual space that drive the neuron.

To clarify this, consider Fig. 2. Fig. 2C shows how an RF for a ganglion cell in the retina is reconstructed in a cat or a monkey. The output layer of

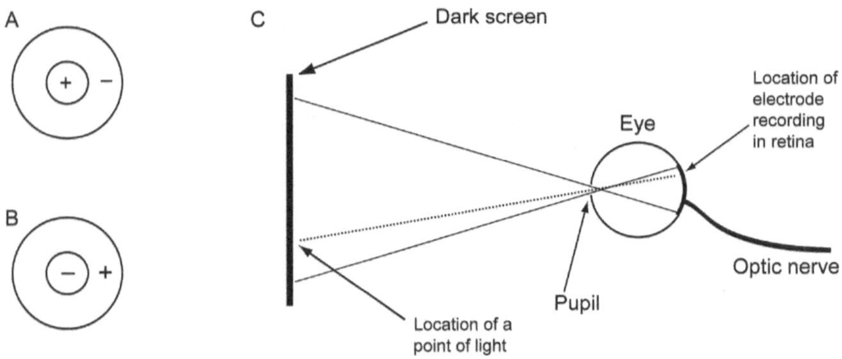

FIG. 2 (A) On-center, off-surround receptive field. (B) Off-center, on-surround receptive field. (C) The point-wise process of reconstructing the receptive field for a neuron in the retina.

the retina is composed of retinal ganglion cells and optic nerve is made of axons from retinal ganglion cells. To reconstruct the RF, a recording electrode is attached to the cell. A point of light is flashed onto different parts of a dark screen, while the eye fixates at the center of the screen. Locations on the screen that affect the firing rate of the cell are part of the cell's RF regardless of whether the light increases or decreases the cell's firing probability. In terms of structural properties, locations that enhance the firing of the cell are said to be excitatory regions of the RF. Areas that diminish the firing of the cell are inhibitory regions. The RFs of retinal ganglion cells have a center-surround structure as shown in Fig. 2A and B. The receptive field comes in two varieties, both of which are circularly symmetric having an excitatory region and an inhibitory region. The width of an RF, as measured by the visual angle to which it is sensitive, varies with position on the retina, giving highest resolution to the fovea (the region on the retina which processes the center of the visual field). Since neurons do not have negative firing rates, pairs of opponent cells, such as on-center/off-surround vs off-center/on-surround, can compensate for this.

After leaving the retina via the optic nerve, the visual pathway passes through the LGN, which is a relay station that recodes the input into a form that the neocortex can understand, and then projects to the visual cortex. Machine-learning-based vision systems rarely, if ever, model the LGN. We have stated earlier that nuclei in the thalamus, such as the LGN, recode the sensory input so that it is understandable by the neocortex. When we get into the first area of the visual cortex, V1, receptive fields can become more complicated. Although there are a number of RF types in V1, the RFs of *simple* and *complex* cells (Hubel and Wiesel, 1962, 1968) have exerted the most influence on the field of machine learning and convolutional networks. The prototypical receptive field of a simple cell resembles a localized, oriented, edge detector and the receptive field of a complex cell resembles the same edge detector with some amount of position invariance. Such cells, whether simple or complex, are said to be orientation selective and there are many theories of how such cells emerge through development. For an influential paper, see Olshausen and Field (1996). The complex cells are downstream from the simple cells. This has been the inspiration of convolutional neural networks beginning with Fukushima (1980) where the motif of alternating simple/complex cell layers is extended deep into a many-layered network.

In modern convolutional networks (ConvNets), the simple cells form a convolutional layer and the complex cells form a subsampling or pooling layer. Layers are said to represent visual feature maps, a term that was imported from cognitive psychology (Triesman, 1977). These maps are also said to be topographic. This concept is most intuitive on the retina where neighboring locations in the visual field project to neighboring locations in the retina. This property is preserved in early areas in the visual cortex with the caveat that the surface of the map can be deformed by stretching and shrinking topologically. For example, the visual field covered by the foveal area on the retina enjoys relatively much more surface area in a V1 map than the visual field in the periphery of the retina. That is, relatively more minicolumns or processing elements are devoted to the fovea than to areas of the peripheral visual field. Topographic deformations are also found in other areas of neocortex, especially the somatosensory and motor cortex that represents input from and output to the body surface. For example, in these areas much more of the cortical surface is devoted to representing and controlling the hands and face than the rest of the body surface.

The prototype feature map is a two-dimensional array, although there are higher dimensional feature maps in both the brain and in ConvNets. Over a small dynamic range, the simple cells are linear and shift invariant, so they can be approximated by a convolution filter kernel. Suppose one convolves a 3×3 kernel with an $n \times n$ image using "valid" mode. The output is then an $n-3+1 \times n-3+1$ image. The kernel can model a center-surround cell, a simple cell, or anything that can be modeled as a linear, shift-invariant filter (complex cells cannot be modeled in this way). The $n-2 \times n-2$ output is a map of features that the filter detects. This might correspond to a cortical area in the brain or a two-dimensional layer of artificial neurons in a ConvNet. This immediately suggests layers of hierarchical processing that is found in convolutional networks and the layers are sometimes called feature maps or image maps. Furthermore, the notion a neural receptive field can be applied to neurons that reside in very deep feature maps. The receptive field is the set of input units that play a role in determining the designated neuron's activity level. The practical use of such a construct is that it can be used to reconstruct the set of input features to which a neuron in a deep layer is sensitive. We will also talk about the immediate RF (local input neighborhood) of a unit, which is the set of neurons in the previous layer that affect the unit's firing properties.

With this background, we can look at Fig. 3. The figure shows the main building block of a state-of-the-art ConvNet. The building block is the pyramid-shaped part of the diagram. The bottom surface of the pyramid is a feature map (S-map) represented by simple cells and the top surface is a feature map (C-map) represented by complex cells. In this particular figure, the S-cells take their input from the retina. As explained above, the S-cells can be modeled using a convolution filter applied to a small neighborhood from the previous layer as their immediate RF input. The weights form the filter and are shared by each S-cell.

The C-cells can be modeled in at least two ways. For the purpose of ConvNets, the modeling takes some form of downsampling or pooling. This gives translation invariance for the S-map features, gives the C-units broader RFs, and reduces the size of the feature map for the C-units. For example, the maximum value in some neighborhood of the S-map could be used at a location in the C-map. In C-maps we also need to be aware of the stride of the units in the

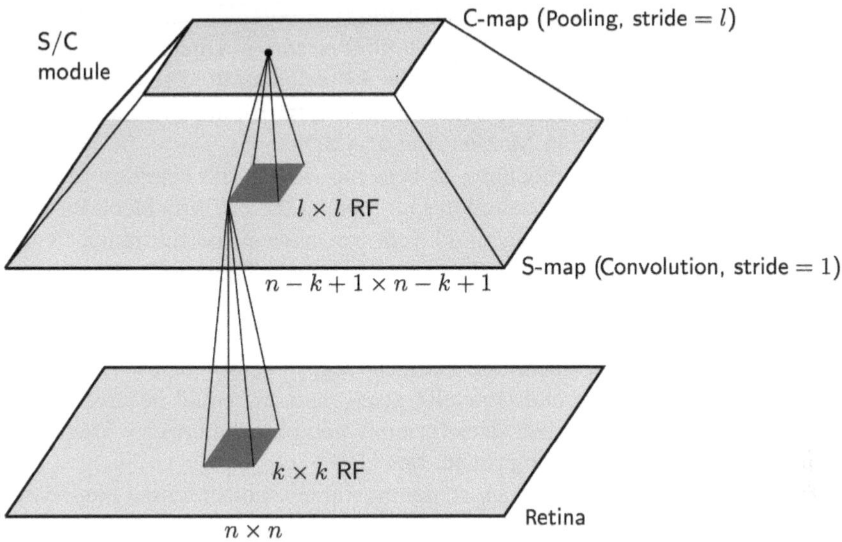

FIG. 3 A simple/complex cell (S/C) module that takes a retina as input. The immediate RFs of units in the S-map and C-map are shown. The C-map of an S/C module can serve as the input to a higher-level S/C module. *Adapted from Lee, H., Grosse, R., Ranganath, R., Ng, A.Y., 2011. Unsupervised learning of hierarchical representations with convolutional deep belief networks. Commun. ACM 54 (10), 97.*

layer. The stride is the distance between immediate RF (input neighborhood) centers. For C-maps, the stride is often equal to the width of the immediate RF. This causes the C-map to partition the space on the S-map. This rule can be broken to obtain what is known as overlapping pooling. We will see an example of this with the description of AlexNet, given in a later section. This approach captures a significant amount of functionality and is compatible with the learning mechanism of modern ConvNets.

Although Fig. 3 only shows one S/C module, the retina can project to many S/C modules, each with different filters, to obtain a large number of different feature maps. For instance, there can be feature maps for orientation selectivity, direction selectivity (direction of movement), and color. In practice, there are many feature maps in each layer of a ConvNet.

A second method to model complex cells has also been used. This method may offer more biological realism in modeling the visual cortex but it is not used in ConvNets. In this approach, complex cells are implemented by pairs of simple cells. The simple cells, respectively, implement sine (odd) and cosine (even) Gabor filter kernels (Adelson and Bergen, 1985). The activation of a complex cell is then given by

$$c_{r,\theta}^{x,y} = \left(o_{r,\theta}^{x,y}\right)^2 + \left(e_{r,\theta}^{x,y}\right)^2 \qquad (1)$$

In the above, c, o, and e are the activations of complex cells, odd Gabors, and even Gabors, respectively. The symbols x and y denote RF centers, r denotes the RF radius, and θ denotes the RF orientation. Theoretically, the complex cell is the sum of two squared quadrature filters. Complex cells of this type have been used in McManus et al. (2008).

Having explained the modeling of both the simple and complex cells, let us resume the topic of the pyramid in Fig. 3 being the building block for deep ConvNets. The input to the pyramid does not have to be the retina. It can instead be the output (C-map) of another pyramid. The pyramids can then be stacked to form deep networks of arbitrary depth, detecting increasingly complex features, until a 1×1 feature map is produced. Because each pyramid uses a subsampling operation, items on deeper feature maps will acquire more position invariance and larger RF sizes. This increased position invariance and larger receptive field sizes for more complex features are also found in the ventral pathway of the primate brain.

In this section, we have discussed the most important structural property of a ConvNet, but we have not yet discussed learning. The technical breakthrough with recent ConvNets is the ability to efficiently train networks with many layers, so-called deep networks. This will be discussed in a later section.

5.4 Optimal Representations with Respect to Natural Image Statistics

It has been stated many places that, from a theoretical viewpoint, simple cells are oriented and resemble a Gabor filter that is bandpass in one direction and low-pass in the orthogonal direction. It has been shown that purely statistical methods can learn representations with very similar properties.

The statistical approaches to analyzing representations in area V1 seek to build theoretically optimal representations for the stimuli that V1 evolved to deal with, namely, natural scenes. With this, the biologically observed representation can be compared with the theoretically optimal representation. An optimal representation reduces the inferential complexity of the representation while maintaining its fidelity (no loss of information). This can be formalized as maximizing the mutual entropy between the input image and the output representation (preserve fidelity), while minimizing the joint entropy among the variables within the representation (maximize independence).

The representations we consider are linear combinations of basis functions and take the following form:

$$I(x,y) = \sum_i a_i \phi_i(x,y)$$

The above states that the gray-scale intensity of an image at point (x,y) is a linear weighted sum of a set of basis functions, ϕ_i, evaluated at (x,y). One can think of the basis functions as simple cell activations in V1. The goal is to find an optimal set of basis functions and, after that, compare them with the biologically observed simple cells to obtain insights into the functional properties of the simple cells.

There have been two prominent approaches toward achieving optimal representations. Both approaches try to preserve representational fidelity. They differ in how they keep the representation simple. The first approach tries to minimize the reconstruction error while achieving a sparse representation, where most of the a_i's in the representation have values close to zero (Olshausen and Field, 1996). The second approach tries to maximize the mutual entropy between the input images and the output representations while minimizing the joint entropy among the a_i's (Bell and Sejnowski, 1997). This is known as independent components analysis (ICA) (Bell and Sejnowski, 1995). Both approaches lead to almost equivalent optimal representations that are quite similar to the simple cells found in V1. In a very detailed comparison, it was shown that ICA-constructed bases can match simple cell receptive fields in orientation tuning, spatial frequency, aspect ratio, and length (van Hateren and Shaaf, 1998). Both approaches consider only a single-layer representation. This is adequate for asking questions about V1. When addressing similar questions for cortical areas later in the processing stream, such as in regard to area V2 which receives inputs from V1, then we must use hierarchical representations.

The sparsity-based approach seems to enjoy more compatibility with deep neural networks, so that is what we will emphasize in this section. To obtain a sparse representation, one can perform gradient descent on a cost function which takes the form below (Olshausen and Field, 1996). As already stated, the cost function tries to make a representation sparse while preserving the reconstruction accuracy:

$$\text{cost} = E_{\text{recon}} - \lambda \cdot M_{\text{sparsity}}$$

The variable E_{recon} is the reconstruction error, λ is a parameter greater than zero that controls the relative importance of the sparsity to the cost, and M_{sparsity} is the sparsity measure. The reconstruction error is given below:

$$E_{\text{recon}} = \sum_{x,y} \left[I(x,y) - \hat{I}(x,y) \right]$$

The expression $\hat{I}(x,y)$ denotes the reconstructed image value at position (x,y), specifically

$$\hat{I}(x,y) = \sum_i a_i \phi_i(x,y)$$

A very simple sparsity measure is given below:

$$M_{\text{sparsity}} = -\sum_i |a_i|$$

The intent is to keep the values of the a_i coefficients near zero. More complex sparsity measures can be found in Olshausen and Field (1996) as well as the details of the gradient descent algorithm to train the ϕ_i.

5.5 Softmax and Winner-Take-All Network

An important building block found in both biologically motivated models and purely artificial models, including ConvNets, is the softmax function. With respect to biology, the softmax function is a very convenient model of a so-called winner-take-all (WTA) network. A WTA is an idealization of a set of excitatory neurons mutually interconnected by inhibitory neurons (reciprocal or lateral connections) as shown in Fig. 4. The mutual inhibition creates a competition that makes it difficult for more than one or a few excitatory neurons to maintain a high activity level. In the figure, the neurons labeled z_k denote units that are interconnected by local, mutual inhibition. The z_k units model idealized pyramidal neurons that receive inputs from outside the WTA and also project to neurons outside the WTA. Such a network can be conveniently modeled by the softmax function (also called the normalized exponential), which captures the WTA functionality while avoiding the network implementation details. The softmax function appears below:

$$p(z_i \text{ fires}) = \frac{\exp(u_i)}{\sum_k \exp(u_k)} \tag{2}$$

In the above, we write $p(z_i \text{fires})$ to signify that the value falls in the range (0,1) and that the values over $p(z_k \text{fires})$ sum to 1, thus creating a foundation to perform probabilistic computations. The notation u_i denotes the membrane potential of neuron or unit, i. This will be used in an expectation maximization, spike-based model discussed later.

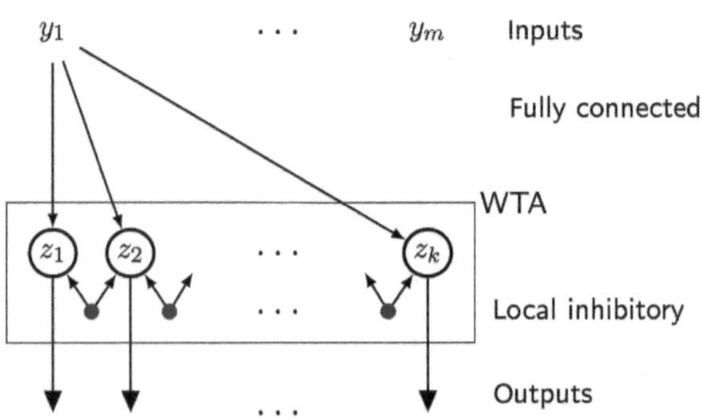

FIG. 4 WTA consisting of K excitatory units, z_k, mutually interconnected by lateral inhibition. In addition, all z-units receive input from external y-units. One or a few z-units will be highly active. The activity of the remaining z-units will over time be suppressed by mutual inhibition.

An approximate *stochastic spiking WTA* can be built from softmax by letting u_i denote the instantaneous membrane potential of excitatory neuron i in the WTA (Nessler et al., 2013). If we assume the spike rate is governed by a Poisson process whose rate parameter is given by the instantaneous membrane potential (taking into account the inhibitory neurons), then the equation above denotes the probability that neuron i fires given some neuron in the set K fires at a given time step. This type of WTA, when combined with a spike-based learning rule, will be used in liquid computing microcircuits and in expectation maximization microcircuits.

5.6 Learning in the spike domain

In the area of ANNs, one of the most common learning algorithms is backpropagation of errors. To date, no one has proposed a plausible implementation of backpropagation in the brain.

In biology, it has been empirically observed that many synapses in the mammalian nervous system adjust their efficacy according to the recent history of timing differences between pre- and postsynaptic spikes (Caporale and Dan, 2008). In brief, if the presynaptic neuron fires slightly before the postsynaptic neuron (e.g., say 15 ms), the synapse is strengthened (LTP) and if the postsynaptic neuron fires slightly before the presynaptic neuron, the synapse is weakened (LTD). This is a form of unsupervised learning that likely explains some of the self-organizing properties of the brain. An idealized plot of the empirically observed functional relationship is shown in Fig. 5 for a single pair of pre- and postsynaptic spikes. The right side of the

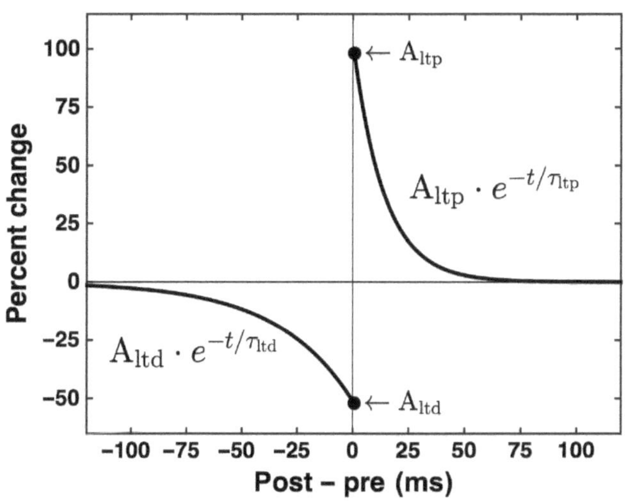

FIG. 5 Idealized plot of experimentally observed STDP. The *graph* shows percentage change in synaptic effectiveness as a function of the relative spike time of pre- and postsynaptic spike events.

graph shows synaptic strengthening, or long-term potentiation (LTP), where the postsynaptic neuron fired after the presynaptic neuron. The left side of the graph shows synaptic weakening, or long-term depression.

The above-mentioned learning is known as spike-timing-dependent plasticity (STDP). In this chapter, we will see several demonstrations of the ability of STDP-type learning rules to form cell assemblies, approximate expectation maximization (EM), and perform unsupervised feature discovery.

5.7 A Spiking EM Classifier

One of the simplest proposed cortical microcircuits is a spiking classifier. A spiking WTA when combined with the appropriate STDP learning rule can be used to build an unsupervised 1-out-of-K classifier. The developers of this model argue that this functionality can be embedded in a cortical microcircuit (Nessler et al., 2013) as a means to explain microcircuit functionality. Additionally, the mechanism within the classifier has the attractive theoretical property that it approximates expectation-based maximization (EM) learning (Nessler et al., 2009).

The inputs to the network are class exemplars whose feature vector components, x_m, consist of samples from a multinomial distribution. For the present discussion, we will assume the samples come from a binomial distribution. These feature vectors need to be somehow represented using spiking neurons. Thus, they are recoded to a population spike code that meets the input requirements of the EM implementation. Specifically, each recoding box has two outputs. If a feature $x_m = 1$, then one of the corresponding y outputs fires at 40 Hz (output produced by a Poisson process) and the other fires at 0 Hz. If $x_m = 0$, then the opposite pattern happens.

The y-units project to the output layer of the classifier which is a K-unit spiking WTA consisting of units z_k. The inputs to the WTA are z-unit membrane potentials, u_k, as discussed previously in terms of Eq. (2). The u_k's are defined by the expression below:

$$u_k(t) = \sum_{i=1}^{n} w_{ki}\widetilde{y}_i(t) + w_{k0} \tag{3}$$

In the above, $\widetilde{y}_i(t)$ denotes the membrane potential of y-unit i at time t. The membrane potential is 1 if the y-unit fired within the previous 10 ms; otherwise it is 0. The weight w_{ki} denotes the strength of the synaptic connection from y-unit i to z-unit k. The weight w_{k0} is the bias weight for z-unit k. The winning z-unit, which is allowed to spike at time step t, is chosen from the softmax distribution given in Eq. (2).

It remains to discuss the learning mechanism. Learning occurs in the w_{ki} weights used in Eq. (3). The weight adjustment rule uses a variant of STDP in which the magnitude of any LTP depends on the current value of the weight. For weights w_{ki}, the learning rule is given below:

$$\Delta w_{ki} = \begin{cases} exp(-w_{ki}) - 1, \text{ if } \tilde{y}_i(t^f) = 1 \\ -1, \text{ otherwise} \end{cases} \tag{4}$$

The above rule is applied to the incoming weights of a z-unit when it fires. If the z-unit does not fire, the weights are not adjusted. The symbol t^f denotes the firing time of the z-unit. Case 1 occurs if and only if y-unit i fires within 10 ms before z-unit k fires. This is the case of LTP. In the second case, the z-unit has fired but the y-unit has not fired within the previous 10 ms. In this case, the weight is decreased by -1. This is LTD (long-term depression). The weight adjustments are also governed by a positive rate parameter whose magnitude decreases as a function of the number of weight adjustments. The bias weights also have a learning rule (Nessler et al., 2009) which we will not discuss in this chapter.

One counterintuitive property of the weights is that they are operate in the negative regime. It can be shown that weight w_{ki} converges to the natural logarithm of $P(\tilde{y}_k = 1 | z_k = 1)$.

In summary, the network has two layers of stochastic neurons that generate Poisson spike trains. The units in the first spiking layer are y-units. These units encode input feature vectors to be classified. The second layer is composed of z-units that represent classification categories after the network has been trained. The layers are fully feedforward connected from layer y to z by weights trained according to an STDP learning rule. In addition to the feedforward connections, the z-units obey a type of WTA discipline that is implemented by the softmax function. The model has been applied to the emergence of orientation selectivity in the visual cortex and to unsupervised classification in the MNIST data set of normalized handwritten digits. In more recent work, the model has been extended with recurrent excitatory synapses (Kappel et al., 2014). This extended motif captures more of the biology of a neocortical microcircuit. This extends the functionality of the previous model so that it can learn to detect sequential structure within spike input sequences and approximate hidden Markov model learning. This model has also been used as a building block in recent versions of the liquid-state machine model, as explained in the next section.

5.8 The Liquid Computing Model

The space of possible representations in the neocortex is larger than the set of possible rate codes or temporal codes. The ubiquitous feedback (or recurrent) connections within a cortical minicolumn form a dynamic system with fading memory for its recent input history. The field of reservoir computing (Lukosevicius and Jaeger, 2009) shows that representations within the brain may take the form of trajectories within a dynamic system. The liquid computing model is a biologically motivated approach within the field of reservoir computing that aims to capture these properties (Maass et al., 2002).

The original liquid computing model does not address the issue of learning power within the minicolumn, but it does address the issue of the representational power of a minicolumn. More recent liquid computing models do address the issue of learning power (Klampfl and Maass, 2013).

A reservoir model consists of three components: (1) a sensory-based time-varying input stream; (2) a recurrent neural network known as a reservoir; and (3) one or more linear readout units that can be trained to extract information from the reservoir. In a liquid computing model, the reservoir is called a liquid (because its dynamics represents transient sequences of network states) and its structure is constrained to model the abstract structure of a cortical microcircuit. The role of the readout units is to demonstrate that various types of information are successfully encoded within the liquid and, because of the linearity of the readout, can be easily decoded. The liquid obtains its representational power by converting the temporal dynamics of the input stream into a high-dimensional spatial pattern embedded in the liquid.

To illustrate, let us examine a particular liquid computing model whose parameters are chosen to fit the somatosensory cortex of the rat (Maass et al., 2002). The model reflects the observation that a minicolumn has many recurrent circuits involving a diverse neural population consisting of 80% excitatory neurons and 20% inhibitory, as previously discussed. This particular model uses 270 LIF neurons with refractory periods of 3 ms for the excitatory and 2 ms for the inhibitory. The time constant is 30 ms for both. Neurons are given spatial locations so that their connection probabilities can be modeled as an exponentially decreasing function of the distance separating them. The connections are kept sparse in order to avoid chaotic dynamics within the networks. The synaptic connections have their own temporal dynamics (dynamic synapses) which we will not discuss. There was no training within the liquid. In general, recurrent neural circuits are difficult to train. The readout neurons could be trained to calculate nonlinear functions of the input and to perform spoken digit classification (Verstraeten et al., 2005), thereby establishing a lower bound on the representation power of the liquid.

Although classically, liquid-state models are not hard-coded to perform a task, more recent models have experimented with adding unsupervised learning to the liquid in order to improve its reservoir properties (Norton and Ventura, 2006). The main reservoir properties are *separation* and *approximation*. Together the properties guarantee the computational power of the corresponding liquid-state models. Informally, separation refers to the difference in trajectories within the liquid that are caused by a difference in input streams. Approximation applies to the readout function and its ability to approximate an arbitrary function.

A recent version (Klampfl and Maass, 2013) of the liquid computing model adds learning and more structured network motifs to the mix. These extensions are based in part on the spiking EM model discussed in the previous section. Specifically, the structured motifs are the stochastic WTA circuits mentioned in the previous section and the learning is the type of STDP learning given in Eq. (3). Recall that the Maass group argued that the spiking EM model could be embedded in a cortical microcircuit. In the present model, the spiking EM model is a building block of a more comprehensive microcircuit. The resulting model is highly consistent with neurophysiological data and can account for the acquisition of dynamic memory traces. One example simulation consists of a recurrent network of 50 WTA circuits distributed over a 10×5 spatial grid. Each WTA consists of K units, where K can be different for each WTA. K was drawn independently from a uniform distribution with minimum and maximum depending on the particular simulation. In one simulation, the range was from 2 to 10 neural units, and in another the range was from 10 to 50. The spacing on the grid was used to control the connection probability between the WTA circuits in a manner similar to that in which connection probabilities between two nearby neurons in the brain vary with their physical separation. Specifically, the probability of a connection between two units is an exponentially decaying function of the spatial distance between the units on the grid.

Input to the network consists of repeated spike trains embedded in Poisson noise. The task of the network is to learn a stereotyped response to the repeated inputs. The response takes the form of activation sequences of neurons recruited to represent that particular input.

5.9 Unsupervised Feature Discovery

We now return to the issue of learning within a spiking ConvNet. A spiking convolutional network, equipped with a form of STDP learning, has been used to perform feature discovery for classification (Masquelier and Thorpe, 2007). It uses three stages of processing: edge extraction, intermediate-level feature discovery, and classification. The modules are described below.

1. The edge-extraction module is hard-wired and does not use learning. It uses four banks of Gabor filters to extract edge segments at four orientations. The spike-time outputs are then processed to give some degree of location invariance.
2. A feature discovery module uses STDP learning. The discovered features are processed to obtain location invariance, similar to what was done in the edge-extraction module.
3. A classifier module such as an SVM or a linear classifier which is trained to use labeled data to make classification decisions using the discovered features as input.

The system has different modes of operation depending on whether it is being trained or being used as a classifier. There are two training modes. The first is *feature acquisition* where the unsupervised STDP rule is used to discover intermediate-level features. In this mode, only the first two modules are used and learning takes place in the feature discovery module. The second learning mode takes place after feature acquisition completes. In this mode, the classifier is trained via supervised learning (feature learning is turned off). Finally, the third mode is performance mode when the system is used to classify novel inputs. In this mode, all learning is turned off and classification accuracy is assessed.

In recent years, the distinction between training and performance has acquired increased practical importance in deep networks. Training a deep convolutional network has become much more compute intensive than deploying it after it has been trained. Tools have become available which make it easy to deploy trained networks which operate in performance mode. Such trained networks can operate on many compute devices, including mobile devices and web pages.

5.9.1 Edge Extraction

Details of the edge-extraction module are shown in Fig. 6. The input to the network is a gray-scale image. The module processes edges at four orientations. In this model, Gabor filters are convolved with the input at the four orientations to yield four orientation-specific edge maps. The receptive field sizes for the Gabor filters are 5×5, yielding four feature maps of size $n-4 \times n-4$. Next, there is a cross-orientation WTA competition to ensure that, at each location, only one edge dominates. For each pixel, only

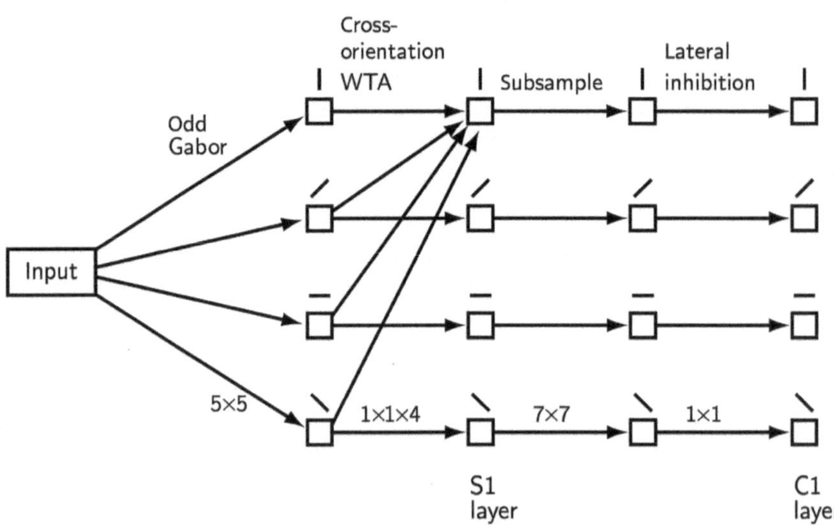

FIG. 6 Information flow within the Masquelier/Thorpe edge detection module. The *input box* represents a two-dimensional image. *Other boxes* represent two-dimensional orientation-specific feature maps. Each layer computes four maps at four orientations. *All arrows* represent image processing steps. The cross-orientation WTA layer is fully connected but not *all arrows* are shown. Receptive field dimensions are shown in the *bottom row*.

one pixel across the four orientations is allowed to emit a spike. This yields four edge maps representing the first simple cell layer, called S1. The outputs of the S1 layer are converted to spike times by a function that ensures that the earliest spike time corresponds to the maximum output.

The model also uses complex cells. The position invariance characteristic of a complex cell is implemented by subsampling within its immediate receptive field neighborhood. Specifically, a complex cell in the model has a 7×7 neighborhood with a stride of 6. Since the model uses spike-time coding, the earliest spike time corresponds to the maximum input. After this, the complex cells undergo local lateral (radial) inhibition. These outputs correspond to the C1 layer depicted in Fig. 6. We have seen that the information flow in the edge-extraction module is primarily feedforward, but there are also lateral interactions within the feedforward flow. These interactions include the WTA competition (S1 layer) and the lateral inhibition (C1) layer.

In the model, the module operates at five spatial scales. Specifically, the network is duplicated at scales: 100%, 71%, 50%, 36%, and 25% of actual size. Cross-scale interactions do not occur within the edge-extraction module. The module simply generates 20 C1 feature maps deriving from four orientations times five spatial scales.

5.9.2 Feature Discovery

The function of the feature discovery module is to acquire domain-relevant features to support the classification task. Let us assume the system is configured to acquire n distinct intermediate-complexity features, such as facial features to support the task of face vs nonface classification. Also, temporarily assume that the network only uses one spatial scale at 100% of actual size. The S2 layer will compute a *separate feature map* for each of the n acquired features. Incoming weights to the S2 units undergo unsupervised learning. An individual unit in an S2 map will represent the feature associated with the map at a specific location in the image. An intermediate-level feature will be composed from the outputs of the four C1 maps provided by the edge-extraction module.

Each S2 unit within an S2 map receives input from a 16×16 neighborhood taken from the four maps computed by the C1 layer corresponding to one processing scale for all four orientations. Thus the immediate receptive field for an S2 unit has dimension $16 \times 16 \times 4$. That is, an intermediate-level feature is made of edges spanning a 16×16 neighborhood across all four orientations. The weight matrix associated with this immediate receptive field is shared by all units in this particular S2 map. This weight-sharing scheme is the mechanism by which all S2 units in the map represent the same feature, but at different locations. These weights are trained using (unsupervised) STDP and this underlies the feature discovery mechanism.

The above description assumed one spatial scale of analysis. However, the network is built to analyze five spatial scales. The outputs of the edge-extraction module for a given spatial scale become the inputs to the feature discovery module for that spatial scale. Although the edge-extraction module analyzed its input at five spatial scales, there were no cross-scale interactions within that module. Cross-scale interactions do not come into play until processing within the S2 layer. These interactions are implemented by cross-scale weight sharing of the $16 \times 16 \times 4$ weight matrix.

To make this more explicit, in the multiscale version of the feature discovery module, there are five S2 maps for each intermediate-complexity feature. Each map reflects a different spatial scale for that feature. However, all five layers share the same weight matrix for that feature. The immediate receptive field associated with the weight matrix covers a larger percentage of the retinotopic map for coarser scale images. The weight matrix can be viewed as a position and scale-invariant prototype for a particular feature.

During feature-acquisition mode, an image from the positive set is presented to the network and some of the S2 weight matrices are adjusted by the STDP rule. If n distinct S2 maps are allocated to acquire n intermediate-complexity features, it is necessary that all n maps acquire different features so that coverage of the feature space is maximized for the resources allocated. This is achieved by allowing competitive interactions between the maps and between the neural units within a map. Before STDP is applied to the weights for an S2 layer, competitive mechanisms are applied to determine which S2 neurons are allowed to control learning. These interactions are summarized below.

1. Within a map, only the earliest firing S2 unit (if it reaches threshold, the S2 unit is an IF neuron) adjusts the shared weights for that map. This is enforced by a 1-WTA mechanism. Since the weights are shared by all S2 units in the map, the effects of the weight adjustments apply to all units within the map, thereby enabling the scale-invariant learning of that feature. The earliest firing unit is associated with the strongest input signal for that pattern and presumably provides the best information for learning the pattern.
2. There is lateral inhibition between different S2 maps. When an S2 cell within a particular map fires, it prevents all other S2 cells at that scale and within an 8×8 radius (even those within a different map) from firing.
3. Finally, there is a 2-WTA competition across maps within a scale. From the first rule listed, only one unit within a map can fire. The current rule states that, within a scale, at most representatives from two separate maps can fire. Winners, again, are determined by their earliest spike times.

The above rules specify when an S2 unit is selected to adjust its weights. Once selected, an idealized STDP rule is used where the weight modification is uninfluenced by the delay magnitude between pre- and postsynaptic spike times. Because of this, only the spike-time rank order between pre- and postsynaptic neurons affects the feature-acquisition phase of learning.

5.9.3 The Weight Modification Rule

This section specifies the weight modification (idealized STDP) rule for an S2 unit that wins the competition. The Masquelier and Thorpe weight modification rule is given below:

$$\Delta w_{ji} = \begin{cases} a^{+} w_{ji}\left(1 - w_{ji}\right), t_j - t_i \geq 0 \\ a^{-} w_{ji}\left(1 - w_{ji}\right), t_i - t_j < 0 \end{cases}$$

In the above, t_i and t_j are the spike times of the pre- and postsynaptic neurons, respectively. The symbols a^{+} and a^{-} denote gain paramaters, with $a^{+} > 0$ and $a^{-} < 0$. Initially, the value of $a^{+} = 2^{-6}$ and is multiplied by 2 every 400 postsynaptic spikes. As the value of a^{+} changes, the ratio of a^{+} to a^{-} is maintained at −4/3. Weight values are constrained to remain between 0 and 1.

5.9.4 C2 Units

There will be one C2 unit for each intermediate-complexity feature. The unit receives input from the five S2 feature maps representing that feature. The spike time of this C2 unit is the spike time of the unit with the earliest spike across all the incoming S2 maps. This yields a feature detector which is globally shift invariant and scale invariant across the five spatial scales. If there are n intermediate-complexity features, there will be n of these C2 units. These n spike-time values are feature vectors which form the input to the categorization module.

5.9.5 Limitations of the Model

In this particular network model, learning occurs only in one layer of the network. This network has not been shown how to scale STDP learning to train multiple layers of a network. This appears to still be an open question. Existing deep spiking convolutional networks are derived from nonspiking convolutional networks (Cao et al., 2015), which are in turn trained by back propagation. In particular, a nonspiking deep convolutional network is trained on some task and then the trained network is translated to a spiking network. The motivation to translate it to a spiking network is so that it can run on low-power hardware. The basis for the translation is the observation that a highly active unit intuitively corresponds to a neuron with a high firing rate. Two recent theoretical attempts to bridge the gap between back propagation and STDP are found in Bengio et al. (2015a,b).

5.9.6 Extensions of the model

The model has been extended to handle a much larger number of C2 features (Kheradpisheh et al., 2016). When used with the 3D-object data set (Savarese and Feifei, 2007) and 400 C2 features, the model achieved state-of-the-art classification performance as compared to the classic HMAX system (Serre et al., 2007). Acquired intermediate-level features were analyzed using dissimilarity matrices, hierarchical clustering, and mutual information.

Other research has explored a probabilistic variant of the STDP rule with the intention of acquiring more insight into why such rules might be effective (Tavanaei and Maida, 2016).

6 LARGE-SCALE CORTICAL SIMULATIONS

Until now, we have only looked at abstract models of cortical circuitry. However, we must also look at the status of highly detailed brain simulations. There are many obstacles to a brute force attempt at reverse engineering the brain via a highly detailed brain simulation.

First, the brain has immense biological detail (Kasthuri et al., 2015) and our knowledge of these details is incomplete. Second, as we learn new biological details, it is unclear how they contribute to brain computations without performing careful simulations. We need a critical mass of biological detail and modeling infrastructure to obtain meaningful results. Adding biological details to assess their impact requires enormous effort in software development and computing power. Third, because of the complex data produced by a large-scale simulation, there is no standardized procedure to distill the simulation results into a tractable abstract model.

6.1 The Blue Brain Digital Reconstruction

To date, the most biologically detailed reconstruction and simulation of a neocortical brain circuit (Markram et al., 2015) comes from the Blue Brain Project. This project represents a long-term multinational effort to build a detailed digital reconstruction and simulation of a small, but essential, piece of neocortical microcircuitry. It includes data on neural morphologies, layer heights, neuronal densities, ratios of excitatory to inhibitory neurons, and so forth. The model has been validated on additional data that were not used to create the reconstruction.

The reconstruction is so detailed, it has been referred to as virtual brain tissue. The reconstruction is modeled after the somatosensory cortex of a juvenile rat. It places 31,000 neurons with high morphological complexity into a small virtual neocortical volume of 0.01 mm^3 spanning all six layers of the neocortex. The simulation contained 310 minicolumns. The simulation uses 37 million dynamic (but nonplastic) synapses. Simulations of this virtual brain tissue can be directly compared with both in vitro (petri dish) and in vivo (living animal) neuroscience experiments using live neural tissue.

Whereas the abstract models we have so far reviewed in this chapter confined themselves to only two types of model neurons (excitatory and inhibitory), the Blue Brain model used 55 layer-specific morphological types in combination with 207 morphoelectrical neuron types using highly accurate compartmental models of their morphology and electrophysiology. They are modeled using Hodgkin-Huxley-type equations (Ermentrout and Terman,

2010) and an extended version of the Neuron simulation tool (Carnevale and Hines, 2005). The simulation runs billions of equations at millisecond resolution using the Blue Brain IV supercomputer (Schurmann et al., 2014). There also exists a collaboration portal for this project (Ramaswamy et al., 2015).

Although the details of the reconstruction were guided by experimental data, the simulations were not designed to predict any specific microcircuit properties. Rather, their purpose was to study emergent phenomena that arose within the simulations without parameter tuning. Several findings were obtained. The most significant was that simulations showed a spectrum of network activity states. Unsurprisingly, a microcircuit can dynamically reconfigure its state "to support diverse information processing strategies" (Markram et al., 2015, p. 456). Sharp transitions between synchronous and asynchronous activity states were observed. Using the reconstruction, it was possible to isolate the mechanism responsible for such transitions. It was found that calcium concentrations influenced synaptic function which, in turned, affected the balance between excitation and inhibition within the microcircuit.

The volume of neocortical tissue simulated seemed to form a critical mass of circuitry. The functional properties of the network could not be obtained with a smaller amount of circuitry. This provides evidence that this network size (310 minicolumns) is the minimum functional unit for cortical processing.

Although this reconstruction does not make a direct contribution to the field of cognitive computing, it does represent a fundamental advance in our understanding of and our ability to model neocortical microcircuits and this, in the long run, may offer deep contributions to artificial intelligence. Simulations of this kind have many long-term goals such as therapeutic drug discovery and simulation of disease processes.

7 HARDWARE SUPPORT FOR BRAIN SIMULATION

Spiking neural networks are amenable to implementations in integrated circuits with extremely low power consumption (Maass, 2015; Merolla et al., 2014). The former authors have built a neurosynaptic chip, called True North, consisting of 1 million spiking neurons running in parallel at a clock speed of 1 kHz. It consumes power at a level which is orders of magnitude lower than the equivalent spiking network running on a conventional machine. It has also been shown that backpropagation-trained ConvNets, discussed below, can be translated into spiking ConvNets implemented in hardware, leading to a two order of magnitude improvement in energy efficiency (Cao et al., 2015).

8 DEEP LEARNING NETWORKS

Deep networks are so called because they have many layers of trainable weights using nonlinear computing elements. Deep learning networks offer state-of-the-art machine learning performance and have spearheaded breakthroughs in speech recognition, visual object recognition, and other areas (LeCun et al.,

2015). One subtype of deep learning network, the deep convolutional network or *ConvNet*, is of special interest in this chapter because of their structural resemblance to the ventral visual pathway of the primate visual system. Even more impressive is that these artificial networks rival the representational power of the actual ventral visual pathway in the brain (Cadieu et al., 2014).

This section will cover stochastic deep belief networks, convolutional deep belief networks, and finally a hybrid convolutional deep belief network.

8.1 Stochastic Deep Belief Networks

Deep belief networks (DBNs) are multilayer stochastic neural networks that are built by stacking restricted Boltzmann machines (RBMs). RBMs are trained in an unsupervised fashion and have a probabilistic semantics. Their relationship to probability theory is based on mathematics adapted from statistical mechanics. An energy value can be assigned to the state of the network, which it uses to represent some information. The Boltzmann distribution is then used to convert the value of the energy function to a probability, thereby assigning a probability to the network state.

DBNs are trained in two phases (Hinton, 2007). The first is called pretraining and uses an unsupervised method based on contrastive divergence. The second is called fine-tuning and uses backpropagation, if supervised training data are available. This section describes the pretraining method.

8.1.1 Restricted Boltzmann Machine

An RBM is a type of stochastic network. In a stochastic network, the units are updated according to a probability function. The RBM is a special case of a Boltzmann machine constrained so that training and probabilistic inference are less computationally intensive. An RBM is a two-layer network. The first layer is the visible layer and the second is the hidden layer. Both layers consist of binary stochastic units. The layers are connected by trainable symmetric weights. The weights are trained using unsupervised learning to model a probability distribution using the method of contrastive divergence, which is an approximation to log likelihood that is less computationally intensive to compute. The absence of lateral connections within the visible and hidden layers is what makes this an RBM in contrast to a full Boltzmann machine.

Let us first look at the probabilistic semantics of the network. Let the vectors $\boldsymbol{v}$ and $\boldsymbol{h}$ denote a particular set of values for the units in the visible and hidden layers of the network, respectively. The hidden units always take on binary values. The visible units can take on real values but, for this exposition, we will restrict ourselves to binary values. This is an example of a state. The probability of a given state is specified by the formula below:

$$P(\boldsymbol{v},\boldsymbol{h}) = \frac{1}{Z} e^{-E(\boldsymbol{v},\boldsymbol{h})}$$

Z denotes a normalizing constant and $E(\boldsymbol{v},\boldsymbol{h})$ denotes the energy associated with state $(\boldsymbol{v},\boldsymbol{h})$, which is defined below:

$$E(\boldsymbol{v},\boldsymbol{h}) = -\sum_i \sum_j v_i w_{ij} h_j - \sum_j b_j h_j - \sum_i c_i v_i$$

v_i and h_j are activations of the visible and hidden units, respectively. w_{ij} are weights from the visible to the hidden units. Recall that these weights are symmetric. b_j and c_i are biases to the hidden and visible units, respectively.

The training procedure is derived from the probabilistic semantics just given. We will present the training procedure without the derivation. The goal of training is to give the hidden layer the ability to reconstruct samples from the input distribution that are applied to the visible layer. The contrastive-divergence learning rule used to update weight w_{ij} is given below:

$$\Delta w_{ij} = \epsilon \cdot \left(\langle v_i h_j \rangle_{\text{data}} - \langle v_i h_j \rangle_{\text{recon}} \right)$$

ϵ is the positive learning rate parameter and the angle brackets indicate the expected value of the data and the reconstruction, respectively. The weights are initialized to 0.

Before proceeding, because there are no lateral connections within either of the layers, the posterior probabilities conditioned on $\boldsymbol{v}$ and $\boldsymbol{h}$, respectively, factorize, as shown below:

$$P(\boldsymbol{h}|\,\boldsymbol{v}) = \prod_j P(h_j|\,\boldsymbol{v})$$

$$P(\boldsymbol{v}|\,\boldsymbol{h}) = \prod_i P(v_i|\,\boldsymbol{h})$$

The above relationships will make our subsequent calculations much simpler.

To compute the expected values for the data, $\langle v_i h_j \rangle_{\text{data}}$, set the v_i's to the values of the pixels in some binary training image. Next, use the formula below to update the hidden units probabilistically. This can be done in parallel over the hidden units because, as we have seen, the hidden unit states are conditionally independent of each other given the state of the visible layer. The formula specifies a stochastic update. Namely, it specifies the conditional probability of any unit, h_j, taking a value of one, where σ denotes the logistic sigmoid function:

$$P(h_j|\,\boldsymbol{v}) = \sigma\left(\sum_i w_{ij} v_i + b_j \right)$$

With the absence of lateral connections in the hidden layer, the states of the hidden units are independent of each other, given the visible state. To obtain data samples to estimate $\langle v_i h_j \rangle_{\text{data}}$, inputs are clamped to the visible layer and a set of hidden unit activations are calculated using the above formula. The resulting value $v_i h_j$ will be a sample.

To obtain samples to estimate $\langle v_i h_j \rangle_{\text{recon}}$, a slightly different method is used. We start by performing the two steps described above. That is, clamp an input image to the visible units and then compute the values or states of the hidden units. Having obtained the hidden unit states, the values of the visible units are reconstructed using the formula below. Again because there are no lateral connections in the visible layer, the states of the visible units are independent of each other, given the hidden state:

$$P(v_i|\boldsymbol{h}) = \sigma\left(\sum_j w_{ij} h_j + c_i\right)$$

Again, the product $v_i h_j$ is used, but with the reconstructed value of v_i instead of the original value that was clampled to the input image. For estimating the reconstruction, we have described one cycle of what is known as block (alternating) Gibbs sampling. The reconstruction estimate improves if more cycles are used. Having obtained expected values for both the data and the reconstruction, we can apply the weight update rule. As training continues, the reconstructed v_i's will tend to improve their match to the clamped v_i's.

8.1.2 Deep Belief Network

RBMs can be used to build and train DBNs that have more representational power than an RBM alone. This is done by stacking the RBMs on top of each other and then training them one layer at a time, starting from the bottom. That is, to train a DBN, we train it layer by layer, starting from the first hidden layer. Once a particular layer has been trained, the weights and biases are frozen for that layer and the next hidden layer above it is trained. The inferred values from the previous hidden layer are used as data to train the next hidden layer. The inferred values are obtained by clamping an input image to the visible layer and performing a feedforward update. The process is repeated until the top layer is trained. After a DBN is trained, we have learned a hierarchy of feature detectors without supervision.

Fig. 7 shows a DBN developed to recognize handwritten digits and associate digit labels with the visual patterns (Hinton et al., 2006). Training is both supervised and unsupervised. Training of the first two hidden layers uses the procedure that we described. The top layer, consisting of 2000 units, receives input from the second hidden layer and also from 10 label units. To train the top layer, the units in the label layer and the units in the second hidden layer are concatenated to give 510 data components.

After training, the network can be run in one of two modes. In recognition mode, we can present an image to the visible layer, run the network, and see what digit label becomes active. Alternatively, in dreaming mode, we can turn a particular digit label on, run the network, and examine the patterns in the visible layer. This will tell us what the network thinks or imagines, say, the digit 6 looks like.

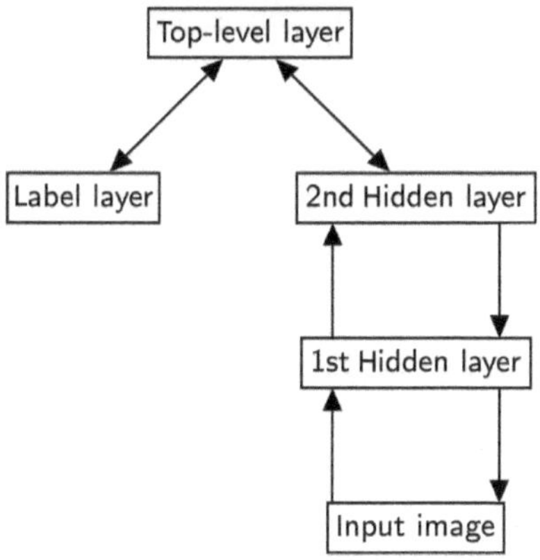

FIG. 7 Hinton et al.'s DBN for MNIST.

8.1.3 Sparse DBNs

Although DBNs already create hierarchical representations, evidence exists that adding sparsity improves their representational effectiveness. We describe a case study that builds a two-layer sparse-variant DBN and serves as a model of cell response properties in cortical area V2 which is one of the output targets of area V1 (Lee et al., 2008). The key insight underlying this model is to add a sparsity component to the RBM cost function in a manner similar to that described in Section 5.4. This yields a cost function for layer, l, which is the sum of two components as shown below:

$$\text{cost}_{\{w_{ij},c_i,b_j\}} = -\sum_{i=1}^{m} \log \sum_{\boldsymbol{h}} P\left(\boldsymbol{v}^{(l)}, \boldsymbol{h}^{(l)}\right) + \lambda \cdot \sum_{j=1}^{n} \left| p - \frac{1}{m} \sum_{l=1}^{m} \mathrm{P}\left(h_j^{(l)} | \boldsymbol{v}^{(l)}\right) \right|^2$$

In the above, m is the training set size, n is the number of hidden units, and p is a target value for the expected activation of the hidden units, sometimes called the target sparsity. This last term, p, is set to a low fixed level such as .005.

The first component of the cost function is the negative log likelihood which can be optimized using the contrastive divergence approximation and the second component is a sparsity regularization term which can be optimized using gradient descent. Training finds parameter values $w_{i,j}$, c_i, and b_j to minimize the cost. Training proceeds layer by layer as with the standard DBN. However, training of the RBM must take into account the sparsity term. The training algorithm for the RBM repeats the two steps below until convergence.

1. Update parameters using contrastive divergence as explained in previous section.
2. Update parameters using gradient descent on the sparsity term.

The results of the simulation are that model V2 neurons are acquired that capture some of the biologically observed response properties of V2 neurons in the monkey (Hegde and Van Essen, 2000). The reader can consult the original work to see the resulting receptive field maps. In contrast to oriented stimuli, the responsiveness to more complex stimuli is strongly characteristic of response types in area V2. The authors do not claim from this that the mechanisms in V2 literally form a sparse RBM. Rather, they note that it may be more indicative of complex shapes such as corners, and junctions being characteristic of the statistics of natural images. That is, there may be several different algorithms that embody the same computational theory of the problem, in the sense of Marr (1982).

8.2 Deep ConvNets and Machine Learning

ConvNets offer a means of training a deep-layered architecture, such as that found in the ventral visual pathway, and prior experience with neural network training shows this is a nontrivial task (Tanaka, 1996). ConvNets are distinguished by the following properties: "local connections, shared weights, pooling, and the use of many layers" (LeCun et al., 2015, p. 439). They require labeled data in large amounts appropriate to the number of trainable parameters in the network, and are normally trained by backpropagation.

There was a breakthrough event in ConvNet machine learning performance in 2012. The event established the dominance of ConvNets to perform visual object classification. In part, the breakthrough was enabled by the availability of a large set of labeled images provided by the large-scale visual object recognition challenge (LSVRC) ImageNet database (Deng et al., 2009). The relevant subset consisted of over 1.2 million high-resolution images with class labels for 1000 categories. An architecture, now known as AlexNet, achieved, by a large margin, the best performance in both the 2010 and 2012 versions of the data set (Krizhevsky et al., 2012).

The network was trained by backpropagation to maximize the multinomial logistic regression objective, had eight trainable layers, and a total of 60 million adjustable parameters. The output layer used a softmax function with a 1000 components to account for the 1000 labeled categories. The images were downsampled to a size of 256×256. Although the input images have dimension 256×256, the input layer has dimension $224 \times 224 \times 3$, including RGB. This apparent discrepancy will be explained when we discuss data augmentation. The full architecture consisted of five convolutional layers followed by three fully connected layers, distributed over two GPUs. The method for using backpropagation (stochastic gradient descent optimization) to train ConvNets was developed by LeCun et al. (1989). When stacked sequentially, these layers of convolutional/pooling modules lead to units that are selective to increasingly complex features, while being invariant to position, scale, and other deformations. Since the network had 60 million trainable parameters, the main challenges were to train the network in a reasonable amount of time and to reduce overfitting of the 1.2 million labeled image training set.

The breakthrough performance of AlexNet is attributable to several specialized techniques that enhance the performance of ConvNets with regard to learning speed, generalization, and the reduction of overfitting. Some of the more essential methods are listed below.

First, the activation function for the units in the network is the rectified linear unit (ReLU), defined below:

$$f(x) = \max(0, x)$$

Its value is 0 for $x < 0$ and it increases linearly for positive x. Its gradient is 0 for $x < 0$ and 1 for $x > 0$. It has been shown on some data sets to speed up backpropagation training by a factor of six in comparison with using sigmoidal units like $\tanh x$ and $\text{logsig}\, x$. It was introduced by Jarrett et al. (2009). In the context of RBMs, a noisy variant of ReLUs for hidden units, $\max(0, x + \mathcal{N}(0, \sigma))$, can also learn better features for object recognition in comparison to the use of binary hidden units, explained in an earlier section (Nair and Hinton, 2010). The noise is sampled from a zero-mean Gaussian distribution with standard deviation as a hyper parameter.

Second, the network was trained on more than one GPU. Training on large DBNs can be impractical without GPU facilitation. Modern GPUs can read and write to each others memory directly. Use of multiple GPUs is now standard in off-the-self training tools with speedups by a factor of 50 being common. AlexNet used two GPUs and took several days to train.

Third, the regularization technique of "dropout" reduces overfitting (Srivasta et al., 2014). The purpose of dropout is to prevent coadaptation of hidden units. Coadaptation refers to the phenomenon that the incoming weights to two separate hidden units become trained to work together. Although this appears desirable, it is better for one hidden unit to work well with a diversity of random hidden units. Dropout is implemented by randomly removing some percentage of hidden units from the network for each mini batch. Removal is applied to each hidden unit with some fixed probability p, of say 50%, and consists of thinning the network by removing the chosen units along with their incoming and outgoing weights. Thus each mini batch trains a different thinned network. The dropout prevents the units from coadapting with each other, thereby allowing them to function more independently. At test time, some sort of averaging scheme must be used over the thinned networks. An approximate, but computationally efficient way to do this, is to use the full network without dropout. However, the weights are multiplied (scaled) with the dropout probability. This technique has been shown to lower the generalization error. The cost of dropout is that it increases the number of training iterations to converge by approximately a factor of 2.

The fourth technique to improve network performance is data augmentation. Although there are about 1 million training images, there are 60 million trainable parameters. Without some form of data augmentation, the network overfits the training data. The training data set size can be increased drastically by introducing label preserving transformations applied to randomly selected image patches. For training, 2048 random patches (translations and reflections) are sampled from each training image and used as input to the network. For testing, a total of 10 reflections and translations on image patches of size 224×224 taken from the test image are used. There are five translations sampled from the center and four corners of each 256×256 image. The predictions of the 10 patches are averaged by the network's softmax layer. AlexNet also uses a second data augmentation scheme which, because of space limitations, we will not discuss.

Fifth, it was found that overlapping pooling aids generalization in AlexNet. Recall from Fig. 3 that pooling layer accepts an immediate RF of size $l \times l$. In many ConvNets, the sampling stride of these RFs is l, so they do not overlap. In AlexNet the size of l was three and the stride was two, so that the immediate pooling RFs had some overlap.

In summary, these networks can be said to provide coarse models of the subset of connections between ventral pathway cortical areas that are feedforward connected. Features of ConvNets that do not have analogs in the brain are the following.

1. They do not have feedback connections.
2. The need for nearly astronomical amounts of labeled training data.
3. The brain does not use the backpropagation algorithm, although it may use some learning algorithm that optimizes a very similar cost function.
4. Within the ConvNet architecture, the combination of convolution/pooling layers is biologically plausible, but there is no clear analog of a cortical microcircuit.
5. The backpropagation-trained ConvNets are not spiking neural networks. However, there has been some work on converting already trained ConvNets into equivalent spiking ConvNets (Cao et al., 2015).

8.3 Convolutional Sparse DBNs

One of the main practical limitations of ConvNets is the need for extremely large amounts of labeled training data. Indeed, a major challenge in training AlexNet was to avoid overfitting despite the availability of over 1 million items of labeled data. DBNs, as we have seen, have the desirable property that feature hierarchies can be trained in an unsupervised fashion. Because of their large number of parameters, they still need large amounts of training data, but they do not need to be labeled training data and, therefore, the data are more readily available.

Up until this point in our exposition, the DBNs, however, did not encode (with shared weights) prior geometric knowledge as was done with ConvNets. This is one of the reasons for the practical success of ConvNets. It has been shown that such knowledge can be incorporated into a DBN, yielding a convolutional DBN (Lee et al., 2011). Unsurprisingly, convolutional DBNs (CDBNs) use convolutional RBMs as their basic building block. The innovations needed to do this are (1) devise an energy function for the convolutional layer so it can be interpreted probabilistically; and (2) similarly provide a probabilistic interpretation for the pooling layer.

To see how these two innovations come into play, let us revisit Fig. 3. As we have seen, this is the architectural motif that underlies the ConvNet. Training will proceed in a bottom-up, stack-based fashion as with the all DBNs. Now, however, a stack element now has two types of layer pairs. From the figure, we see that there are two pair types: retina/S-map and S-map/C-map. To convert this to DBN friendly terminology, we would say visible/hidden and hidden/pooling.

The first step is to extend the RBM energy function to use shared weights so that it is compatible with a ConvNet architecture. The new energy function is then used to obtain new conditional distributions by marginalizing on $\boldsymbol{v}$ and $\boldsymbol{h}$, respectively. Block Gibbs sampling can then be used as before to compute the gradient according to the formulas below (Lee et al., 2011):

$$P\left(h_{xy}^{k}=1\,|\,\boldsymbol{v}\right)=\sigma\left(\left(\tilde{W}^{k}*v\right)_{xy}+b_{k}\right)$$

$$P\left(v_{xy}=1\,|\,\boldsymbol{h}\right)=\sigma\left(\left(\sum_{k}W^{k}*h^{k}\right)+c\right)$$

Fig. 8 makes explicit that a retinal or visible layer can project to more than one S-map or module. In the above formulas, k denotes the module under consideration. W^k denotes the weight matrix for module k, $\tilde{W}$ means that the array is flipped horizontally and vertically, and the asterisk indicates convolution.

Next, we provide a probabilistic interpretation for a pooling unit in a C-map. We enforce the following constraint: at most one of the S-cells in

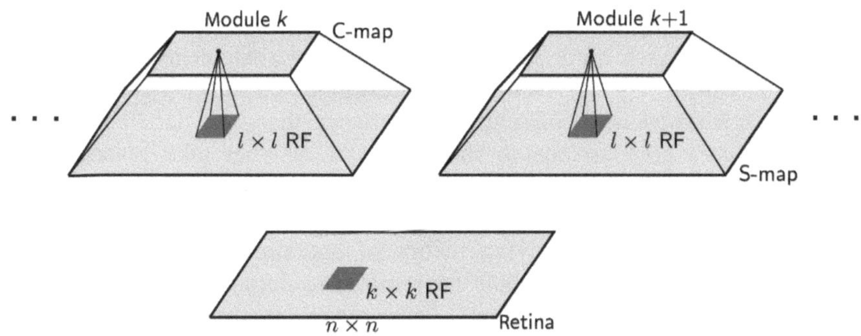

FIG. 8 A retinal neighborhood can project to K different module (feature) maps each with its own weight matrix. Lee et al. (2011) used a three-layer system with many modules in each layer.

the immediate RF of a pooling unit may have value 1. The pooling unit is on if and only if exactly one of the S-cells is on. With this constraint, we can treat the collection of l^2+1 units as a softmax random variable. This allows us to build an energy function for this pair of layers and then appropriate conditional probability update formulas (details in Lee et al., 2011). Training also includes sparsity regularization as defined in Section 8.1.3.

The reader can consult the original work to see visualizations of trained weight matrices of the second and third layers of the CBDN for the Caltech-101 data set that consists of faces, cares, airplanes, and motorbikes. The network successfully learned an object part hierarchy as can be seen by inspecting the visualized bases for the respective layers.

9 SUMMARY AND CONCLUSION

This chapter offered a perspective on the state of the art of spiking neural network models and deep learning models as providing insight into the workings of the cortical microcircuits and visual areas of the brain. We have summarized models that attempt to provide insight into cortical microcircuits and also into computations that may occur within the ventral visual pathway.

In regard to cortical microcircuits, we have outlined how they might be modeled as spiking networks that use STDP to perform EM classifications and how liquid-state machines that encode rich representations of input activity that can be read out by simple linear classifiers. We have seen that STDP can be incorporated within a liquid-state machine to allow it to acquire stereotyped responses to repeated input. In contrast to these idealized models, we have also looked at a highly detailed biological simulation of 310 microcolumns in adolescent rat somatosensory cortex. There is still a large gap between matching up the idealized models with the biologically detailed models.

We have looked at the problem of unsupervised feature learning in spiking neural networks. Although successful when a single learning layer is used, it has not been yet shown to scale to multilayer networks. It still remains an open question as to how to train a deep spiking network.

Learning is much more successful using more abstract models of deep neural networks. By far, the most successful performancewise is the backpropagation-trained convolutional network that uses labeled data. It suggests direct comparisons to the computations that take place in the ventral visual pathway. Perhaps more interesting are the probabilistic DBNs that are built by stacking RBMs. They have been extended to create convolutional DBNs. Their underlying theory of operation is more developed. Their units have two states, which have more similarity to spiking neurons, and they have feedback connections. These models might serve as a guide to the development of stochastic spiking neural networks to perform similar tasks.

Although we have emphasized the workings of the brain, the recent success of ConvNets may provide a means to bridge the signal-to-symbol divide (Murphy, 2012, p. 1007). This is more commonly known as the symbol-grounding problem (Harnad, 1990) and is concerned with how symbols get their representational authority. Without this, symbol-based cognitive systems operating in the real world, in contrast to a formalized artificial world, do not have a robust and reliable semantics.

REFERENCES

Adelson, E.H., Bergen, J.R., 1985. Spatiotemporal energy models for the perception of motion. J. Opt. Soc. Am. 2, 284–299.

Allman, J., 1991. Evolution of neocortex. In: Jones, E.G., Peters, A. (Eds.), Cerebral Cortex, vol. 8a. Plenum Publishing, New York, pp. 269–283.

Azevedo, F., Carvalho, L., Grinberg, L., Farfel, J., Ferretti, R., Leite, R., et al., 2009. Equal numbers of neuronal and nonneuronal cells make the human brain an isometrically scaled-up primate brain. J. Comp. Neurol. 513, 532–541.

Bell, A.J., Sejnowski, T.J., 1995. An information-maximization approach to blind separation and blind deconvolution. Neural Comput. 6, 1129–1159.

Bell, A.J., Sejnowski, T.J., 1997. The "independent" components of natural scenes are edge filters. Vis. Res. 37, 3327–3338.

Bengio, Y., LeCun, Y., 2007. Scaling learning algorithms toward AI. In: Bottou, L., Chapelle, O., DeCoste, D., Weston, J. (Eds.), Large-Scale Kernel Machines. MIT Press, Cambridge, pp. 321–360.

Bengio, Y., Fischer, A., Mesnard, T., Zhang, S., Wu, Y., 2015a. From STDP towards biologically plausible deep learning. In: Deep Learning Workshop, International Conference on Machine Learning (ICML) 2015, Lille.

Bengio, Y., Lee, D., Bornschein, J., Lin, Z., 2015b. Towards biologically plausible deep learning. arXiv:1502.04156.

Bishop, C., 2007. Pattern Recognition and Machine Learning. Springer, New York.

Buxhoeveden, D., Casanova, M., 2002. The minicolumn hypothesis in neuroscience. Brain 125, 935–951.

Cadieu, C.F., Hong, H., Yamins, D., Pinto, N., Ardila, D., Solomon, E., et al., 2014. Deep neural networks rival the representation of primate IT cortex for core visual object recognition. PLoS Comput. Biol. 10, e1003963.

Cao, Y., Chen, Y., Khosla, D., 2015. Spiking deep convolutional networks for energy efficient hardware. Int. J. Comput. Vis. 113, 54–66.

Caporale, N., Dan, Y., 2008. Spike timing-dependent plasticity: a Hebbian learning rule. Annu. Rev. Neurosci. 31, 25–46.

Carnevale, N., Hines, M., 2005. The NEURON Book. Cambridge University Press, Cambridge.

Deng, J., Dong, R., Socher, R., Li, L., Li, K., Fei-Fei, L., 2009. ImageNet: a large-scale hierarchical image database. In: IEEE Conference on Computer Vision and Pattern Recognition. IEEE, pp. 248–255.

Domingos, P., 2015. The Master Algorithm: How the Quest for the Ultimate Learning Machine Will Remake Our World. Basic Books, New York.

Ermentrout, G.B., Terman, D.H., 2010. Mathematical Foundations of Neuroscience. Springer, New York.

Fukushima, K., 1980. Neocognitron: a self-organizing neural network model for a mechanism of pattern recognition unaffected by shift in position. Biol. Cybern. 36, 193–202.

Gerstner, W., Kreiter, A., Markram, H., Hertz, A., 1997. Neural codes: firing rates and beyond. Proc. Natl. Acad. Sci. U.S.A. 94, 12740–12741.

Gurstner, W., Sprekeler, H., Deco, G., 2012. Theory and simulation in neuroscience. Science 338, 60–65.

Harnad, S., 1990. The symbol grounding problem. Phys. D 42, 335–346.

Harris, K., Shepard, G., 2015. The neocortical circuit: themes and variations. Nat. Neurosci. 18, 170–181.

Haykin, S.O., 2008. Neural Networks and Learning Machines, third ed. Prentice Hall, Upper Saddle River, NJ.

Hegde, J., Van Essen, D.C., 2000. Selectivity for complex shapes in primate area V2. J. Neurosci. 20, RC61–RC66.

Hinton, G.E., 2007. To recognize shapes, first learn to generate images. In: Cisek, P., Drew, T., Kalaska, J. (Eds.), Computational Neuroscience: Theoretical Insights into Brain Function. Elsevier, Amsterdam, pp. 535–548.

Hinton, G.E., Osindero, S., Teh, Y., 2006. A fast learning algorithm for deep belief nets. Neural Comput. 18, 1527–1554.

Hubel, W.H., Wiesel, T.N., 1962. Receptive fields, binocular interaction, and functional architecture in the cat's visual cortex. J. Physiol. 160, 106–154.

Hubel, W.H., Wiesel, T.N., 1968. Receptive fields and functional architecture of monkey striate cortex. J. Physiol. 195, 215–243.

Hung, C.P., Kreiman, G., Poggio, T., DiCarlo, J.J., 2005. Fast readout of object identity from macaque inferior temporal cortex. Science 310, 863–866.

Izhikevich, E.M., 2004. Which model to use for cortical spiking neurons? IEEE Trans. Neural Netw. 15, 1063–1070.

Jarrett, K., Kavukcuoglu, K., Ranzato, M., LeCun, Y., 2009. What is the best multi-stage architecture for object recognition? In: Proceedings of the International Conference on Computer Vision (ICCV'09). IEEE.

Jolivet, R., Rauch, A., Lüscher, H., Gerstner, W., 2006. Predicting spike timing of neocortical pyramidal neurons by simple threshold models. J. Comput. Neurosci. 21, 35–49.

Kappel, D., Nessler, B., Maass, W., 2014. STDP installs in winner-take-all circuits an online approximation to hidden Markov learning. PLoS Comput. Biol. 10 (3), 1–22.

Kasthuri, N., Hayworth, K.J., Berger, D.R., Schalek, R.L., Conchello, J.A., Knowles-Barley, S., et al., 2015. Saturated reconstruction of a volume of neocortex. Cell 162 (3), 648–661.

Kheradpisheh, S.R., Ganjtabesh, M., Masquelier, T., 2016. Bio-inspired unsupervised learning of visual features leads to robust invariant object recognition. Neurocomputing 205, 382–392.

Klampfl, S., Maass, W., 2013. Emergence of dynamic memory traces in cortical microcircuit models through STDP. J. Neurosci. 28, 11515–11529.

Krizhevsky, A., Sutskever, I., Hinton, G., 2012. ImageNet classification with deep convolutional neural networks. In: Pereira, F., Burges, C., Bottou, L., Weinberger, K. (Eds.), Advances in Neural Information Processing Systems, vol. 25. Curran Associates, Red Hook, NY, pp. 1097–1105.

Krüger, N., Janssen, P., Kalkan, S., Lappe, M., Leonardis, A., Piater, J., et al., 2012. Deep hierarchies in the primate visual cortex: what can we learn for computer vision. IEEE Trans. Pattern Anal. Mach. Intell. 35 (8), 1847–1871.

Kurzweil, R., 2013. How to Create a Mind. Penguin Books, New York.

LeCun, Y., Boser, B., Denter, J.S., Henderson, D., Howard, R.E., Hubbard, W., et al., 1989. Handwritten digit recognition with a back-propagation network. In: Proceedings of the Advances in Neural Information Processing Systems, pp. 396–404.

LeCun, Y., Bengio, Y., Hinton, G., 2015. Deep learning. Nature 521, 436–444.

Lee, H., Ekanadham, C., Ng, A.Y., 2008. Sparse deep belief net model for visual area V2. In: Platt, J.C., Koller, D., Singer, Y., Roweis, S.T. (Eds.), Advances in Neural Information Processing Systems, 20. Curran Associates, Red Hook, NY, pp. 873–880.

Lee, H., Grosse, R., Ranganath, R., Ng, A.Y., 2011. Unsupervised learning of hierarchical representations with convolutional deep belief networks. Commun. ACM 54 (10), 95–103.

Lukosevicius, M., Jaeger, H., 2009. Reservoir computing approaches to recurrent neural network training. Comput. Sci. Rev. 3 (3), 127–149.

Maass, W., 1996. Lower bounds for the computational power of networks of spiking neurons. Neural Comput. 8 (1), 1–40.

Maass, W., 2000. On the computational power of winner-take-all. Neural Comput. 12 (11), 2519–2535.

Maass, W., 2015. To spike or not to spike: that is the question. Proc. IEEE 103 (12), 2219–2224.

Maass, W., Natschlager, T., Markram, H., 2002. Real-time computing without stable states. Neural Comput. 14 (11), 2531–2560.

Markram, H., Muller, E., Ramaswamy, S., Reimann, M.W., Abdellah, M., Sanchez, C.A., et al., 2015. Reconstruction and simulation of neocortical microcircuity. Cell 163, 456–492.

Marr, D., 1982. Vision: A Computational Investigation into the Human Representation and Processing of Visual Information. Freeman, New York.

Masquelier, T., Thorpe, S.J., 2007. Unsupervised learning of visual features through spike-timing-dependent plasticity. PLoS Comput. Biol. 3 (2), 247–257.

Masquelier, T., Thorpe, S.J., 2010. Learning to recognize objects using waves of spikes and spike-timing-dependent plasticity. In: International Joint Conference on Neural Networks. IEEE, Barcelona.

McManus, J., Ullman, S., Gilbert, C., 2008. A computational model of perceptual fill-in following retinal degeneration. J. Neurophysiol. 99, 2086–2100.

Merolla, P., Authur, J., Alvarez-Icaza, R., Cassidy, A., Sawada, J., Akopyan, F., et al., 2014. A million spiking-neuron integrated circuit with a scalable communication network and interface. Science 345 (6197), 668–673.

Modha, D.S., Singh, R., 2010. Network architecture of the long-distance pathways in the brain. Proc. Natl. Acad. Sci. U.S.A. 107 (30), 13485–13490.

Modha, D., Ananthanarayanan, R., Esser, S., Ndirango, A., Sherbondy, A., Singh, R., 2011. Cognitive computing. Commun. ACM 54 (8), 62–71.

Mountcastle, V.B., 1997. The columnar organization of the neocortex. Brain 120, 701–722.

Mountcastle, V.B., 1998. Perceptual Neuroscience: The Cerebral Cortex. Harvard University Press, Cambridge, MA.

Murphy, K.P., 2012. Machine Learning: A Probabilistic Perspective. MIT Press, Cambridge, MA.

Nair, V., Hinton, G.E., 2010. Rectified linear units improve restricted Boltzmann machines. In: Proceedings of the 27th International Conference on Machine Learning, Haifa.

Nessler, B., Pfeiffer, M., Maass, W., 2009. STDP enables spiking neurons to detect hidden causes of their inputs. In: Bengio, Y., Schuurmans, D., Lafferty, J., Williams, C. (Eds.), Advances in Neural Information Processing Systems, 22. Curran Associates, Red Hook, NY, pp. 1357–1365.

Nessler, B., Pfeiffer, M., Buesing, L., Maass, W., 2013. Bayesian computation emerges in generic cortical microcircuits through spike-timing-dependent plasticity. PLoS Comput. Biol. 9 (4), 1–30.

Norton, D., Ventura, D., 2006. Preparing more effective liquid state machines using Hebbian learning. In: 2006 International Joint Conference on Neural Networks. IEEE, Vancouver, pp. 8359–8364.

Olshausen, B.A., Field, D.J., 1996. Emergence of simple-cell receptive field properties by learning a sparse code for natural images. Nature 381, 607–609.

Ramaswamy, S., Courcol, J., Abdellah, M., Adaszewski, S., Antille, N., Arsever, S., et al., 2015. The neocortical microcircuit collaboration portal: a resource for rat somatosensory cortex. Front. Neural Circuits 9 (44).

Savarese, S., Feifei, L., 2007. 3D Generic object categorization, localization, and pose estimation. In: Proceedings of the IEEE 11th International Conference on Computer Vision (ICCV), pp. 1–7.

Schmidhuber, J., 2015. Deep learning in neural networks: an overview. Neural Netw. 61, 85–117.

Schurmann, F., Delalondre, F., Kumbhar, P.S., Biddiscombe, J., Gila, M., Tacchella, D., et al., 2014. Rebasing I/O for scientific computing: leveraging storage class memory in an IBM Bluegene/Q Supercomputer. In: Kunkel, J.M., Ludwig, T., Meuer, H.W. (Eds.), Supercomputing. Springer International Publishing, Heidelberg, pp. 331–347.

Serre, T., Wolf, L., Bileschi, S., Riesenhuber, M., Poggio, T., 2007. Robust object recognition with cortex-like mechanisms. IEEE Trans. Pattern Anal. Mach. Intell. 29 (3), 411–426.

Shadlen, M., 2006. Rate versus temporal coding models. Encyclopedia of Cognitive Science. Macmillan, London, UK.

Sharma, J., Angelucci, A., Sur, M., 2000. Induction of visual orientation modules in auditory cortex. Nature 404, 841–847.

Sherman, S.M., Guillery, R.W., 2004. Thalamus. In: Shepard, G.M. (Ed.), The Synaptic Organization of the Brain, fifth ed. Oxford University Press, Oxford, pp. 311–359.

Srivasta, N., Hinton, G., Krizhevsky, A., Sutskever, A., Salakhutdinov, R., 2014. Dropout: a simple way to prevent neural networks from overfitting. J. Mach. Learn. Res. 15, 1929–1958.

Tanaka, K., 1996. Inferotemporal cortex and object vision. Annu. Rev. Neurosci. 19, 109–139.

Tavanaei, A., Maida, A.S., 2016. Acquisition of visual features through probabilistic spike-timing-dependent plasticity. In: International Joint Conference on Neural Networks (IJCNN), Vancouver, pp. 307–314.

Triesman, A., 1977. Focused attention in the perception and retrieval of multidimensional stimuli. Percept. Psychophys. 22 (1), 1–11.

van Hateren, J.H., Shaaf, A., 1998. Independent component filters of natural images compared with simple cells in primary visual cortex. Proc. R. Soc. Lond. B 265, 359–366.

Verstraeten, D., Schrauwen, B., Stroobandt, D., Van Campenhout, J., 2005. Isolated word recognition with the liquid state machine: a case study. Inf. Process. Lett. 95, 521–528.

Wolpert, D.H., 1996. The lack of a priori distinction between learning algorithms. Neural Comput. 8 (7), 1341–1390.

Section B

Complex Analytics and Machine Learning

Chapter 3

Visual Analytic Decision-Making Environments for Large-Scale Time-Evolving Graphs

S.R. Venna, R.N. Gottumukkala[1] and V.V. Raghavan
University of Louisiana at Lafayette, Lafayette, LA, United States
[1]*Corresponding author: e-mail: raju@louisiana.edu*

ABSTRACT

Decision makers in multiple domains are increasingly looking for ways to improve the understanding of real-world phenomena through data collected from Internet devices, including low-cost sensors, smart phones, and online activity. Examples include detecting environmental changes, understanding the impacts of adverse manmade and natural disasters, and obtaining marketing intelligence on company profiles. Real-world observations from streaming data sources such as sensors, click streams, and social media are becoming increasingly common in many domains, including cybersecurity, infectious disease surveillance, social community networks, and web-based recommendation systems. An emerging approach is to leverage graph-based modeling of these events, to understand ongoing trends and predict future ones. Data scientists are faced with the challenge of analyzing large-scale graphs that are changing dynamically, while existing tools and metaphors for data collection, processing, storage, and analysis are not suitable for handling large-scale evolutionary graphs. This chapter describes visual analytics as a cognitive computing approach to improve decision making with large-scale dynamic graphs. We provide a conceptual introduction to time-varying graphs and various components of the visual analytics that affect the performance of decision support systems, including data management, analytics, visualization, and visual interaction. We provide techniques to improve the performance of each of these components in an integrated visual analytics framework. We also describe a visual graph analytics sandbox architecture and sample applications implemented within it.

Keywords: Visual analytics, Time-evolving graphs, Data streams, Graph visualization, Data warehouse

Handbook of Statistics, Vol. 35. http://dx.doi.org/10.1016/bs.host.2016.07.002

1 INTRODUCTION

Large-scale graphs with billions of nodes and edges created from real-world observations are emerging in multiple domains and disciplines—these include social community networks, infrastructure networks, epidemiology networks, IP traffic networks, etc. There is a lot of interest by researchers, business, and government organizations to get real-time visibility into the health state of communities, infrastructures, and communication networks, and to discover emerging trends and anomalies. This in turn would help to both defend against and mitigate the impact of adverse events such as natural and manmade disasters, cyberattacks, and infectious disease outbreaks. Graphs are a powerful and intuitive tool to model and study the relationships between various objects. However, sense making from big graphs generated from streaming data sources such as sensor networks, click streams, and social media is a significant big data challenge given the size, complexity, and dynamic nature of these graphs. Existing data warehouse architectures and business intelligence tools to collect, store, process, mine, and visualize information are designed for relational data bases and do not scale well for NoSQL data types such as graphs. Various graph databases (Neo4j (Miller, 2013), Titan (Titan, 2016a), AllegroGraph (AllegroGraph, 2016), and ArangoDB (ArangoDB, 2016)) are emerging that offer optimized storage and query performance on large and complex graphs.

Dynamic graphs have a temporal dimension that is lost when graphs are aggregated or when only a snapshot of a graph is analyzed. The dynamics and evolution of these graphs in terms of their growth can be better understood by introducing the time dimension. This gives visibility into the dynamic properties of graph elements (nodes, edges, and subgraph) in terms of when and where elements appear, their duration, and their community interaction. Time-varying (or time-evolving) graphs have been proposed as a model to represent the time dimension in highly dynamic graphs. The evolutionary aspect of graphs has been studied in many applications (Aggarwal and Subbian, 2014; Casteigts et al., 2012; Kostakos, 2009) in terms of how the properties of graphs grow or shrink. The application of studying these properties has become a recent subject of interest. Notable examples include understanding co-authorships and citation networks to predict future collaborations, understanding evolution of word co-occurrence graphs generated from social media to understand the evolution of topics and events of interest, detecting malware and attack graphs from Internet traffic, understanding disease spread in cancer cells from gene networks containing information on protein and DNA information, etc. It is not possible to study such properties of real-world networks by ignoring the temporal dimension.

Decision-making environments provide computerized assistance to analysts for rapidly extracting relevant insights from data. The overall process of decision making involves knowledge discovery, visual presentation of knowledge, and human cognition for obtaining relevant insights. Knowledge discovery involves tools and processes for data collection, processing, management, and mining for information. Visualization involves generating visual representations for extracted knowledge through a human computer interface. The desired result of this process is for decision makers to gain knowledge and obtain insights from it. In his famous article on human computer symbiosis (Licklider, 1960), Licklider highlighted that humans and machines should blend together seamlessly so that they complement each other in the decision-making process.

As graphs grow in size, number, and complexity, so does the cognitive load for observing relevant events or trends of interest. Given massive size and complexity of dynamic or time-evolving graphs, such graphs are often summarized or reduced in size to adapt to the capability of existing tools. Most of the existing graph mechanisms (definition, metrics, and models) and decision support tools are primarily defined for static graphs and cannot translate to capture the dynamic properties for time-varying graphs. The complexity of time-evolving graphs exceeds the representational capabilities of key-value stores, column stores, relational SQL, or document stores. Also, the computational and visualization requirements to manage these stores exceed the capabilities of commercially available tools. Moreover, new interaction techniques have to be developed to visually portray and manage these graphs. Research is therefore needed in all aspects of decision support tools for managing these graphs—how graphs are stored, managed, processed, and visualized, and the best mechanisms to interact with these graphs. Most of the existing graph analysis tools such as Gephi (Gephi, 2016), GraphLab (Dato (GraphLab), 2016), and Cytoscape (Cytoscape, 2016) are designed for small graph data with limited capability for interactive analysis. Palantir (2016) is another graph-based decision support tool that discovers trends, relationships, and anomalies, including predictive analytics. These tools rely on visualization and visual interaction metaphors that are unsuitable for handling complex evolving graphs.

Here are some representative domains where decision makers need to manage time-evolving graphs:

- *Homeland security and emergency management*: Sensor streams obtained from critical infrastructures (transportation, food supply, energy grids, water supply, and communication networks) and their physical or logical interdependencies can be modeled as graphs (Deshpande et al., 2004; Rinaldi et al., 2001). With growing monitoring technologies such as low-cost sensors, the dynamics of these graphs are always changing due to changes in the environment (Galati et al., 2013). Emergency responders and critical infrastructure analysts need situational awareness and interactive capabilities to monitor, analyze, and visualize the status and health of these dynamic networks, eg, to analyze the movement of people, vehicles, or commodities.

- *Computer networks*: Network traffic-related events including peer-to-peer connections, call data from telecommunication networks, email communications can be modeled as graphs (Dong et al., 2011; Iliofotou et al., 2011; Nguyen et al., 2011; Sperotto et al., 2010). The change in these networks in terms of how new nodes and links get added or removed is evident given the dynamics of new connections, interactions, or subnetworks. Monitoring this change and evolutionary patterns in the networks is important, especially when system administrators and security officers need to detect emerging anomalies such as attack graphs (Noel and Jajodia, 2009; Noel et al., 2015) or malware propagation (Chau et al., 2010; Chen and Ji, 2005).
- *Social networks*: Web data such as click streams and social media can be leveraged to generate social network graphs that represent friendships, shared communities, message interactions, product recommendations, or other forms of relationship (Kumar et al., 2010). These graphs are used by recommendation engines to recommend products or page links, to detect emerging events in social media, etc. There are few graph visualization tools that let marketing and product supply chain experts take advantage of these time-varying graphs to make decisions based on evolutionary patterns of graphs.
- *Biological networks*: The interaction or correlation of behaviors in biological networks such as protein–protein interactions and infectious disease networks can be modeled as time-varying graphs. The protein–protein interaction graph can be used to understand the evolution of proteins with respect to different drugs. Infectious disease networks that represent people-to-people contacts can help in understanding the evolutionary patterns of a disease in different sociodemographic and environmental conditions, to help public health officials better track and predict the evolution of infectious diseases across different spatiotemporal regions (Masuda and Holme, 2013; Nepusz et al., 2012).

In order to overcome the limitation of existing decision support tools, a new generation of tools is needed that enables the synergistic collaboration between computational power and human intuition to support insight-driven knowledge discovery. These tools would provide a suite of tightly integrated cognitive decision-making capabilities, comprised of (1) a data management environment to handle large-scale graphs, (2) an analytics engine that supports temporal and nontemporal graph analytics and provides support for automated analysis with minimal configuration, and (3) an intuitive visualization and human interaction component.

Visual analytics is an interdisciplinary field that integrates the following areas: *analytical reasoning* approaches that let users obtain deep insights that directly support assessment; *planning and decision making*, *visual representations*, and *interaction techniques* that exploit the human eye's broad bandwidth pathway into the mind to let users see, explore, and understand large amounts of information simultaneously; *data representations and transformations* that convert all types of conflicting and dynamic data in ways that support visualization and analysis; and techniques to support *production, presentation, and dissemination* of analytical results to communicate information in an appropriate context to a variety of audiences (Andrienko and Andrienko, 2012). Fig. 1 shows a high-level mind map of the different components and their subtasks involved in a typical visual analytics system. While the size and complexity of data is increasing rapidly, existing tools for knowledge discovery are becoming usable for humans to find patterns, make predictions, and enhance understanding of large-scale dynamic graphs—a rapidly growing area in the big data era.

This chapter presents visual analytics as a mechanism to bring closer collaboration between humans and computers to improve the overall performance of decision-making environments for analyzing large-scale time-evolving graphs. We discuss the fundamental characteristics of time-evolving graphs, and state-of-the-art tools available for data management, analytics, visualization, and human–computer interaction (HCI). We also present our ongoing research in developing a visual analytics sandbox, along with some applications. Finally, we discuss some open challenges to building visual analytics capabilities into decision-making environments.

2 VISUAL ANALYTICS AS AN APPROACH TO COGNITIVE COMPUTING

Graph-based reasoning and decision support tools provide an advantage of performing what-if scenario analysis. However, the complex nature of graphs presents numerous challenging for decision making, such as visual layouts, navigation, and selecting the right mining techniques. Decision making involves a systematic process of data cleaning, transformation, model configuration, analysis, and discovery. Designing decision-making environments that work in symbiosis with humans requires an understanding of human thought processes, incorporating these processes in a model to automate tasks and reduce human burden, and making the decision-making process more efficient. Cognitive computing involves simulation of human thought in a computerized model to automate some of the activities performed by the human brain.

According to a recent definition by Briody, cognitive computing refers to systems that work synergistically with humans in terms of learning collaboratively, reasoning with purpose, and interacting naturally with humans (Briody, 2015; Chen et al., 2016; Kelly, 2015). In the context of decision-making environments, cognitive systems should support:

- Deeper engagement with humans for understanding the context of data analysis in terms of relevance and computational constraints, and making recommendations on data sample (size and time period), models, and visual representation.

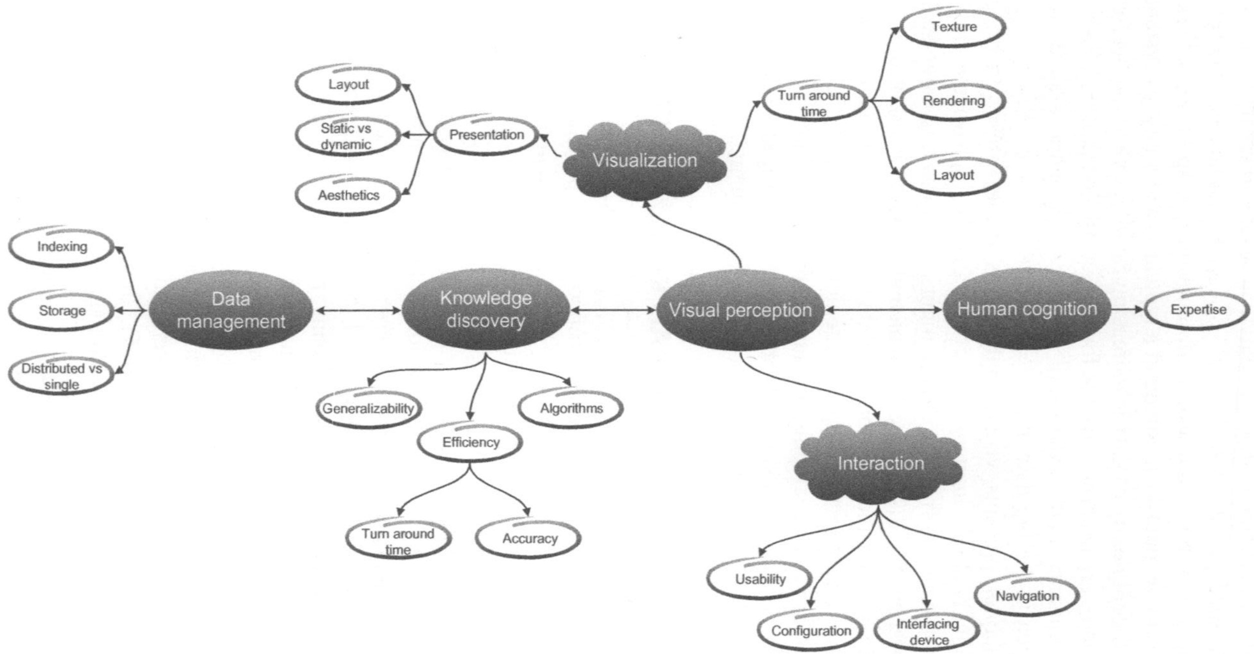

FIG. 1 A high-level mind map of tasks involved in a visual analytics system.

- Automation of data collection, preprocessing, transformation, integration, and information extraction based on the decision goals and application domain.
- Discovery and exploration of insights gained from knowledge through what-if scenario analysis.
- Integration of multiple components of the decision-making environment to improve their collective capabilities.

2.1 Visual Analytics: A Definition

The primary goal of visual analytics is to make the way that users understand and process the analytic discourse simple, transparent, interactive, and efficient (Keim et al., 2010). Visual devices and interactive techniques can be combined with data analytics, for faster and more efficient analysis of larger amounts of data. Visual analytics researchers are trying to exploit and combine the strengths of human and electronic data processing (Andrienko and Andrienko, 2012). This demands close collaborations between workers from multiple disciplines including visualization experts, statistics, data scientists, GIS experts, etc. According to Thomas, 2005, visual analytics is an interdisciplinary field that integrates the following areas: analytical reasoning approaches that let users obtain deep insights that directly support assessment, planning, and decision making; visual representations and interaction techniques that exploit the human eye's broad bandwidth pathway into the mind to let users see, explore, and understand large amounts of information simultaneously; data representations and transformations that convert all types of conflicting and dynamic data in ways that support visualization and analysis; and techniques to support production, presentation, and dissemination of analytical results to communicate information in an appropriate context to a variety of audiences. Fig. 2 shows the visual analytics process as an integrated framework that has data, models, knowledge, and visualization interaction process interacting with each other. The capabilities and targets of visual analytics systems have been significantly expanding both in terms of data and domain applications (Wong et al., 2012).

3 TIME-EVOLVING GRAPHS

3.1 Definition

A graph is a pair $G=(V, E)$ satisfying $E \subseteq V \times V$, where V is the set of vertices or nodes and E is the set of edges or links between any two vertices and can also be referred to as a static graph. A time-evolving graph is characterized by edges and nodes that appear and disappear over time. A time-evolving graph G is defined as $G=(V, E, T)$, where V is the set of vertices (or nodes), E is the

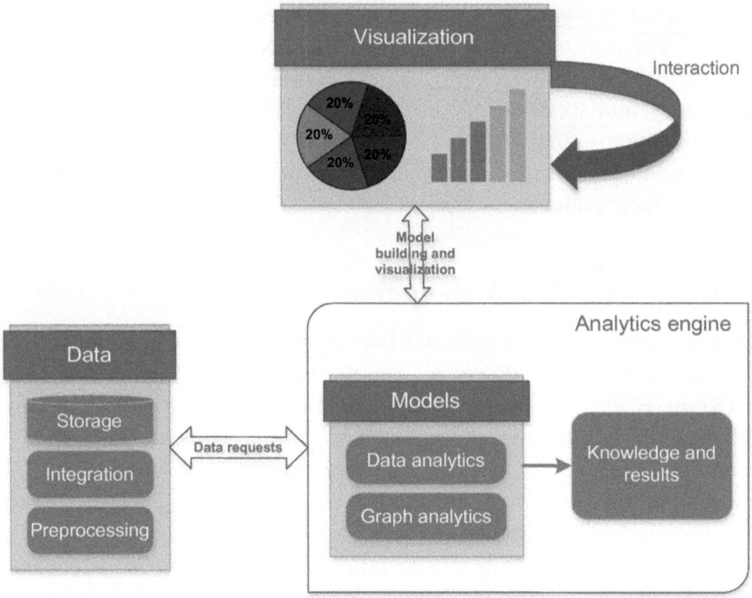

FIG. 2 Visual analytics as an integrated framework.

set of edges, and T is the set of time instances. Also, $E \subseteq (V \times T \times V \times T)$ is the set of edges. An edge $e \in E$ is defined by $e = (v_1, t_a, v_2, t_b)$, where $v_1, v_2 \in V$ are the origin and destination nodes and $t_a, t_b \in T$ are origin and destination time instances, respectively. The edge e is a directed edge from node v_1 at time t_a to node v_2 at time t_b. An undirected edge can be represented when E has both (v_1, t_a, v_2, t_b) and (v_1, t_b, v_2, t_a). This definition was introduced in Wehmuth et al. (2015). Evolution can represent the variation of availability of a node, a edge, or a graph. If the time ordering of the edges in a time-evolving graph is ignored, we consider them as concurrent in time and we get its aggregated static graph. Detailed definition and evolutionary properties of time-varying graphs are discussed in Kostakos (2009) and Wehmuth et al. (2015).

3.2 Properties of Time-Evolving Graphs

Computing regular network properties such as degree, density, diameter, and modularity on a static graph is straightforward, but in the case of time-evolving graphs, these can be studied as evolving parameters at different levels of granularities across the time stamps. The lowest level of granularity is the graphs individual snapshot at a unit time, and the highest is the aggregated graph of all snapshots (the static graph), and in between is the aggregation of snapshots between selected time intervals of convenience

(Santoro et al., 2011). Depending on the parameter and application domain different granularities should be chosen to extract meaningful summaries. Based on the analysis task we are performing, evolutionary properties of time-evolving graphs can be labeled as edge centric, vertex centric, or graph centric. Edge centric evolution refers to the changes of availability and latency of an edge over time, node centric evolution looks into the changes in a node and its neighborhood, and graph centric properties deal with the entire graph. Different evolutionary properties of time-evolving graphs are discussed later.

3.2.1 Journey

A journey in a time-evolving graph is a walk of edge sequences with ascending time stamps. A journey has departure and arrival times and is thought of as paths between source and destination node pairs. A journey can have both temporal and topological lengths. Temporal length refers to the total time duration of the journey—the time difference between its arrival and departure times—whereas topological length refers to the number of hops or edges between source and destination. Direct journeys are those paths where the next edge follows on directly; otherwise, it is an indirect edge (Casteigts et al., 2014). Minimum topological length is labeled as shortest path distance, and temporal distance can be labeled as foremost, fastest, or latest distances.

3.2.2 Structural Properties

There are many structural properties associated with time-evolving graphs. A frequently used indicator for quantifying the graph structure is *density*, which is a measure of how close a graph is to being complete. The density D for an undirected graph $G=(V, E)$ is computed as:

$$D=\frac{2\times|E|}{|V|\times(|V|-1)}$$

In a time-evolving graph, the temporal density is the trend of densities computed at different snapshots of the graph. *Eccentricity* of a node is the largest distance between this node and any other node from this graph. *Diameter* is the maximum eccentricity of any node in the graph, whereas *radius* is the minimum eccentricity of any node. Similar to density, evolution of eccentricity and diameter can be observed as a trend across different snapshots. *Modularity* quantifies the quality of a particular division of a network by comparing the fraction of edges within the division with such fraction when random connections between the nodes are made. Fig. 3 is a comparative chart showing the evolution of degree and edge weight of selected nodes against the whole graph; it can be observed that something interesting happens with the selected nodes after a certain time (about 13:36). *Conductance*

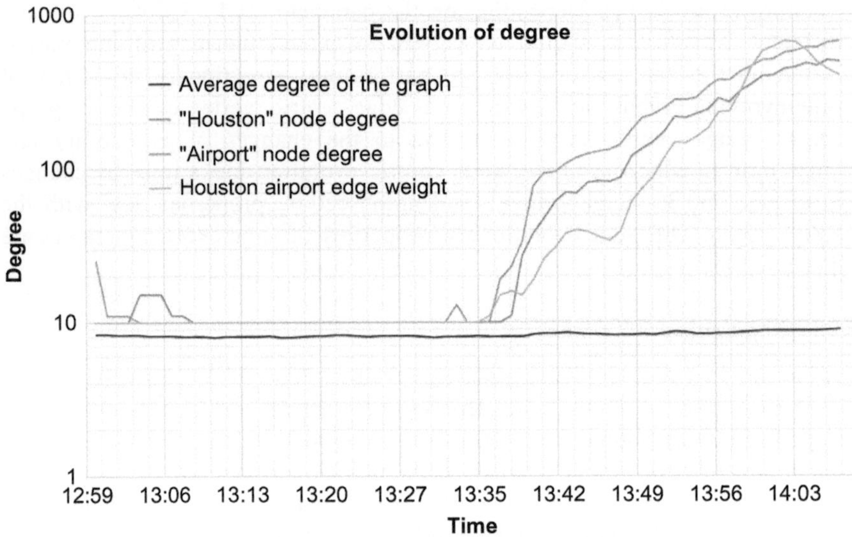

FIG. 3 A sample chart showing evolution of degree and edge weights.

measures how firmly connected a graph is, and it characterizes the time of convergence of a random walk toward its uniform distribution. In general, nodes in a graph tend to create tightly connected groups or clusters with high intradensity; *clustering coefficient* of a node quantifies how close to a clique its neighborhood is. An increase in clustering coefficient of node in a time-evolving graph captures the formation of a tightly connected cluster or community.

3.2.3 Reachability

Reachability between two nodes in a time-evolving graph is possible when the second node can be reached directly through a single edge or a sequence of time-ordered edges. A *walk or temporal walk* in a time-evolving graph between two nodes is defined as a sequence of edges with increasing sequence of times. A *temporal path* is a walk during which each node can be visited at most once. Two nodes in a time-evolving graphs are said to be *temporally connected* if a temporal path exists between them and this relation is not symmetric. Fig. 4 shows the evolution of average shortest path from a selected node. It can be observed that after some time the selected node becomes very active, leading to a reduction in its average shortest path. Two nodes A and B are called *strongly connected* if and only if A is temporally connected to B and B is temporally connected to A. Nodes A and B are *weakly connected* when there is a temporal path between A and B, and between B and A when directionality in the time-evolving graph is discarded. Both strong connectedness and weak connectedness are reflexive and symmetric but not transitive.

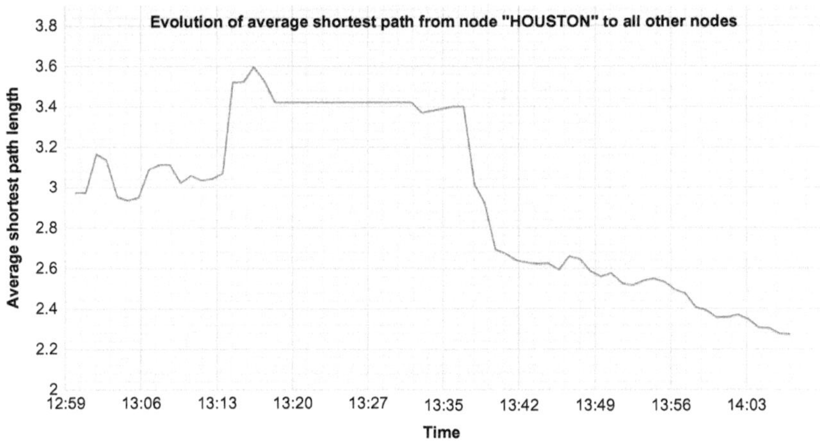

FIG. 4 Chart showing the evolution of average shortest path from a selected node.

A *strongly connected component* of a time-evolving graph is defined as a set of nodes in which each node from the set is strongly connected to the remaining nodes in the set (Nicosia et al., 2012). If a node can reach all other nodes and be reached back, then that node is said to be *round connected.* If a subgraph stays connected across time, then that subgraph is said to possess *recurrent connectivity.*

3.2.4 Centrality Measures

Centrality indicators were first introduced in social network analysis to identify influential nodes in a graph. Simple *degree centrality* of a node is the number of edges incident on it. *Betweenness centrality* of a node is defined as the fraction of all-pairs shortest paths that pass through the node. *Closeness centrality* of a node is measured as its average shortest distance to all other nodes in the graph (Newman and Girvan, 2004; Nicosia et al., 2013; Tang et al., 2010).

4 VISUAL ANALYTICS AS A FRAMEWORK FOR TIME-EVOLVING GRAPHS

4.1 Visual Analytics Frameworks

Implementation of a visual analytics framework in a decision-making environment requires tighter integration and optimization of the various components shown in Fig. 5, namely: (1) an efficient data model and memory management to store large data volumes and run different graph processing and mining algorithms, (2) interaction techniques based on various visualization interfaces to interact, manipulate, and follow the graphs with respect to its dynamics, and (3) an integration framework that facilitates seamless interaction of graph data sets. Not all graph operations can be performed on the

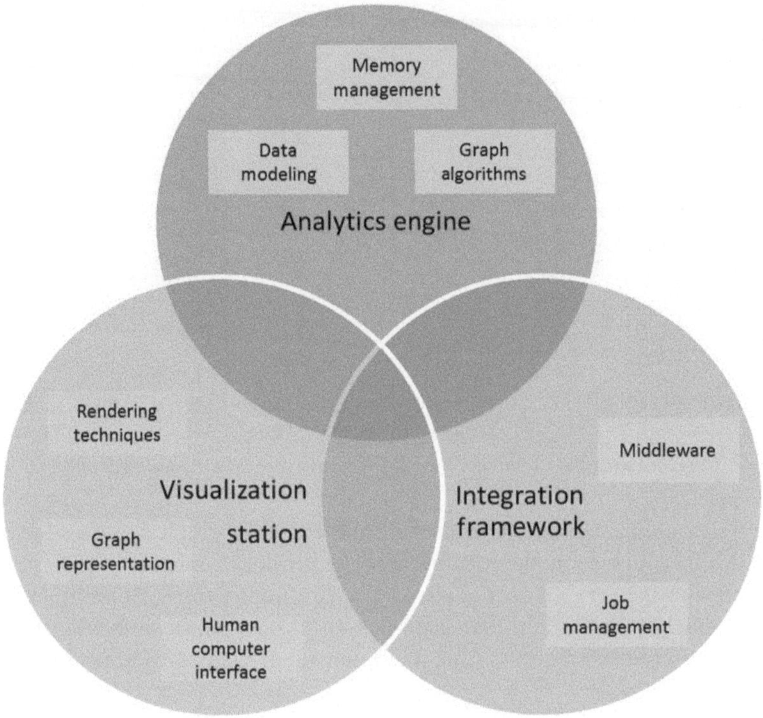

FIG. 5 Visual analytics as an integration framework with various components.

visualization system; hence, there should be seamless communication between the visualization system and the analytics server. The middleware (integration framework) serves as the key interface between the visualization system and the analytics server. The middleware takes care of management and prioritization of various jobs, and translation of user actions into analytical queries.

Visual analytics has become a buzz word in the business intelligence domain. Many leading companies in statistical analysis for business are pursuing novel ways to improve their data presentation, through new products such as SAS Visual Analytics (SAS® Visual Analytics, 2016) and IBM Many Eyes (IBM Many Eyes, 2016). Moreover, several new business intelligence tools such as Tableau (Tableau, 2016), Birst (Birst, 2016), and Google fusion tables (Google Fusion Tables, 2016) also provide various interactive visualization capabilities. While these tools provide some basic visualization and interaction capabilities for users to interact with the data, these tools are far from promoting analytics discourse with the visualization environment. The overarching vision of visual analytics is to provide technology that combines the strengths of human and electronic processing (Keim et al., 2010; Thomas, 2005). A recent comparative study on the landscape of current business intelligence platforms and research in visual analytics capabilities of business intelligence systems (Von Landesberger et al., 2011) highlights the capabilities and limitations with respect to individual components, ie, data management, automated analysis, visualization, and system architecture. Another survey paper on visual analytics (Beck et al., 2014) also highlights the state of the art in visual analytics and the challenges in individual research areas. However, the visual analytics solutions actually lie in the integration of various research areas and optimization of data management, analytics, visualization, and human interaction modules. All these business intelligence tools have visual analytics capabilities added into the existing platform; hence, they offer limited flexibility to support interactive visual analytics of complex, and real-time data sets. Bridging these disciplines into an integrated framework offers new opportunities and challenges for researchers to experiment with different visual analytics components to improve the overall end-user experience to manipulate the information.

4.2 Data Management

The most straightforward way to store a graph is in the form of an adjacency list or adjacency matrix. There are multiple ways to store a time-evolving graph while preserving its temporal structure. Choosing the right data model depends on the nature of the data, the type of graph (strongly connected vs weakly connected, sparse or dense graphs, etc.), and the targeted data processing and analytical tasks.

The simplest approach is to store a different snapshot of the graph for each time instance as in Fig. 6A (Tang et al., 2010). This model consumes a lot of memory and works only when it is not necessary to capture relationships between nodes across time stamps. Also, running certain queries across time stamps is inefficient. Alternative ways include creating a single graph for all time stamps and storing the time information on the edge as an attribute. This can be accomplished in two different ways: as a simple list of time stamps as shown in Fig. 6B (Holme and Saramäki, 2012) or by specifying limits when edges are persistent during sequences of time stamps as in Fig. 6C (Miller, 2013). For example, in Fig. 6B, the edge between node A and node B is available at time stamps T1, T3, and T5. In Fig. 6C, the edge from node A to node B is available between time stamps T1 and T4. One of the limitations of these models is that the relationship between the nodes across time stamps cannot be stored. One way of storing more complex time-evolving graphs, where there is a possibility of having edges across nodes from different time stamps, is creating duplicates of nodes for each time stamp in which it is present and adding edges between required nodes as shown in Fig. 7 (Kostakos, 2009).

One of the most important elements of a graph database is its data model, which is the data structures for schema and instances modeled as graphs (or generalizations of them), to support efficient storage and to query, index, or

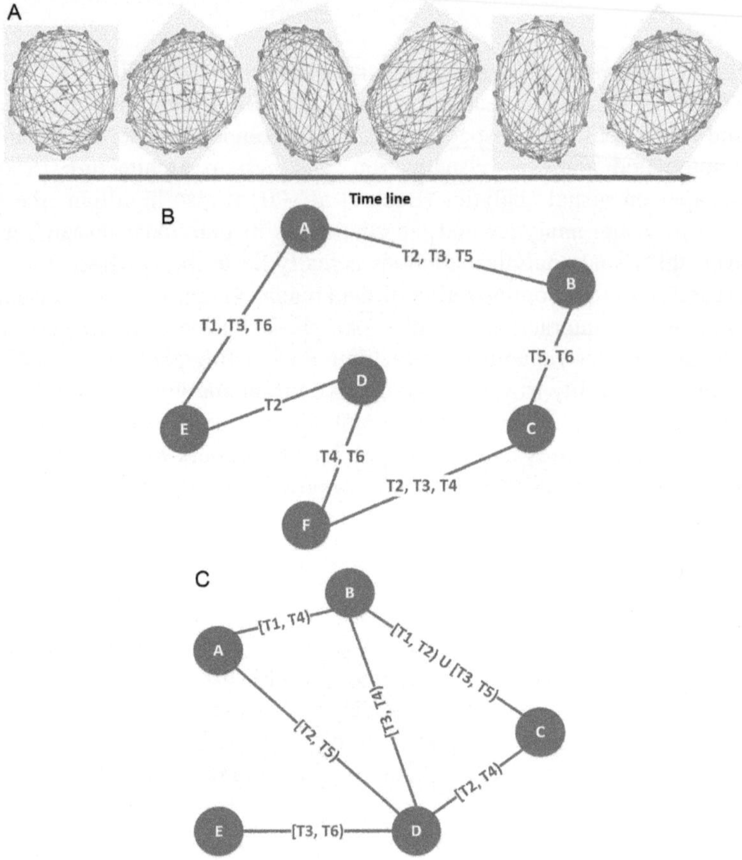

FIG. 6 Three ways of storing a time-evolving graph. (A) One snapshot of the graph for each time instance; (B) list of time stamps labeled on the edge; and (C) the range of time stamps labeled on the edge.

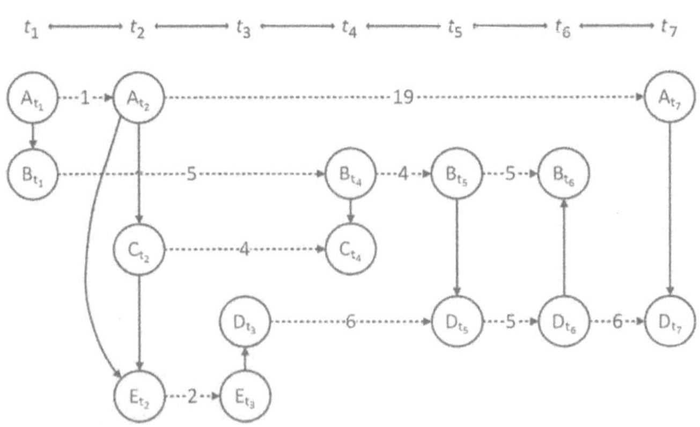

FIG. 7 Graph representation of time-evolving graph where edges are present across nodes from different time stamps.

aggregate data. The data can be centralized or distributed, either storing graphs in main memory or storing them on the disk for retrieval on demand.

Graph access patterns have poor spatial memory locality, which results in large amounts of random memory access. High-throughput processing of massive graphs that might not fit in main memory requires efficient memory management that includes caching strategies to write unused data to disk and indexing mechanisms for efficient retrieval of these graphs. Storing and managing graph data on disks suffers from poor I/O latency, but it is not possible to store the entire graph in memory. Solid-state drives offer low latency storage for both caching and long-term storage. Most graph databases provide basic cache management and indexing schemes. These are not generally optimal for all types of graphs or graph operations, so strategies are needed for storing the graph in single vs distributed nodes, and considerations for the dynamic nature of the data (bursty or highly dynamic), the graph topology, the type of processing needed, etc.

When processing or analyzing any data, important aspects to be considered are volume (size of the data), variety (complexity), and velocity (speed of the data coming in). Even though relational DBMSs have matured over the past four decades, they are still limited to handling relatively small and less complex data. Fig. 8 (Shao et al., 2012) compares different data management schemes available between size and complexity. Map/reduce file systems or key-value stores are best suited for data with low complexity and high volume, whereas in-memory key-value stores or graph databases can handle high complex data of moderately less sizes as they are mostly dependent on the amount of primary memory.

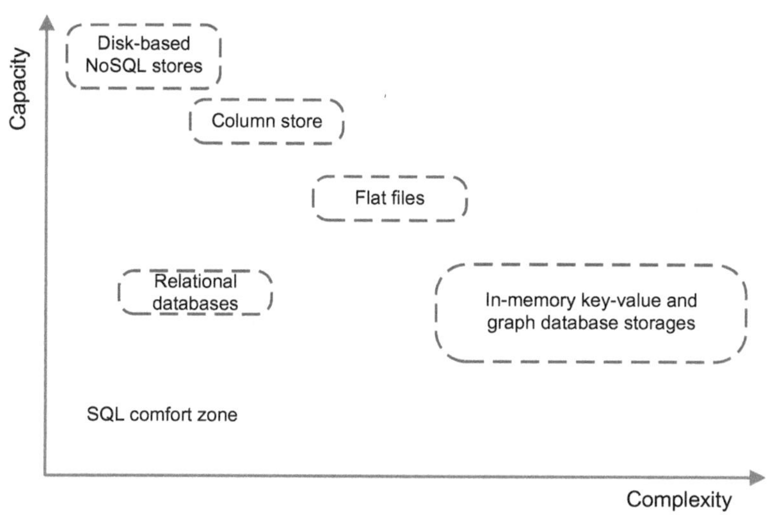

FIG. 8 Data capacity vs complexity scaling.

For example, one of the most widely used graph databases, Neo4j (Miller, 2013), stores a graph on disk and retrieves it into the main memory for computation. FlockDB from Twitter (FlockDB, 2016), RDF-based AllegroGraph (Aasman, 2006), Titan (Titan, 2016b), and Objectivity's InfiniteGraph (InfiniteGraph, 2012) are well-known distributed databases than can support storing node or edge labels as temporal attributes. The choice of graph database depends on the requirements of the application and graph type. This includes storing features (main memory, external storage, indexing) and the graph structures to store temporal attributes (either on nodes, edges, or graphs) for efficient retrieval.

The high complexity of graph data is determined by the number of nodes, their relationships (or edges), and different properties associated with them. Time-evolving graphs introduce a new dimension, making them even more complex. Time-evolving graphs can be modeled as regular static graphs by associating time with their respective entities. With varying pros and cons, each of the graph systems has its own way of storage and retrieval of time-evolving graphs. Figs. 9 and 10 visualize two different ways of storing the same graph using Neo4j. The simple example data used here is a meeting schedule for five persons across 3 days. Fig. 9 represents an accumulated graph that was created when the time is stored as an edge property. Fig. 10 represents the same data as multiple snapshots of the graphs at each time stamp when the time is stored as a node property. With Neo4j each of these storage schemes has pros and cons, because in Neo4J indexing can be done

FIG. 9 Data stored as an accumulated graph when time is stored as an edge property.

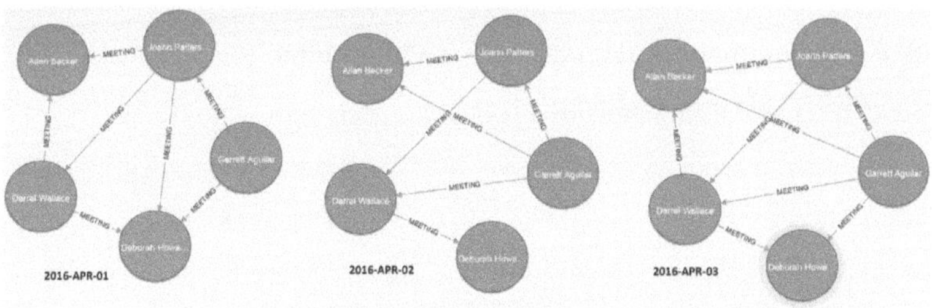

FIG. 10 Data stored as multiple graph snapshots when time is stored as node property.

only on the properties of nodes but not those on edges. Also, the second representation takes more disk space as each of the nodes and its properties are replicated at each snapshot (compared to the first representation).

4.3 Graph Analytics

There is a great demand for close to real-time analysis of massive graphs—given the demand in several real-time applications such as online recommendations for click stream processing, fraud detection, and analysis of cyberattack graphs. The performance of a graph analytics engine is affected by three main factors: the graph data model, the memory management/caching scheme, and the graph analytics algorithms.

The performance of graph mining algorithms depends on both the algorithm and the data management techniques discussed in Section 4.2. The complexity of these algorithms also depends on the speed of evolution of these graphs, eg, slowly evolving graphs in which substantial changes occur on a timescale of days or weeks (web-based networks, citation graphs, etc.). In streaming or fast-evolving graphs, overall graph structure changes in a matter of seconds, eg, social media graphs, transportation networks, etc. (Aggarwal and Subbian, 2014; Kostakos, 2009; Pienta et al., 2015; Tse and Tsui, 2002). General graphs mining topics have been studied extensively in books and surveys (Aggarwal and Wang, 2010; Aggarwal et al., 2011). Specific graph mining topics such as graph classification (Bhagat et al., 2011; Tsuda and Saigo, 2010), link prediction (Al Hasan and Zaki, 2011), and influence analysis (Sun and Tang, 2011) are well discussed. A recent survey (Aggarwal and Subbian, 2014) covers literature on clustering, classification, influence analysis, and link prediction on both slowly evolving networks and streaming networks. Table 1 shows a summary of graph analysis operations. The basic graph operations for identifying temporal network topology and structure, and reachability analysis have been discussed in Section 3. Here we discuss graph mining algorithms with respect to predicting network topological properties, outlier detection, and community detection.

Predicting network topologies: This involves predicting node labels or links based on historical graph data. The node classification problem involves

TABLE 1 A Summary of Graph Analysis Operations

Type of Temporal Characteristic	Graph Operations
Temporal network topology and structure	Degree, connectivity, density
Reachability analysis	Paths, walks, trails
Predicting network topological properties	Link prediction and classification
Detecting outliers	Node or edge clustering
Node neighborhoods and communities	Persistent patterns and motifs

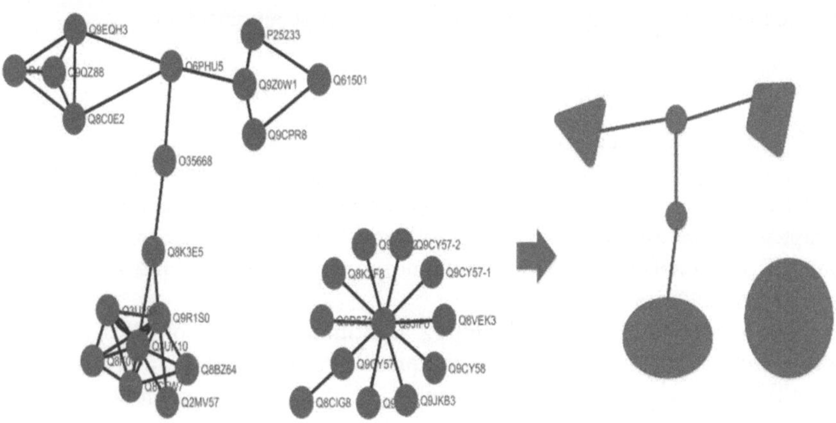

FIG. 11 An example graph summarization.

predicting the labels of unlabeled nodes in a time-evolving graph based on the existing node labels.

4.3.1 Graph Summarization

It is difficult to understand the information encoded within a large graph or to get visual insights. Therefore, graph summarization is a critical task for better understanding of the internal characteristics of large graphs. Graph summarization aims to generate a compressed version of the original graph without losing important information, where each node in the summarized graph is a super node consisting of one or more nodes from the original graph exclusively (Navlakha et al., 2009) as shown in Fig. 11 (Dunne and Shneiderman, 2013). Summarization can be done through simple statistics extracted from graph properties like degree distributions, hop lengths, and clustering coefficients (Chakrabarti et al., 2007; Zhou et al., 2009), graph isomorphism and other frequent graph patterns, and graph aggregation or partitioning to identify subgraph communities (Louati et al., 2011; Tian et al., 2008; Williams et al., 2007).

4.4 Visualization

4.4.1 Visualizing Time-Varying Graphs

The overall goal of visualization is to enable users to obtain insights from data. Given the scalability and dynamic nature of time-evolving graphs, visualization needs to take into account how much information can be perceived and understood, computed, and displayed. These include the graph topology to be projected (graph representation) into 2D or 3D space using different layout schemes, understanding what interaction and human computer interface tools are best suitable, and how to render data efficiently on display screens. These factors are common for visualization of large-scale multivariate graphs.

In the case of visualizing time-evolving graphs, analysts need to understand temporal changes, their topological structure, or graph characteristics. The most common way to represent dynamic graphs are animated diagrams or static graphs with a timeline.

4.4.2 Animation-Based Visualization

When no interaction or manipulation of graphs is desired, a simplest way to show temporal evolution is through animation. A graph is constructed by creating an animation from a series of graphs at different time stamps. An initial super graph layout is created to have a consistent layout for graphs in multiple time stamps. Generally, a super graph is constructed using the graphs from considered time stamps, and a single graph layout is computed as shown in Fig. 12 (Beck et al., 2014). There are several variations of animation-based approaches that represent time transitions using color coding, shape, or layout techniques covered in Beck et al. (2014).

4.4.3 Timeline-Based Visualization

Another (and the most common) way to display temporal evolution is by projecting time into space dimension. This can be done multiple ways by juxtaposed node–link presentation over time as in Fig. 13A, or superimposed nodes and links where layers are used to represent time stamps as in Fig. 13B. The temporal changes in graphs may also be represented by matrix-based approaches, which are better suitable for more readability. The matrix

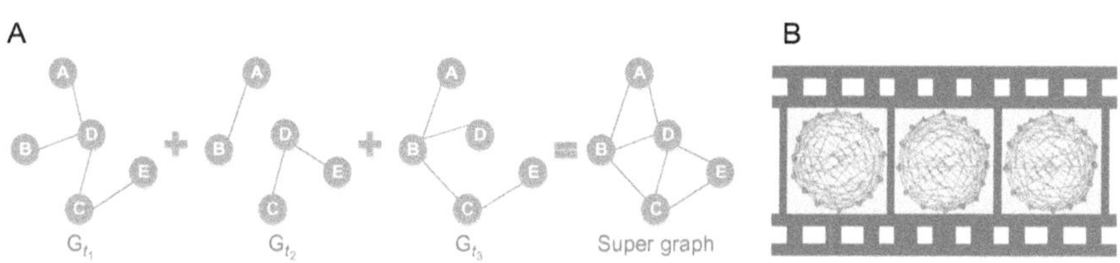

FIG. 12 (A) Constructing a super graph from three sequential dynamic graphs and (B) a simple animation for a small time-evolving graph.

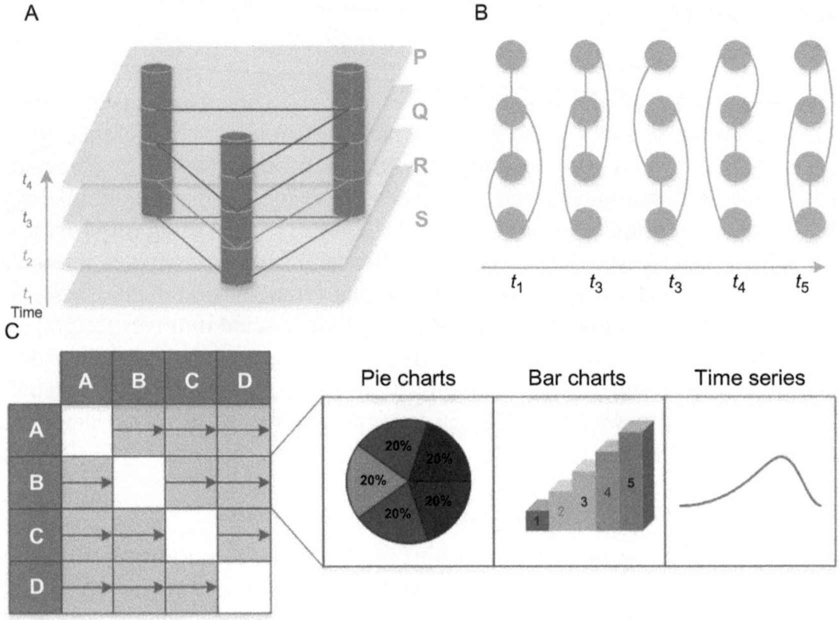

FIG. 13 (A) Juxtaposed node–link-based timeline presentation. (B) Superimposed node–link approach with layers representing time steps. (C) Intracell timelines in a matrix representation.

notation provides the ability to encode dynamic changes in the cell and an edge using colors, and charts. For example, Fig. 13C uses superimposed node–link approach to show different forms of intracell timeline representation.

4.5 Visual Representation and Interaction

For visualizing time-evolving graphs, it is important to choose a good visual representation to show them in a presentable and understandable format. These visual representations should reduce visual clutter and minimize temporal aliases for node positions across time and maximize readability and scalability. Selecting a visual representation for time-evolving graphs is restricted by the data at hand, size of the graph, amount of data to visualize, purpose of the visualization, etc. Some of the visual representations are limited by graph layouts as it is difficult to find automatic layouts for static graphs, and to do that for every time stamp is an enormous task. Other extensions to the animation-based and timeline-based visualization techniques include 3D visualization (Gaertler and Wagner, 2005; Kumar and Garland, 2006), hybrid representations combining animation, and timeline drawings. Several application-specific visualizations are available in the literature, including time line trees (Burch et al., 2008), tree maps (Hao et al., 2005), icicle plots (Tekušová and Schreck, 2008), node–link diagrams with time series (Saraiya et al., 2005), and time arc trees (Greilich et al., 2009).

HCI enables users to interactively browse the data set to discover hidden insights. An effective HCI is equally important as visual representation for a good visual analytics framework. These HCIs should enable the user to have control over what and how they want to see and to define the flow and parameters of decision informatics. Recent studies (Heer and Shneiderman, 2012; Kerren and Schreiber, 2012; Yi et al., 2007) provide taxonomies for visual interaction techniques to help better understand and improve visual analytics designs. Interactions with the visual representations are divided into three high-level categories:

- *Data and view specifications*: An HCI should allow the user to reconfigure views based on attributes of interest, to filter portions of the graph, and to derive simple analytics using statistical computations.
- *View manipulations*: A user should be able to select, highlight, and bookmark portions of the graph either by manual selection or through search criterion, and to navigate and explore over graphs using zooming, magic and fish eyed lenses, panning, etc. The HCI should allow the user to coordinate and organize multiple views for easy comparisons of results from different interactions.
- *Process and provenance*: Visual analytics systems should record different interactions for fast recall or revisiting of past analyses. They should also support multiuser collaboration, reporting, and sharing of views, interactions, and results.

The other important aspect of visualization is rendering the graph to display large-scale data sets. GPU-based rendering is becoming increasingly common. Gephi provides a time-sliding-based tool to navigate a time-varying graph. There are several other rendering techniques for multivariate graphs that can be applied for time-varying graphs, graph visualization libraries and tools, and network visualization tools available for use. The choice of tools depends on the size, scale, the nature of the graphs, the type of analysis (flow-based, relationships, clusters, cliques), and the platform for visualization (desktop or web browser). Some of the widely used visualization desktop visualization tools are Gephi (Gephi, 2016), Cytoscape (Cytoscape, 2016), Palantir (Palantir, 2016), ComVis (Matković et al., 2008), and Dato (GraphLab) (Dato (GraphLab), 2016). There are also several web-based visualization libraries that include D3.js (D3.js, 2016), Sigma.js (Sigma.js, 2016), and Vivagraph.js (VivaGraph.js, 2016). More detailed lists of visualization tools are available in kdnuggets (2016) and blog (2016).

5 VISUAL ANALYTICS SANDBOX: AN IMPLEMENTATION ARCHITECTURE

5.1 Implementation Architecture

As shown in Fig. 14, our visual analytics sandbox environment is a reference architecture that integrates various components of visual analytics. This

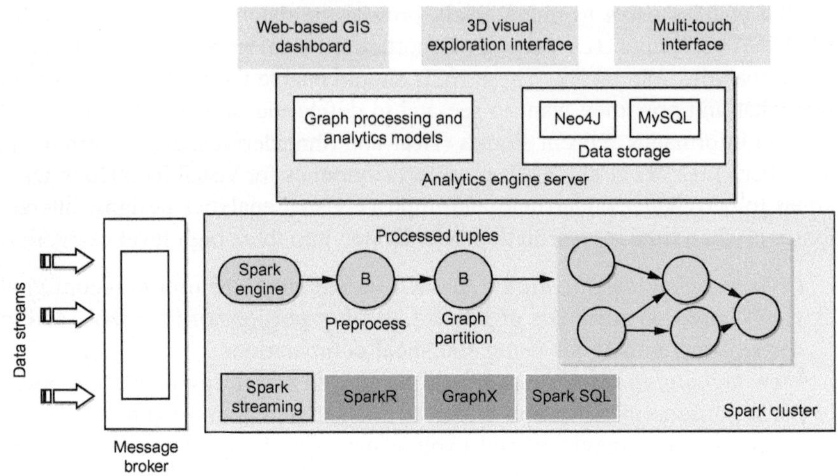

FIG. 14 Tasks performed by different components of the visual analytics sandbox.

includes the data management module with preprocessing tools for graph construction and graph storage and indexing, the analytics engine server that integrates the graph data management components and analytics modules, and visualization and visual interaction components.

The end-to-end system implementation architecture consumes data streams, constructs graphs, and updates the time-evolving graph stored in the graph database based on new incoming data streams. The dynamically updated time-evolving graphs can be accessed by a browser, a 3D environment, or a multitouch interface. The message broker receives data streams from multiple real-time sources, integrates these streams, and sends them to an Apache Spark cluster (Apache Spark, 2016). The Spark cluster does initial preprocessing in terms of extracting relevant information and reducing the dimension of the graph. A graph is constructed for every time window. The transformed graph is loaded into an in-memory graph database. The temporal information about nodes and edges are updated in the new transformed graph. The graph can be queried from a visual interface. The queries include basic node- and edge-based statistics, to mining motifs, cliques, and other persistent graph patterns. The visual interface has various libraries for multiple devices (Table 2).

5.2 Case Study: Emerging Event Detection for Social Media

Various graph-based techniques have been applied on social media data for extracting useful information. These include emerging event detection (Cataldi et al., 2010), community detection (Papadopoulos et al., 2012), opinion mining (Pang and Lee, 2008), and link analysis (Roth et al., 2010; Tang et al., 2011). A social network based on connections between different entities

TABLE 2 Components of the Visual Analytics Sandbox Implementation

Component	Purpose	Methods
Data broker	Integration and distribution of different data streams	Social media streams, Internet traffic streams, sensor network streams, etc.
Online data preparation	Collect data and prepare for preprocessing	Data collection, integration, normalization, representation (schema), etc.
Distributed preprocessing	Improve quality of the data	Data cleaning, correction, transformation, dimensionality reduction, etc.
Batch processing	Generate and clean graphs	Graph generation, pruning, clustering, transformation, etc.
Analytics engine	Graph processing	Graph querying—topology, paths, walks, persistent patterns, motifs, link classification, and prediction
Visual processing	Prepare graphs for visualization	Layout computation and visual representation
Visual interface	User interaction	Web based, 3D exploration, and multitouch interfaces

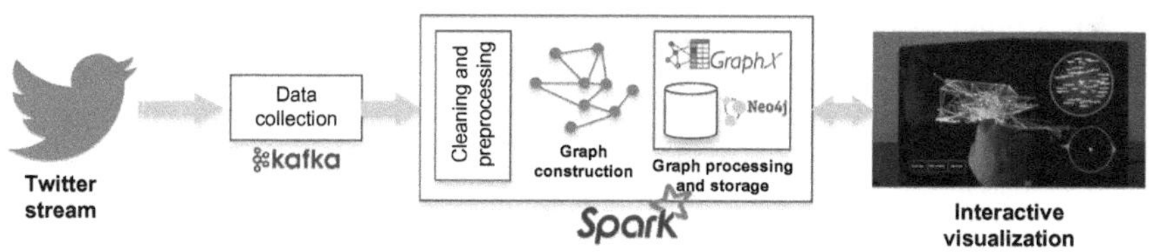

FIG. 15 Work flow schematic for the event detection system.

can be used for identifying and analyzing communities of nodes that are densely connected.

Social networks or graphs constructed from word co-occurrences, user interactions, or friendship networks are temporal in nature. There is ongoing research in understanding the evolution of these temporal social networks, but most of these studies are offline or stop-gap analyses. Tracking the evolution of these graphs is a challenge, because existing infrastructures and tools cannot handle the big data life cycle efficiently from data ingest to knowledge.

The visual analytics sandbox generates word co-occurrence graphs from Twitter streams for detecting emerging events typically within 4–8 min of the initiation of the event. As shown in Fig. 15, the data management work flow consists of a message broker that collects data from the Twitter API, a distributed stream processing tool that extracts relevant information from tweets to construct a graph, a graph database that stores the data, and a visualization interface to interact with the data. Apache Kafka (2016), a low latency message broker, is used for collecting the data from twitter streams and distributing it to multiple consumers. Kafka, compared to other messaging systems, guarantees better throughput and fault tolerance, low latency, built-in partitioning, and replication. The data is sent to Spark, which performs distributed processing of these data streams for extracting word co-occurrences through a sequence of steps such as cleaning and other preprocessing before constructing the word co-occurrence graphs. Spark is a fast in-memory data processing engine with high data parallelism and complete fault tolerance. GraphX (2016) on top of Spark is used to perform some common graph operations; with GraphX it is easy to integrate or join multiple data sources as it is built on top of Spark (Gonzalez et al., 2014). Graph data is stored into Neo4j, a highly scalable ACID-compliant graph database. When the stored data is queried for, Neo4j can perform standard operations like retrieval (nodes, relationships, and subgraphs), filtering, and traversals (paths, loops, and communities). Otherwise, the subgraphs are loaded into a Spark environment where GraphX or other graph libraries perform more complex operations such as graph clustering and reduction, and finding cliques.

The detection analysis task is accomplished by conducting the following four steps. First, a fast and efficient divergence model is used to identify unusual activity in the usage of words. Second, we reduce the co-occurrence graph around those words with unusual activity. Third, candidate events are extracted from the graph using a combination of fast and efficient graph-pruning techniques and a graph-clustering method as shown in Fig. 16.

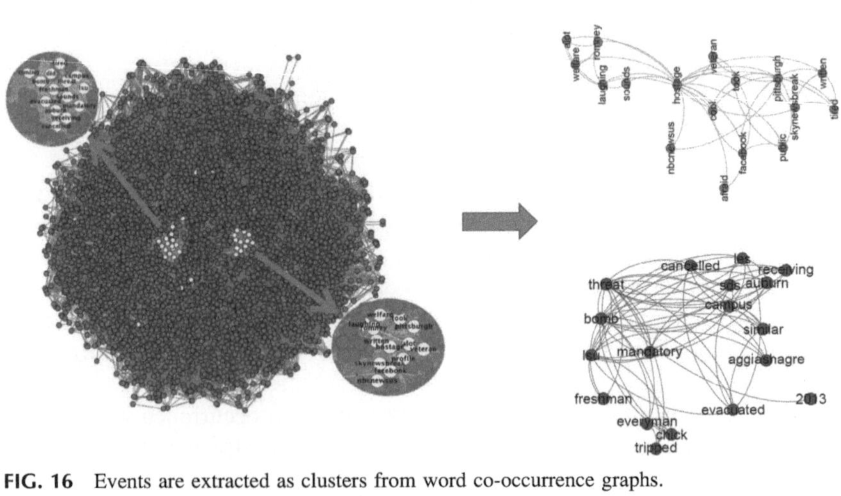

FIG. 16 Events are extracted as clusters from word co-occurrence graphs.

Fourth, spurious clusters (nonevents) are eliminated via an event evolution model, which requires candidate events to be discussed for certain duration of time before being considered a real event. Evaluation of our approach, compared to similar work (Cataldi et al., 2010), shows that the proposed method detects a greater percentage of known true events and a greater number of true events. Moreover, events are detected earlier.

The graphs and analyses are accessible from the visualization node. A multi-touch user interface was developed for exploring and analyzing the temporal graphs interactively. This visual analytics framework has been implemented as a client–server model where the Spark analytics node acts as the server with visualization node acting as the client. Using the visual interface, the end user can interactively explore the temporal graphs through filtering, reduction, zooming, etc. and also can query the graph for structural properties, request more complex analyses. These requests are sent back to the server node for on-demand processing.

5.3 Case Study: Real-Time Information-Based Evacuation Decision Support System for Emergency Management

The dynamics of evacuation are complex due to the number of people, vehicles, road networks, and the uncertainty and perception of the event, and humans decide when and where to evacuate.

The transportation network in this case can be represented as a graph where links represent road networks and nodes represent intersections, origins, and destinations. The visual analytics sandbox provides a platform for consuming and processing real-time information from multiple sources such as traffic sensors, and social media. Fig. 17 shows a detailed description of system architecture. The information processing and synthesis component extracts data from various heterogeneous sources into origin–destination matrices and evacuation network configuration. The origin–destination matrix represents the number of people evacuating from each origin to multiple destinations. The evacuation network configuration represents the state of evacuation highway network in terms of their most recent capacities, taking into consideration the contraflow lanes, road closures, and recent traffic conditions.

Evacuation traffic prediction model: Evacuation is typically modeled as a network optimization problem, where the goal is to select optimal routes from a set of candidate nodes within an existing evacuation network. The selection of optimal routes involves determining the potential evacuation routes based on the location of origin and destination points, the capacity of highways, and the evacuee traffic patterns, that is, how, when, and where people would evacuate. The objective will be to minimize the maximum of the individual remaining times to the destinations, given the evacuee configuration and times to destination at the present time step. The constraints will be on fuel

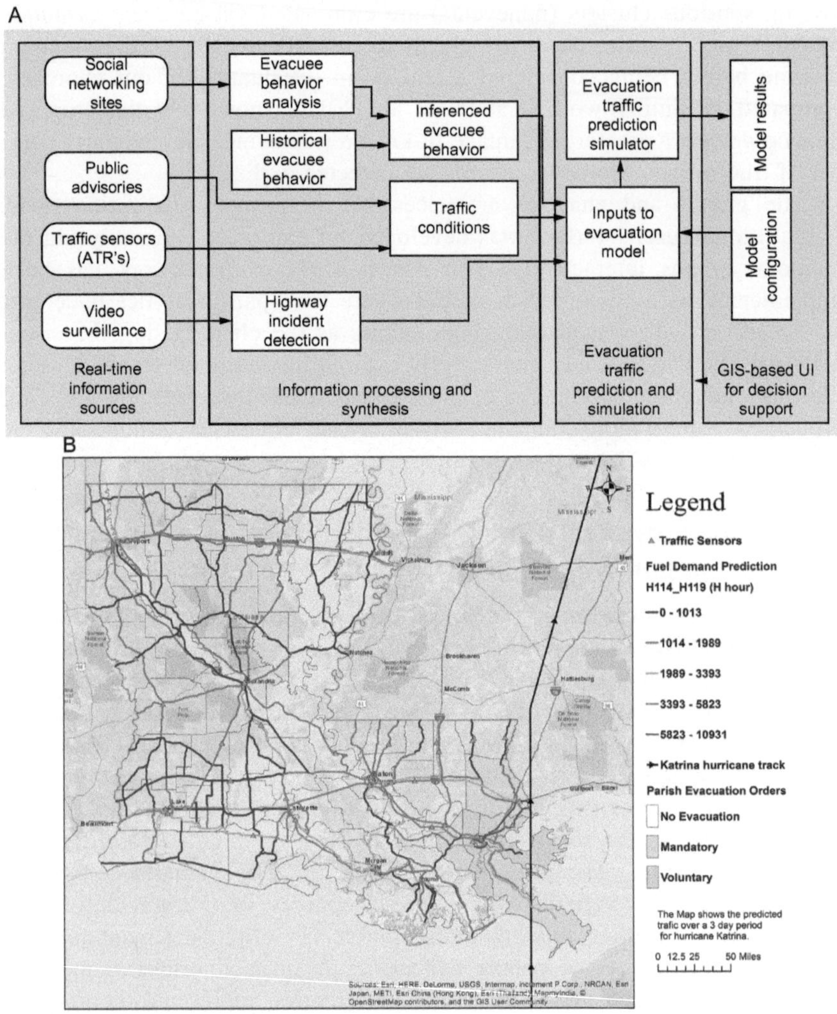

FIG. 17 (A) Data flow diagram showing real-time information-driven evacuation model (Gottumukkala et al., 2012). (B) The map shows the traffic predictions along the evacuation highways for hurricane Gustav.

availability at the nodes at the next time step, traffic capacities along the edges, and the condition that no time to destination increases. Since this is a local-in-time model with reinitialization at each time step, it can be relatively simple, with robustness supplied by the constant data input. The model will compute traffic, fuel, and route capacities (adjusted, e.g., through contraflow) for the next time step. The model would be adjusted with the most recently available data before the next time step. An example objective constructed with these considerations is of the form

$$\min(\varnothing(f)) = \max\{T_i^{t+\Delta t}\}$$

where $T_i^{t+\Delta t}$ is the sum of total remaining travel times of all population at node i to reach their destination at time t. This objective and the constraints will be computed in terms of variables $x_{i,j,l}^{t}$, representing the number of people to be traveling from location i to location j, with the ultimate final destination l in the next time step. Since the $x_{i,j,l}$ can be preidentified to equal zero at a given time step, the resulting linear program is extremely sparse, reducing the computational time. Furthermore, contrast to the evacuation model proposed in Lu et al. (2005), our optimization model will not be done before the evacuation and over the entire evacuation period (that is, over multiple time steps), but during the evacuation, over a single time step, leading to a much smaller problem. The idea is that we are not optimizing the total evacuation event, but are minimizing the maximum delay in reaching the goal over each individual period of time, with adjustment from new observational data at each new time period.

5.4 Case Study: Forecasting Influenza

Forecasting the dynamics of influenza and influenza-like illnesses (ILI) helps in detecting outbreaks well in advance and is useful in public health resource allocations. A handful of literature exists for understanding the spatiotemporal dynamics of ILI outbreaks, and each of these models has different targets and goals for predicting peak volumes and the time of peak surge. Given the benefits of timely forecasts of ILI outbreaks, it is still a huge challenge for developing robust and efficient prediction models. There are several classes of flu prediction models such as ARIMA-based time series models, seasonality-based analog methods, and compartmental models based on disease states SIR (susceptible–infectious–recovered) (Nsoesie et al., 2014). Forecasting ILI volumes using just the historical data is not efficient as it does not follow a simple analog curve even though there is some observable seasonality present because there is some randomness involved. That randomness is generated by a lot of external factors like weather conditions, current ILI situation in nearby or connected locations, etc. It has been accepted and discussed in the literature that these factors correlate or affect ILI volumes (Sloan et al., 2011). Recently, some models have been developed to incorporate some of the variables (Dugas et al., 2013), but still it assumes these correlations are constant or linear. In general, with varying climatic conditions, a lot of things change like travel patterns, virus transmission rates, etc. With this in mind we developed a time-varying graph-based model for forecasting ILI volumes. Broad set surveillance information sources such as Google Flu Trends data and other environmental variables like temperature, precipitation, humidity, etc. are integrated to achieve more sophisticated and accurate analysis.

As shown in Fig. 18, each city is a node with a set of different time series and each of those nodes is connected with each other, and the dynamics of the

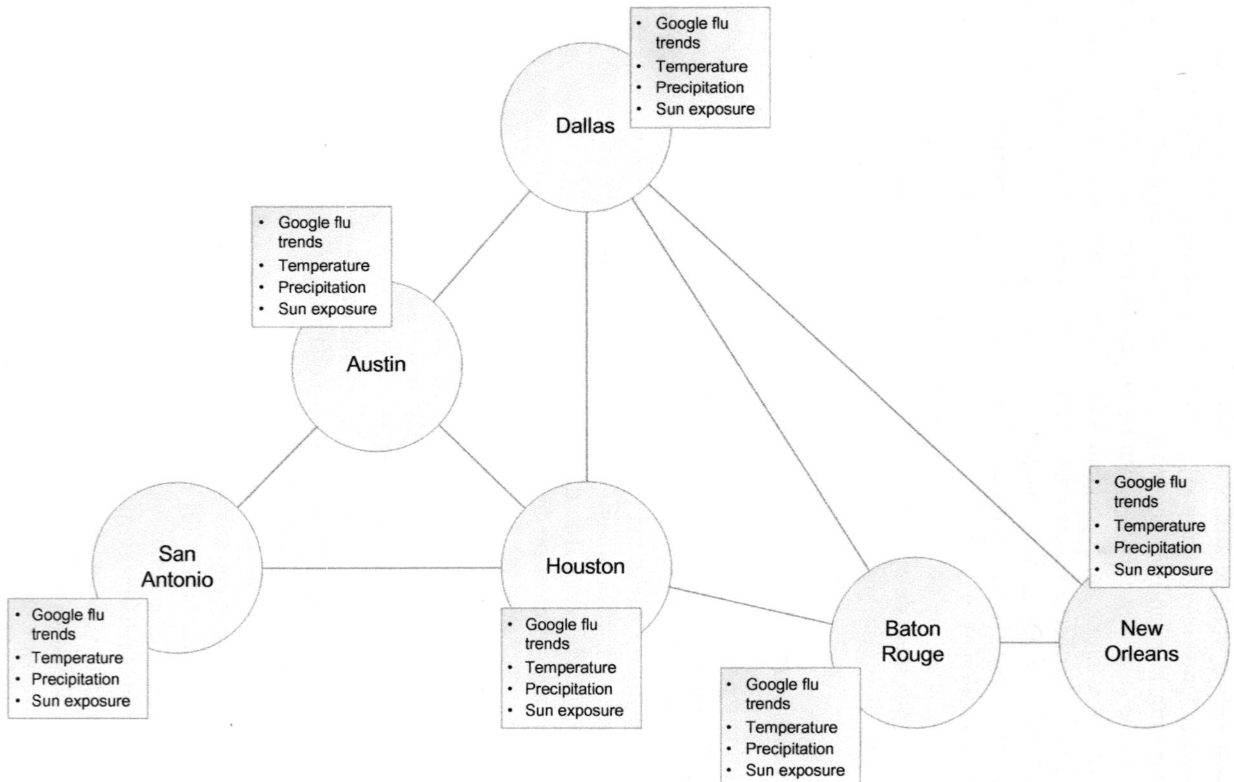

FIG. 18 A graph representation of the data for flu prediction.

graph changes based on the correlations among different time series within a node and also between different nodes. Our model generates situational correlations by transforming these numeric time series into symbolic time series and makes predictions in two stages: in stage 1 a local forecast for each city is made by using variables from each city or location and then in stage 2 the local forecast is adjusted by using the variables information from all those cities within a selected distance to this city, and then the impact of the flu spread pattern happening at those cities is computed to make an adjustment to the forecast for that city. The results of the proposed method after stage 1 are comparable or better than the state-of-the-art method (Dugas et al., 2013), and the results after stage 2 are better than both the stage 1 and the state-of-the-art results.

6 CONCLUSION AND FUTURE RESEARCH

Extracting insights from large dynamic graphs cannot be carried out with traditional data analysis and visualization metaphors. Cognitive computing is an emerging paradigm about building computational artifacts that mimic human intelligence. In this chapter, we present visual analytics as a mechanism for improving decision-making performance of humans. Visual analytics is a synergistic engineering of human and machine capabilities to view and interpret information to combine approaches for data management, data mining, visualization, and HCI to amplify the performance of decision makers. The work presented in this paper provides visual analytics as a framework for improving decision-making performance to handle large-scale time-evolving graphs. In order for data scientists and researchers to develop next generation decision support tools, this book chapter covers the key components including the definition, properties, data management, graph mining, visualization, and visual interaction techniques for time-varying graphs. The chapter also discusses a sandbox implementation and implementation of sandbox for various projects.

There are many challenges for visual analytics in the context of time-evolving graphs. These include:

- *Scalability*: Given that real-world graphs are hard to distribute unlike relational databases, the growing big graph data introduces challenges for pre-processing, graph loading, mining, and visualization. The layout algorithms for visualizing the large-scale graphs need to take the dynamic evolution of these graphs—which is a hard problem because it is hard to estimate the layout of the graphs in domains such as social media where topic and event evolution are so rapid and unpredictable. In order to address scalability challenges, better user controlled graph simplification approaches are needed for filtering, sampling, and aggregation of the graphs.
- *Interactive graph analysis*: Most of the existing graph processing and visual analytics techniques employ black box techniques where the user

has very limited knowledge or no control over the analysis process or what factors contributed to the end result. Systems should be designed to allow the user to guide and control the parameters during the analysis.

- *In situ analysis*: Traditional approaches for storing the data into secondary storage and analyzing later are not feasible with large-scale time-evolving graphs, especially for fast-evolving graphs. Visual analytics systems should explore the idea of in situ analysis and process the data as much as it can while the data is still in memory. A major challenge to address with in situ analysis is to effectively share computing resources and collaborate with overall visual analytics process flow between multiple entities and other user interactions.
- *Visualization*: In addition to addressing the scalability challenges of visualization (rending, layout algorithm, resolution), there is a need to investigate better layout algorithms for different types of time-varying graphs (dense, sparse, slow evolving, fast evolving).
- *Parallel algorithms*: To be on pace with the ever increasing size of graphs and their evolution speed, parallel processing should be explored. As computing resources are getting cheaper and equipped with multiple cores, it is necessary to redesign most of the graph processing and visualization algorithms to support parallel processing.
- *Adaptability*: Designing graph visual analytics frameworks that adapt fast across different application domains is necessary as each application has specific analysis focus and data type. Building such an integrated unified visual analytics framework for time-evolving graphs is a difficult task.
- *Visual analytics APIs*: Lack of resource libraries supporting integrated visual analytics for time-evolving graphs hinders the rapid application development in this scenario. Most of the graph algorithms are designed to support static graphs and some of these have limitations while adapting for time-evolving graphs, and prove to be expensive when they have to be developed from scratch.

ACKNOWLEDGMENTS

This material is based upon work supported by NSF Grant Nos. 1429526 and 1160958. We would also like to acknowledge the following people for their help in contributing data and other discussions. These include Dr. Christoph Borst, Mr. Nicholas Lipari, Mr. Satya Katragadda, and Dr. Ryan Benton.

REFERENCES

Aasman, J., 2006. Allegro Graph: RDF Triple Database. Oakland Franz Incorporated, Cidade.

Aggarwal, C., Subbian, K., 2014. Evolutionary network analysis: a survey. ACM Comput. Surv. 47 (1), 10.

Aggarwal, C.C., Wang, H. (Eds.), 2010. Managing and Mining Graph Data, vol. 40. Springer, New York.

Aggarwal, C.C., Zhao, Y., Yu, P.S., 2011. Outlier detection in graph streams. In: 2011 IEEE 27th International Conference on Data Engineering (ICDE), April. IEEE, Hannover, Germany, pp. 399–409.

Al Hasan, M., Zaki, M.J., 2011. A survey of link prediction in social networks. In: Aggarwal, C.C. (Ed.), Social Network Data Analytics. Springer, Boston, MA, USA, pp. 243–275.

AllegroGraph: http://franz.com/agraph/allegrograph/ (accessed 28 June 2016).

Andrienko, N., Andrienko, G., 2012. Visual analytics of movement: an overview of methods, tools and procedures. Inf. Vis. 12, 3–24. http://dx.doi.org/10.1177/1473871612457601.

Apache Kafka: http://kafka.apache.org/ (accessed 28 June 2016).

Apache Spark™: http://spark.apache.org/ (accessed 28 June 2016).

ArangoDB: https://www.arangodb.com/ (accessed 28 June 2016).

Beck, F., Burch, M., Diehl, S., Weiskopf, D., 2014. The state of the art in visualizing dynamic graphs. In: EuroVis STAR.

Bhagat, S., Cormode, G., Muthukrishnan, S., 2011. Node classification in social networks. In: Aggarwal, C.C. (Ed.), Social Network Data Analytics. Springer, Boston, MA, USA, pp. 115–148.

Birst: https://www.birst.com/ (accessed 28 June 2016).

https://blog.profitbricks.com/39-data-visualization-tools-for-big-data/ (accessed 28 June 2016).

Briody, D., 2015. New vocabulary: cognitive computing. Think Leaders https://www.think-exchange.com/how-to/cognitive-computing-new-vocabulary (accessed 28 June 2016).

Burch, M., Beck, F., Diehl, S., 2008. Timeline trees: visualizing sequences of transactions in information hierarchies. In: Proceedings of the Working Conference on Advanced Visual Interfaces, May. ACM, Napoli, Italy, pp. 75–82.

Casteigts, A., Flocchini, P., Quattrociocchi, W., Santoro, N., 2012. Time-varying graphs and dynamic networks. Int. J. Parallel Emerg. Distrib. Syst. 27 (5), 387–408.

Casteigts, A., Flocchini, P., Mans, B., Santoro, N., 2014. Measuring temporal lags in delay-tolerant networks. IEEE Trans. Comput. 63 (2), 397–410.

Cataldi, M., Di Caro, L., Schifanella, C., 2010. Emerging topic detection on twitter based on temporal and social terms evaluation. In: Proceedings of the Tenth International Workshop on Multimedia Data Mining, July. ACM, Washington, DC, p. 4.

Chakrabarti, D., Faloutsos, C., Zhan, Y., 2007. Visualization of large networks with min-cut plots, A-plots and R-MAT. Int. J. Hum. Comput. Stud. 65 (5), 434–445.

Chau, D.H., Nachenberg, C., Wilhelm, J., Wright, A., Faloutsos, C., 2010. Polonium: tera-scale graph mining for malware detection. ACM SIGKDD Conference on Knowledge Discovery and Data Mining Washington, DC, USA.

Chen, Z., Ji, C., 2005. Spatial-temporal modeling of malware propagation in networks. IEEE Trans. Neural Netw. 16 (5), 1291–1303.

Chen, Y., Argentinis, J.E., Weber, G., 2016. IBM Watson: how cognitive computing can be applied to big data challenges in life sciences research. Clin. Ther. 38 (4), 688–701.

Cytoscape: http://www.cytoscape.org/ (accessed 28 June 2016).

D3.js: http://d3js.org/ (accessed 28 June 2016).

Dato (GraphLab): https://dato.com (accessed 28 June 2016).

Deshpande, A., Guestrin, C., Madden, S.R., Hellerstein, J.M., Hong, W., 2004. Model-driven data acquisition in sensor networks. Proceedings of the Thirtieth International Conference on Very Large Databases, vol. 30. VLDB Endowment, Toronto, Canada, pp. 588–599.

Dong, W., Lepri, B., Pentland, A.S., 2011. Modeling the co-evolution of behaviors and social relationships using mobile phone data. In: Proceedings of the 10th International Conference on Mobile and Ubiquitous Multimedia, December. ACM, Beijing, China, pp. 134–143.

Dugas, A.F., Jalalpour, M., Gel, Y., Levin, S., Torcaso, F., Igusa, T., Rothman, R.E., 2013. Influenza forecasting with Google flu trends. PLoS One 8 (2), e56176.

Dunne, C., Shneiderman, B., 2013. Motif simplification: improving network visualization readability with fan, connector, and clique glyphs. In: Proceedings of the SIGCHI Conference on Human Factors in Computing Systems, April. ACM, Paris, France, pp. 3247–3256.

FlockDB: https://github.com/twitter/flockdb (accessed 28 June 2016).

Gaertler, M., Wagner, D., 2005. A hybrid model for drawing dynamic and evolving graphs. In: Healy, P., Nikolov, N.S. (Eds.), Graph Drawing. Springer, Berlin, Heidelberg, pp. 189–200.

Galati, A., Vukadinovic, V., Olivares, M., Mangold, S., 2013. Analyzing temporal metrics of public transportation for designing scalable delay-tolerant networks. In: Proceedings of the 8th ACM Workshop on Performance Monitoring and Measurement of Heterogeneous Wireless and Wired Networks, November. ACM, Barcelona, Spain, pp. 37–44.

Gephi: http://gephi.github.io/ (accessed 28 June 2016).

Gonzalez, J.E., Xin, R.S., Dave, A., Crankshaw, D., Franklin, M.J., Stoica, I., 2014. Graphx: graph processing in a distributed dataflow framework. In: 11th USENIX Symposium on Operating Systems Design and Implementation (OSDI 14), pp. 599–613.

Google Fusion Tables: https://sites.google.com/site/fusiontablestalks/ (accessed 28 June 2016).

Gottumukkala, R., Zachary, J., Kearfott, B., Kolluru, R., 2012. Real-time information driven decision support system for evacuation planning. In: 2012 IEEE International Multi-Disciplinary Conference on Cognitive Methods in Situation Awareness and Decision Support (CogSIMA), March. IEEE, New Orleans, LA, pp. 206–209.

GraphX: http://spark.apache.org/graphx/ (accessed 28 June 2016).

Greilich, M., Burch, M., Diehl, S., 2009. Visualizing the evolution of compound digraphs with TimeArcTrees. Comput. Graph. Forum 28 (3), 975–982. Blackwell Publishing Ltd.

Hao, M.C., Dayal, U., Keim, D.A., Schreck, T., 2005. Importance-driven visualization layouts for large time series data. In: IEEE Symposium on Information Visualization. INFOVIS 2005, October. IEEE, Minneapolis, MN, pp. 203–210.

Heer, J., Shneiderman, B., 2012. Interactive dynamics for visual analysis. Queue 10 (2), 30.

Holme, P., Saramäki, J., 2012. Temporal networks. Phys. Rep. 519 (3), 97–125.

IBM Many Eyes: http://www.ibm.com/analytics/us/en/technology/advanced-analytics/ (accessed 28 June 2016).

Iliofotou, M., Kim, H.C., Faloutsos, M., Mitzenmacher, M., Pappu, P., Varghese, G., 2011. Graption: a graph-based P2P traffic classification framework for the internet backbone. Comput. Netw. 55 (8), 1909–1920.

InfiniteGraph: The Distributed Graph Database, white paper from Objectivity Inc., 2012. http://www.objectivity.com/wp-content/uploads/Objectivity_WP_IG_Distr_Benchmark.pdf (accessed 28 June 2016).

http://www.kdnuggets.com/2015/06/top-30-social-network-analysis-visualization-tools.html (accessed 28 June 2016).

Keim, D.A., Kohlhammer, J., Ellis, G., Mansmann, F. (Eds.), 2010. Mastering the Information Age-Solving Problems with Visual Analytics. Eurographics, Goslar, Germany.

Kelly, J.E., 2015. Computing, cognition and the future of knowing. IBM Archives. http://www.research.ibm.com/software/IBMResearch/multimedia/Computing_Cognition_WhitePaper.pdf.

Kerren, A., Schreiber, F., 2012. Toward the role of interaction in visual analytics. In: Proceedings of the Winter Simulation Conference, December. IEEE, Berlin, Germany, p. 420.

Kostakos, V., 2009. Temporal graphs. Phys. A Stat. Mech. Appl. 388 (6), 1007–1023.

Kumar, G., Garland, M., 2006. Visual exploration of complex time-varying graphs. IEEE Trans. Vis. Comput. Graph. 12 (5), 805–812.

Kumar, R., Novak, J., Tomkins, A., 2010. Structure and evolution of online social networks. In: Yu, S.P., Han, J., Faloutsos, C. (Eds.), Link Mining: Models, Algorithms, and Applications. Springer, New York, pp. 337–357.

Licklider, J.C., 1960. Man-computer symbiosis. IRE Trans. Hum. Factors Electron. 1, 4–11.

Louati, A., Aufaure, M.A., Lechevallier, Y., Chatenay-Malabry, F., 2011. Graph aggregation: application to social networks. In: Rong, G., Yves, L., Gilbert, S., Wang, H. (Eds.), HDSDA. Hermann, Toulouse, France, pp. 157–177. https://hal.inria.fr/hal-00838649/.

Lu, Q., George, B., Shekhar, S., 2005. Capacity constrained routing algorithms for evacuation planning: a summary of results. In: Advances in Spatial and Temporal Databases. Springer, Berlin, Heidelberg, pp. 291–307.

Masuda, N., Holme, P., 2013. Predicting and controlling infectious disease epidemics using temporal networks. F1000 Prime Rep. 5, 6.

Matković, K., Freiler, W., Gračanin, D., Hauser, H., 2008. ComVis: a coordinated multiple views system for prototyping new visualization technology. In: 12th International Conference on Information Visualisation, IV'08. IEEE, London, England, pp. 215–220.

Miller, J.J., 2013. Graph database applications and concepts with Neo4j. In: Proceedings of the Southern Association for Information Systems Conference, Atlanta, GA, USA, vol. 2324.

Navlakha, S., Schatz, M.C., Kingsford, C., 2009. Revealing biological modules via graph summarization. J. Comput. Biol. 16 (2), 253–264.

Nepusz, T., Yu, H., Paccanaro, A., 2012. Detecting overlapping protein complexes in protein-protein interaction networks. Nat. Methods 9 (5), 471–472.

Newman, M.E., Girvan, M., 2004. Finding and evaluating community structure in networks. Phys. Rev. E 69 (2), 026113.

Nguyen, N.P., Dinh, T.N., Xuan, Y., Thai, M.T., 2011. Adaptive algorithms for detecting community structure in dynamic social networks. In: INFOCOM, 2011 Proceedings IEEE, April. IEEE, Shanghai, China, pp. 2282–2290.

Nicosia, V., Tang, J., Musolesi, M., Russo, G., Mascolo, C., Latora, V., 2012. Components in time-varying graphs. Chaos 22 (2), 023101.

Nicosia, V., Tang, J., Mascolo, C., Musolesi, M., Russo, G., Latora, V., 2013. Graph metrics for temporal networks. In: Holme, P., Saramäki, J. (Eds.), Temporal Networks. Springer, Berlin, Heidelberg, pp. 15–40.

Noel, S.E., Jajodia, S., 2009. George Mason Intellectual Properties, Inc., Attack graph aggregation. U.S. Patent 7,627,900.

Noel, S., Harley, E., Tam, K.H., Gyor, G., 2015. Big-data architecture for cyber attack graphs: representing security relationships in NoSQL graph databases. In: IEEE Symposium on Technologies for Homeland Security (HST), Boston, MA.

Nsoesie, E.O., Brownstein, J.S., Ramakrishnan, N., Marathe, M.V., 2014. A systematic review of studies on forecasting the dynamics of influenza outbreaks. Influenza Other Respir. Viruses 8 (3), 309–316.

Palantir: https://www.palantir.com/ (accessed 28 June 2016).

Pang, B., Lee, L., 2008. Opinion mining and sentiment analysis. Found. Trends Inf. Ret. 2 (1–2), 1–135.

Papadopoulos, S., Kompatsiaris, Y., Vakali, A., Spyridonos, P., 2012. Community detection in social media. Data Min. Knowl. Disc. 24 (3), 515–554.

Pienta, R., Abello, J., Kahng, M., Chau, D.H., 2015. Scalable graph exploration and visualization: Sensemaking challenges and opportunities. In: 2015 International Conference on Big Data and Smart Computing (BigComp), February. IEEE, Jeju, South Korea, pp. 271–278.

Rinaldi, S.M., Peerenboom, J.P., Kelly, T.K., 2001. Identifying, understanding, and analyzing critical infrastructure interdependencies. IEEE Control. Syst. 21 (6), 11–25.

Roth, M., Ben-David, A., Deutscher, D., Flysher, G., Horn, I., Leichtberg, A., Leiser, N., Matias, Y., Merom, R., 2010. Suggesting friends using the implicit social graph. In: Proceedings of the 16th ACM SIGKDD International Conference on Knowledge Discovery and Data Mining, July. ACM, Washington, DC, pp. 233–242.

Santoro, N., Quattrociocchi, W., Flocchini, P., Casteigts, A., Amblard, F., 2011. Time-varying graphs and social network analysis: temporal indicators and metrics. arXiv. preprint arXiv:1102.0629.

Saraiya, P., Lee, P., North, C., 2005. Visualization of graphs with associated time series data. In: IEEE Symposium on Information Visualization. INFOVIS 2005, October. IEEE, Minneapolis, MN, pp. 225–232.

SAS® Visual Analytics: http://www.sas.com/en_id/software/business-intelligence/visual-analytics.html (accessed 28 June 2016).

Shao, B., Wang, H., Xiao, Y., 2012. Managing and mining large graphs: systems and implementations. In: Proceedings of the 2012 ACM SIGMOD International Conference on Management of Data, May. ACM, Scottsdale, AZ, pp. 589–592.

Sigma.js: http://sigmajs.org/ (accessed 28 June 2016).

Sloan, C., Moore, M.L., Hartert, T., 2011. Impact of pollution, climate, and sociodemographic factors on spatiotemporal dynamics of seasonal respiratory viruses. Clin. Transl. Sci. 4 (1), 48–54.

Sperotto, A., Schaffrath, G., Sadre, R., Morariu, C., Pras, A., Stiller, B., 2010. An overview of IP flow-based intrusion detection. IEEE Commun. Surv. Tutorials 12 (3), 343–356.

Sun, J., Tang, J., 2011. A survey of models and algorithms for social influence analysis. In: Aggarwal, C.C. (Ed.), Social Network Data Analytics. Springer, Boston, MA, USA, pp. 177–214.

Tableau: http://www.tableau.com/ (accessed 28 June 2016).

Tang, J., Musolesi, M., Mascolo, C., Latora, V., Nicosia, V., 2010. Analysing information flows and key mediators through temporal centrality metrics. In: Proceedings of the 3rd Workshop on Social Network Systems, April. ACM, Paris, France, p. 3.

Tang, W., Zhuang, H., Tang, J., 2011. Learning to infer social ties in large networks. In: Machine Learning and Knowledge Discovery in Databases. Springer, Berlin, Heidelberg, pp. 381–397.

Tekušová, T., Schreck, T., 2008. Visualizing time-dependent data in multivariate hierarchic plots-design and evaluation of an economic application. In: 12th International Conference on Information Visualisation, IV'08, July. IEEE, London, England, pp. 143–150.

Thomas, J.J., 2005. Illuminating the Path: [The Research and Development Agenda for Visual Analytics]. In: IEEE Computer Society, Los Alamitos, CA.

Tian, Y., Hankins, R.A., Patel, J.M., 2008. Efficient aggregation for graph summarization. In: Proceedings of the 2008 ACM SIGMOD International Conference on Management of Data, June. ACM, Vancouver, Canada, pp. 567–580.

Titan: http://thinkaurelius.github.io/titan/ (accessed 28 June 2016).

Titan: Distributed Graph Database. http://thinkaurelius.github.io/titan/ (accessed 28 June 2016).

Tse, Y.K., Tsui, A.K.C., 2002. A multivariate generalized autoregressive conditional heteroscedasticity model with time-varying correlations. J. Bus. Econ. Stat. 20 (3), 351–362.

Tsuda, K., Saigo, H., 2010. Graph classification. In: Aggarwal, C.C., Wang, H. (Eds.), Managing and Mining Graph Data. Springer, Boston, MA, USA, pp. 337–363.

VivaGraph.js: https://github.com/anvaka/VivaGraphJS (accessed 28 June 2016).

Von Landesberger, T., Kuijper, A., Schreck, T., Kohlhammer, J., van Wijk, J.J., Fekete, J.D., Fellner, D.W., 2011. Visual analysis of large graphs: state-of-the-art and future research challenges. Comput. Graph. Forum 30 (6), 1719–1749. Blackwell Publishing Ltd.

Wehmuth, K., Ziviani, A., Fleury, E., 2015. A unifying model for representing time-varying graphs. In: IEEE International Conference on Data Science and Advanced Analytics (DSAA), 2015, 36678, October. IEEE, Paris, France, pp. 1–10.

Williams, D.W., Huan, J., Wang, W., 2007. Graph database indexing using structured graph decomposition. In: IEEE 23rd International Conference on Data Engineering (ICDE 2007), April. IEEE, Istanbul, Turkey, pp. 976–985.

Wong, P.C., Shen, H.W., Johnson, C.R., Chen, C., Ross, R.B., 2012. The top 10 challenges in extreme-scale visual analytics. IEEE Comput. Graph. Appl. 32 (4), 63.

Yi, J.S., ah Kang, Y., Stasko, J.T., Jacko, J.A., 2007. Toward a deeper understanding of the role of interaction in information visualization. IEEE Trans. Vis. Comput. Graph. 13 (6), 1224–1231.

Zhou, Y., Cheng, H., Yu, J.X., 2009. Graph clustering based on structural/attribute similarities. Proc. VLDB Endowment 2 (1), 718–729.

Chapter 4

CyGraph: Graph-Based Analytics and Visualization for Cybersecurity

S. Noel[1], E. Harley, K.H. Tam, M. Limiero and M. Share
The MITRE Corporation, McLean, VA, United States
[1]Corresponding author: e-mail: snoel@mitre.org

ABSTRACT
This chapter describes *CyGraph*, a system for improving network security posture, maintaining situational awareness in the face of cyberattacks, and focusing on protection of mission-critical assets. CyGraph adopts a unified graph-based cybersecurity model relevant to potential and actual cyberattacks, defenses, and mission impacts. It captures incremental attack vulnerability, security events, and mission dependencies within a network environment, builds a predictive model of possible attack paths and critical vulnerabilities, and correlates events to known vulnerability paths. It also includes dependencies among mission requirements and network assets, for analysis in the context of mission assurance. The resulting knowledge graph captures the complex relationships among entities in the cybersecurity domain. CyGraph brings together isolated data and events into an overall picture for decision support and situational awareness. It prioritizes exposed vulnerabilities, mapped to potential threats, in the context of mission-critical assets. In the face of actual attacks, it correlates intrusion alerts to known vulnerability paths and suggests best courses of action for responding to attacks. For postattack forensics, it shows vulnerable paths that may warrant deeper inspection. CyGraph also supports CyQL (CyGraph Query Language), a domain-specific query language for expressing graph patterns of interest, with interactive visualization of query results. To help manage visual complexity, CyGraph supports the separation of graph models into interdependent layers. For time-dependent graph models, it provides dynamic visualization of evolving graph state. CyGraph also integrates with third-party tools for visualizing graph state changes (e.g., driven by simulations). Furthermore, it has capabilities for synthesizing graph models with particular statistical properties.

Keywords: Cybersecurity modeling, Situational awareness, Mission assurance attack graphs, NoSQL graph databases, Graph visualization

Handbook of Statistics, Vol. 35. http://dx.doi.org/10.1016/bs.host.2016.07.001

1 INTRODUCTION

Cyberattacks and defenses against them are conducted in complex environments, with numerous factors contributing to attack success and mission impacts. Network topology, host configurations, vulnerabilities, firewall settings, intrusion detection systems, mission dependencies, and many other elements can play parts. To go beyond rudimentary assessments of security posture, organizations need to merge isolated data into higher-level knowledge of network-wide attack vulnerability and mission readiness in the face of cyber threats.

Network environments are always changing, with machines added and removed, patches applied, applications installed, firewall rules changed, etc., all with potential impact on security posture. Intrusion alerts and antivirus warnings need attention, and even seemingly benign events such as logins, service connections, and file share accesses may be associated with adversary activity.

The problem is often not lack of available information, but rather the ability to assemble disparate pieces of information into an overall picture for situational awareness, optimal courses of action, and maintaining mission readiness. Security analysts and operators can be overwhelmed by a variety of consoles from multiple tools; each tool provides only a limited view of one aspect of the overall space under consideration. Tools such as security information and event management (SIEM) can help by normalizing data and bringing it together under a common framework. But the data still remain as individual pieces of information, rather than a comprehensive model of network-wide vulnerability paths, adversary activities, and potential mission impacts.

Our goal is to maximize the ability to discover potential threats and mission impacts, while minimizing the time needed for organizing multiple disparate data sources into meaningful relationships. For example, in the well-publicized Target retailer data breach (Harris and Perlroth, 2014), it was revealed that cyber defenders were actually aware of an alert for a particular aspect of the attack, but decided that it was a false positive. We could surmise that if those defenders understood the potential downstream ramifications of that alert, they would have considered it much more carefully, preformed additional investigations, etc. The goal is to provide the higher-order correlations that defenders need for truly informed decisions.

For the Target data breach, the attack began with a compromise within a partner (contractor) network. A common way for this to happen is through Trojan malware. Alerts for such malware are happening with high frequency in many environments and are often considered a low business risk (i.e., mainly a risk for individual clients). However, in the case of the Target breach, the infected host in the contractor became a launching point into the Target network. Several other steps were part of the breach, in which the attackers incrementally increased their scope of control, until they met their attack goals (exfiltrating large-scale credit card data).

The key lesson is that there were multiple attack steps, with multiple corresponding opportunities for detection. However, such alerts and other indicators occur within a large background of event noise. Since it is not practical for human defenders to consider all the possible multistep inferences, this needs to be automated. Also, defenders can make even more informed decisions (and reduce numbers of truly critical incidents to consider) by focusing such inference on mission-critical network assets. This can also be done preemptively, to discover and reduce such critical vulnerability paths.

To help address these challenges, we introduce *CyGraph*, a tool for cyber warfare analytics, visualization, and knowledge management. CyGraph brings together isolated data and events into an ongoing overall picture for decision support and situational awareness. It prioritizes exposed vulnerabilities, mapped to potential threats, in the context of mission-critical assets. In the face of actual attacks, it correlates intrusion alerts to known vulnerability paths and suggests best courses of action for responding to attacks. For post-attack forensics, it shows vulnerable paths that may warrant deeper inspection.

CyGraph builds an attack graph model that maps the potential attack paths through a network. This includes any network attributes that potentially contribute to attack success, such as network topology, firewall rules, host configurations, and vulnerabilities. The dynamically evolving attack graph provides the context for reacting appropriately to attacks and protecting mission-critical assets. CyGraph then ingests network events such as intrusion detection alerts and other sensor outputs, including packet capture. It also incorporates mission dependencies, showing how mission objectives, tasks, and information depend on cyber assets.

CyGraph fuses information from a variety of data sources to build its unified graph-based model. As shown in Fig. 1, this is a layered model, which includes the comprehensive information needed for making informed judgments about mission readiness in the face of cyber warfare.

The *network infrastructure layer* captures how the network is segmented and organized topologically, the locations of sensors, etc. The *cyber posture layer* considers elements within the network infrastructure that might impact cyberattacks/defenses, e.g., host configurations, vulnerabilities, services, shared resources, firewall policies, etc. The *cyber threats layer* describes potential adversary threats, for application against the defensive posture. This includes threat intelligence (e.g., shared among trusted partners) as well as event streams of alerts and other behavioral indicators. Finally, the *mission dependencies layer* captures dependencies among various mission components (from high-level objectives to tasks that support objectives to information required for task, etc.), as well as the particular cyber assets that support the mission components.

CyGraph has the potential for dramatically shortening the analytical cycle. It provides the network-specific context needed for mapping cyber threats to specific network environments, reducing false alarms, and suggesting optimal

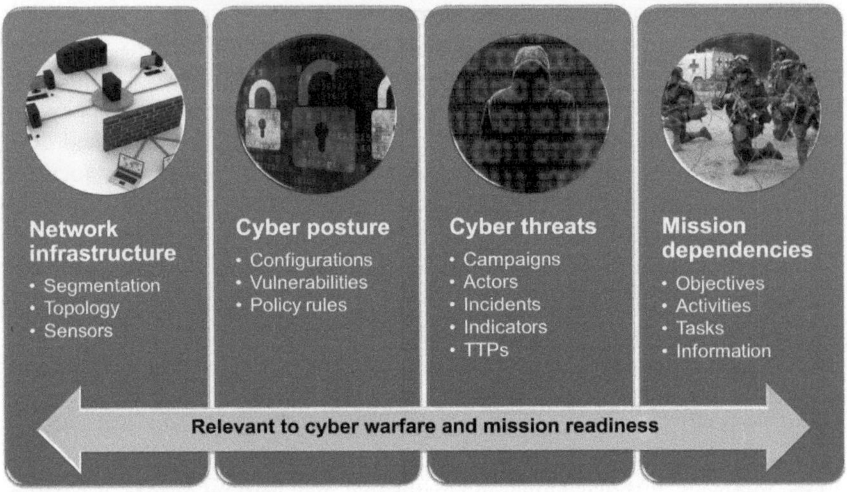

FIG. 1 CyGraph knowledge stack.

attack responses. It helps prioritize exposed vulnerabilities, alone and in combination, with focus on protecting mission-critical assets against potential threat sources. It also provides the context for correlating intrusion alerts and other kinds of network events, matching them to known vulnerability paths. This in turn suggests best courses of action for responding to attacks. Specifically, for postattack situational awareness, CyGraph shows possible paths leading up to the current attack locus (backward looking) as well as potential paths for the attacker to advance the attack (forward looking). It also provides a comprehensive framework for computing a variety of metrics for tracking security readiness over time.

CyGraph provides comprehensive query capabilities over its graph knowledge base, including a query language specific to its knowledge domain. This supports a range of cyber analysis tasks, such as mapping an attacker's potential reach and combining isolated alerts into coordinated multistep attack campaigns. CyGraph also provides a variety of interactive visualization capabilities for portraying complex graph query results.

Section 2 discusses previous work related to the CyGraph system. Section 3 then describes CyGraph in more detail. In Section 4, we examine a number of example applications of CyGraph. Section 5 then summarizes this chapter.

2 RELATED WORK

Traditionally, the development of cybersecurity models and analytics has been hampered by a lack of information sharing. MITRE's *Making Security Measurable* (Martin, 2008; The MITRE Corporation, 2013) is a collection of collaborative initiatives for shared information, languages, and processes for cybersecurity that has helped in that regard. These initiatives encompass software assurance, threat analysis, vulnerability management, malware protection, intrusion detection, incident coordination, and other areas of security. Collaborators include teams from US Department of Homeland Security (DHS), Defense Information Systems Agency (DISA), National Institute of Standards and Technology (NIST), Internet Engineering Task Force (IETF), and many others from government, academia, and industry (The MITRE Corporation, 2014). These standardization efforts facilitate information sharing needed for building CyGraph knowledge graphs. Also helpful are a variety of ontologies and taxonomies that have been proposed for the cybersecurity domain, e.g., Iannacone et al. (2014).

On the product side, various tools for attack graph analysis are available, both Government Off-The-Shelf (GOTS) and Commercial Off-The-Shelf (COTS). GOTS examples include TVA (Jajodia et al., 2005) (commercialized as Cauldron (NSA, 2009)) and NetSPA (Artz, 2002) (commercialized as CyberAnalytix, 2008). Other COTS tools include Skybox (2016) and RedSeal (2016). Attack graphs have also been generated by logic programming (Ou et al., 2005) and model checking (Sheyner and Wing, 2004). More comprehensive reviews of previous work in attack graphs and other graph-based cybersecurity models are given in Lippmann and Ingols (2005) and Schweitzer (2013).

Previous approaches and tools for attack graph analysis have generally employed specialized data structures and algorithms designed for solving specific problems, e.g., attack reachability (Ingols et al., 2006), data aggregation (Noel and Jajodia, 2009a), network hardening (Albanese et al., 2012), sensor placement (Noel and Jajodia, 2007), alert correlation (Ning and Xu, 2004), security posture metrics (Noel and Jajodia, 2014), or risk of unknown (zero-day) vulnerabilities (Wang et al., 2013). Flexibility and extensibility in the face of evolving network environments and adversary threats have not been first-class design criteria. For example, TVA/Cauldron lacks a database persistence layer, and changing the model (new vulnerability scans, firewall rule changes, etc.) requires the entire attack graph to be rebuilt. Such tools are usually implemented with custom code that is difficult to extend as new data sources, model abstractions, analytic techniques, and visualization capabilities are introduced.

Relational database representations have been proposed for attack graphs (Wang et al., 2006). While this has the advantage of a standard model for data representation and queries, the relational model is not the best match for graph problems, especially for evolving network environments and analytic requirements. Extending a relational model requires schema redesign, database reloading, etc. Many graph operations are difficult to express in Structured Query Language (SQL). Moreover, graph traversal in relational databases requires computationally expensive self-join operations.

A class of NoSQL databases known as graph databases has emerged, which are optimized for graph operations. In CyGraph, we employ the Neo4j graph database (Neo4j, 2016). Neo4j represents node adjacency via direct pointers, which avoids expensive join operations for graph traversal. Neo4j has demonstrated graph traversal performance orders of magnitude better than relational databases (Baas, 2012; Batra and Tyagi, 2012). Query execution times depend only on the size of the traversed subgraph, independent of the size of the overall graph (Robinson et al., 2015).

Many of the problems in big data are amenable to established methods of high-performance computing (HPC), so that boundaries between big data and HPC are blurring (Nadkarni and Vesset, 2014). Graph data structures lack the spatial locality implicit in traditional HPC architectures (e.g., for data arrays). To address this, Cray has developed the Urika-GD™ appliance (Cray, 2014), which has specialized hardware and software for high-performance large-scale graph analytics.

3 DESCRIPTION OF CyGraph

CyGraph is a comprehensive, scalable, high-performance system for analyzing and reasoning about network attack relationships. It correlates data from numerous sources (topology, vulnerabilities, client/server configurations, firewall rules, events, etc.) into a common, normalized model and builds a persistent graph data store representing network attack relationships and associated network data. CyGraph supports queries that identify key vulnerabilities, suggest optimal mitigation strategies, map host-to-host trust relationships, show downstream/upstream paths for attack response, etc. The system includes components to compute analytical graph-theoretic measures such as centrality, degree, connectivity, and diameter. It also provides interactive visualization capabilities for conveying complex dependency relationships.

Section 3.1 describes the overall architecture of the CyGraph tool. Section 3.2 explains in detail about the variety of data sources that CyGraph ingests for building its cyber graph models. Section 3.3 describes how CyGraph leverages NoSQL graph databases within an integrated environment for big data analytics, for synthesizing its graph knowledge base from raw data. Section 3.4 examines details of CyGraph's distributed client–server implementation. In Section 3.5, we introduce a domain-specific query language for CyGraph, which provides a level of abstraction that hides lower-level details of the underlying CyGraph data model. Section 3.6 examines a variety of interactive visualization capabilities in CyGraph that help analysts better understand and communicate knowledge base query results.

3.1 CyGraph Architecture

Fig. 2 is a high-level view of the CyGraph architecture. The architecture includes REpresentational State Transfer (REST) web services for ingest, transform, and analytics (queries and visualization). CyGraph ingests data

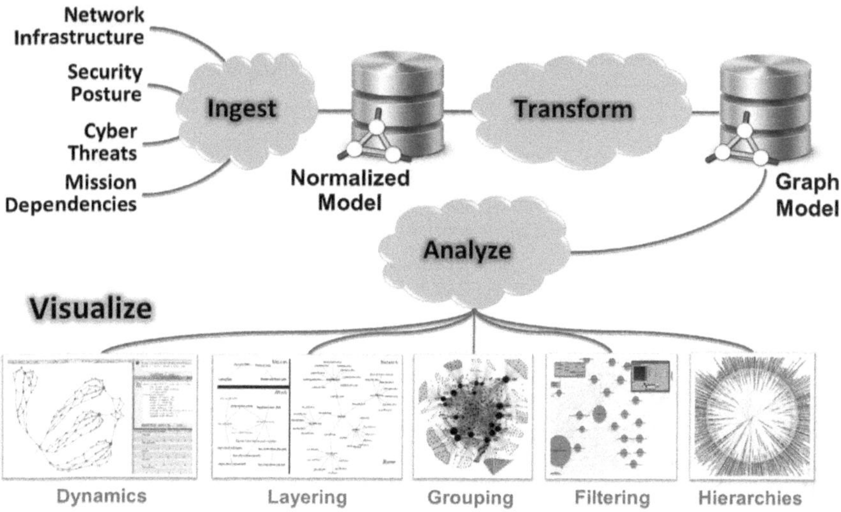

FIG. 2 CyGraph architecture.

from a variety of sources at all layers of its cyber knowledge stack, mapping source-specific data to a common normalized data model. It then transforms the isolated elements of the normalized model into a graph model that captures relevant relationships for the cybersecurity and mission dependence domains. CyGraph also provides a variety of client-side analytic and visual capabilities, including graph dynamics, layering, grouping, filtering, and hierarchical views.

In the CyGraph architecture, the cybersecurity model schema is free to evolve with the available data sources and desired analytics, rather than being fixed at design time. The data model is based on a flexible property-graph formulation implemented in Neo4j. Model extensions are simply the creation of additional of nodes, relationships, and properties in the property-graph data model and require no schema changes or other database renormalizing. Graph pattern-matching queries are expressed in either native Neoj4 query language (Cypher) or our domain-specific CyGraph Query Language (CyQL), which CyGraph compiles to native Cypher.

3.2 CyGraph Data Sources

In the CyGraph architecture, the ingest service provides a standard format for input data, which is processed by source-specific adapters. Thus, data "in the wild" are mapped to the layered CyGraph data model. The network infrastructure layer of this model captures the configuration and policy aspects of the network environment, which forms the basis for modeling security posture (potential vulnerability paths). The cyber threats layer captures events and indicators of actual cyberattacks, which are correlated with elements at the

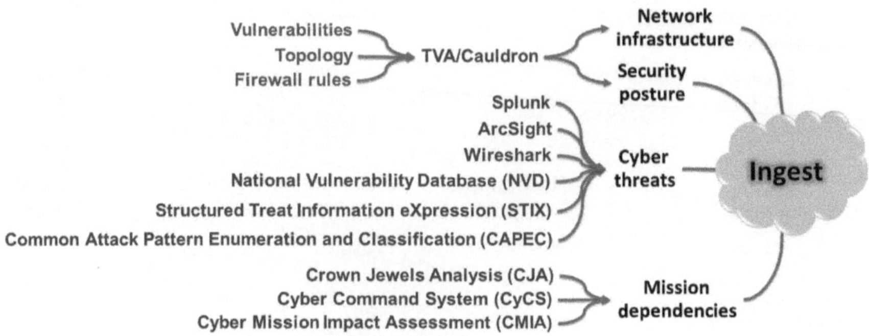

FIG. 3 Example sources for CyGraph data ingest.

lower levels, providing context for cyber events and supporting predictions of subsequent attack spread. The mission dependencies layer shows how cyber activities relate to mission elements.

Fig. 3 shows example data sources for building CyGraph models. These provide elements of CyGraph's cybersecurity/mission knowledge stack. In this way, CyGraph leverages existing tools and data sources for building its rich knowledge graph.

First we consider data sources pertaining to network infrastructure and cyber posture, i.e., network topology, firewall rules, and host vulnerabilities. In CyGraph, we leverage TVA/Cauldron (Jajodia et al., 2011; Noel and Jajodia, 2009b; Noel et al., 2002, 2009; O'Hare et al., 2008), a tool developed at George Mason University for building and analyzing network attack graphs. TVA/Cauldron imports scan results from various vulnerability scanner products. It also parses firewall rules (access control lists) from various firewall vendors. It then analyzes host vulnerabilities, firewall rules, and network topology (subnets, routes, and firewall locations) to enumerate attacker reachability to vulnerable hosts. This in turn provides a model for network infrastructure and security posture in CyGraph. The security posture layer of CyGraph's knowledge stack supports prioritization and optimization of proactive security measures in advance of attack. This layer also provides context for responding to attacks.

In addition to source data about network infrastructure and cyber posture, CyGraph ingests various data sources for cyber threats, both potential (threat intelligence) and actual (cyberattack events). The Splunk log analysis tool (Zadrozny and Kodali, 2013) indexes data from network- and host-based sensors, e.g., intrusion detection systems and other specialized tools providing live threat indicators. CyGraph also processes packet capture data via Wireshark (Sanders, 2011), e.g., for analyzing general traffic patterns. Cyber threat intelligence sources for CyGraph include the National Vulnerability Database (NVD) (2016), Structured Threat Information eXpression (STIX™) (STIX, 2016), and Common Attack Pattern Enumeration and Classification

(CAPEC™) (The MITRE Corporation, 2016). Threat Assessment and Remediation Analysis (TARA) (Wynn et al., 2011) provides a structured methodology for elements of both cyber posture and threats.

Through its mission dependencies layer, CyGraph supports analysis of potential and actual impact of cyberattacks and defenses on organizational functions (missions). To populate this layer, CyGraph leverages established methodologies and tools (Noel and Heinbockel, 2015). This includes Crown Jewels Analysis (CJA) (The MITRE Corporation, 2009), which is a structured methodology for identifying mission-critical cyber assets. It also includes Cyber Command System (CyCS) (The MITRE Corporation, 2016), a tool that captures hierarchical dependencies among mission components, and maps mission operations to the network operations that support them. CyGraph can also leverage Cyber Mission Impact Assessment (CMIA) (Musman et al., 2011), which evaluates time-dependent effects of cyberattacks on mission effectiveness and performance.

In the CyGraph framework, the data model is schema free, so that the model is decoupled from the storage implementation. This provides flexibility in data sources, and how the data are transformed (cast as a graph) determines a particular instantiated CyGraph model. User queries must match a given instantiation. This means that there is a knowledge engineering phase in developing source-specific adapters for populating CyGraph instances, and for formulating relevant queries according to the graph model.

3.3 Big Data Analytics in CyGraph

Big data analytics involve a process of continual discovery. CyGraph discovers interrelationships relevant to attacker progress through a network and corresponding mission impact. This includes vulnerabilities in the usual sense, as well as other attack relationships that enable attacker progress, such as remote desktop and stored credentials. Network events (alerts, flows, etc.) are mapped to these attack relationships, providing context for correlating otherwise isolated events. This shows the next steps that an adversary can take, for optimal response by defenders.

This kind of complex, interconnected, unpredictable data is best captured in a graph model (data structure). Relational databases work well for referencing discrete data items and fixed relationship patterns, e.g., bank customers and their accounts. But the relational model has difficulties when the relationships themselves are variable, especially as in cyber security.

Graph databases are a class of NoSQL database (Gudivada et al., 2016) that embraces graphs as the underlying model for data representation and storage (Angles and Gutierrez, 2008). They often employ semantic (pattern-matching) query languages, which allows retrieval of both explicitly defined information as well as information that can be implied (e.g., through graph traversal). They are applicable when the information about data relationships is as important (or even more important) as the data themselves. A survey of popularity trends for various classes of databases (Andlinger, 2015) shows that graph databases have had a fivefold increase in popularity over the last 2 years (ending March 2015), a much higher growth rate than any other class of database.

Within the graph database class, Neo4j dominates popularity (DB-Engines, 2015). Neo4j is open source, with commercial licensing for an enterprise version also available (Neo4j, 2016). Unlike most NoSQL databases, Neo4j enforces ACID (atomicity, consistency, isolation, durability) transaction properties usually associated with relational database systems.

Graph databases represent node adjacency via direct pointers. This avoids expensive join operations or other index lookups for graph traversal. Graph databases have been shown to be orders of magnitude faster than relational databases for graph traversal, especially deeper traversals (Vukotic et al., 2015). In Neo4j, graph traversal speed depends only on the size of the query result actually traversed, independent of the total size of the graph.

CyGraph is deployed in the MITRE Cyber Analytic Virtual Environment (CAVE), shown in Fig. 4. CAVE provides an integrated, scalable, fault-tolerant, and managed virtual environment for big data analytics. It hosts a suite of cyber data repositories, knowledge bases, and analysis engines, as well as capabilities for querying and visualization.

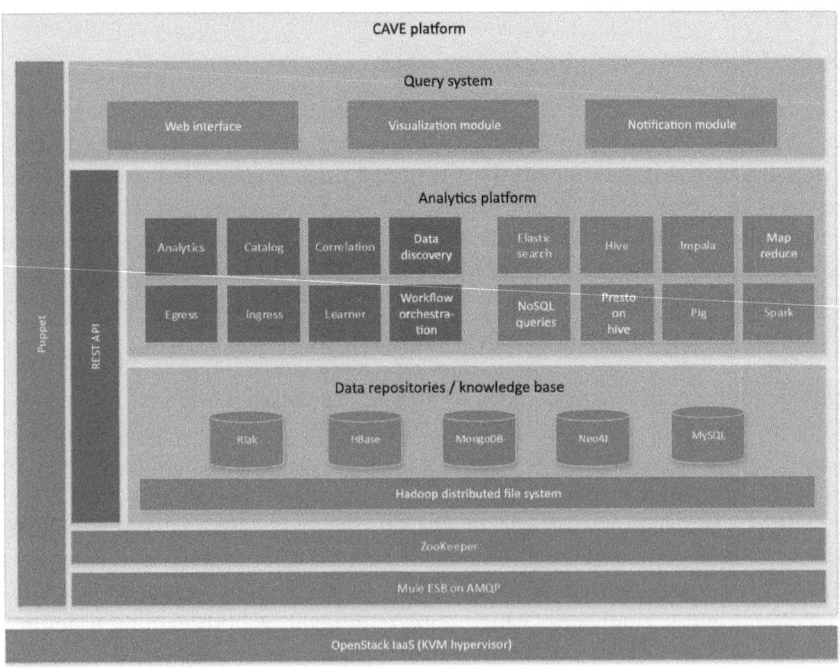

FIG. 4 CAVE stack for big data analytics.

CyGraph synthesizes new network, cyber, and mission knowledge, which is stored in CAVE's knowledge base. In the analytics layer, CyGraph provides graph-based data cataloging, correlation, analytics, and queries. At the top of this big data analytics stack, CyGraph provides novel forms of interactive visualization.

3.4 CyGraph Client–Server

The CyGraph architecture is based on a distributed client–server model, partitioning tasks between service providers (servers) and service requesters (clients). CyGraph clients and servers can be deployed on separate hosts (real and/or virtual) or can be configured to run on a single (local) host. Typically, CyGraph services (middle-tier server and backend database) are implemented as separate virtual machine instances, i.e., in VMware (2016). Such a VMware instance can also be migrated to Amazon Elastic Compute Cloud (Amazon EC2) (Amazon Web Services, 2016) for web-scale cloud computing.

Here are the functional roles for each tier of the CyGraph client–server architecture:

CyGraph Client: The CyGraph Client is a graphical user interface for posing CyGraph queries and visualizing query results. The client communicates with the CyGraph Server through RESTful web application program interface (API) calls. The predominant CyGraph Client is implemented in Java as a desktop application. Because of the decoupled client–server architecture, other clients can be implemented on other platforms, e.g., a web browser.

CyGraph Server: The CyGraph Server acts as a middle-tier intermediary between the CyGraph Client and the CyGraph Database. It provides a layer of abstraction that gives a common service interface, regardless of the how the database backend (and its native query language) is implemented. The server handles the interpretation of CyGraph domain-specific query language (CyQL) into native database queries. The server also houses a library of commonly issued queries, to capture domain knowledge, streamline the analytic workflow, and help ease the learning curve for new analysts.

CyGraph Database: The CyGraph Database stores the graph data (nodes, relationships, and properties). It processes queries from CyGraph Server, in the native language of the database implementation (Neo4j).

Fig. 5 shows the key components for the implementation of each tier of the CyGraph client–server architecture. A key library for CyGraph Client is GraphStream (Dutot et al., 2007), which provides basic functions for graph visualization, styling, and user interaction. CyGraph Server leverages the Spring web model-view-controller (MVC) framework (Yates et al., 2006), which separates the representation of information from how that information

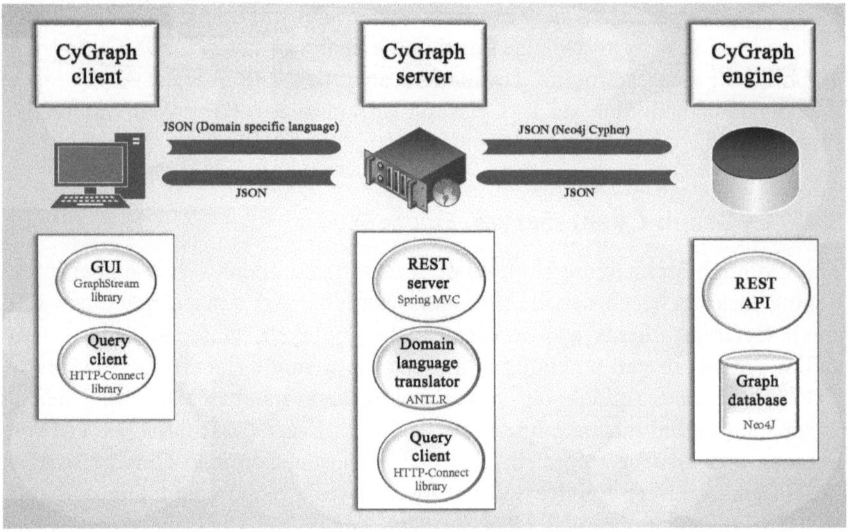

FIG. 5 Components of CyGraph client–server architecture.

is presented to the user. CyGraph Server also applies ANTLR (ANother Tool for Language Recognition) (Parr, 2013) for translating queries in the CyGraph domain-specific query language (CyQL) to corresponding queries in a native graph database query language. CyGraph Client and CyGraph Server also leverage Jersey RESTful web services (Gulabani, 2014) for connecting to CyGraph services via Hypertext Transfer Protocol (HTTP).

In the CyGraph architecture, the analyst formulates a graph pattern-matching query in CyGraph Client (in either domain-specific CyQL or native Neo4j Cypher query language) and submits it. CyGraph Client wraps the query in a JSON message and sends it to the corresponding service (CyQL or Cypher) on CyGraph Server. CyGraph Server processes the query and sends it to CyGraph Database. For a CyQL query, CyGraph Server translates (compiles) it to the corresponding Cypher query; otherwise, it just forwards the native Cypher query.

In response, CyGraph Database (backed by Neo4j) executes the Cypher query and returns the results (matched subgraph) to CyGraph Server as Neo4j-formatted JSON. CyGraph Server parses the Neo4j JSON, builds the corresponding vendor-neutral CyGraph model JSON, and sends that to CyGraph Client for rendering as dynamic graph visualization.

The loosely coupled modular design of CyGraph makes it straightforward to develop alternative implementations for the various tiers of the client–server architecture. For example, CyGraph clients provide alternative graph visualization capabilities in web browsers through standards such as HTML5, JavaScript, Cascading Style Sheets (CSS), and Scalable Vector Graphics (SVG), by leveraging D3.js (Bostock et al., 2011), vis.js (vis.js, 2016), and FoamTree (Carrot Search, 2016). Client-side code also exports CyGraph models into Graphviz.

3.5 CyQL: CyGraph Domain-Specific Query Language

Graph queries in CyGraph are specifications for matching subgraph patterns of interest, written in declarative query language (Harper, 2013). In declarative (nonprocedural) languages, one specifies what needs to be done (i.e., match a particular graph pattern) rather than exactly how to do it (checking properties, traversing edges, etc.).

We provide an additional layer of abstraction by defining a domain-specific language (DSL) for the CyGraph data model, which we call *CyQL*. There are a number of advantages for doing this. A DSL increases the clarity of analytic queries against our cyber data model, especially as the model becomes more complex. CyQL does this by encoding cyber semantics into the query language itself, encapsulating and hiding many of the constraints that must be expressed in the native graph database queries.

While a general-purpose graph query language is broadly applicable across all domains, CyQL is specialized to the application domain within the scope of CyGraph. This in turn helps reduce the learning curve and increase the productivity of security analysts and content developers using CyGraph.

The additional layer of abstraction provided by CyQL also allows CyGraph to support multiple backend data engine implementations, each with their own native query language. This means that users and application software that integrate with CyGraph interact with a single (domain-specific) language, which is independent of the particular native query language that implements the DSL.

Section 3.5.1 describes CyQL in more detail. Section 3.5.2 examines a number of example CyQL queries to see the language in practice.

3.5.1 Description of CyQL

Fig. 6 shows an instance of the underlying CyQL data model, expressed as a graph of entity (node) and relationship (edge) types. Here, properties (name–value pairs) for nodes and edges are omitted for clarity. It is interesting that these four data model areas (mission readiness, network infrastructure, cyber threats, and cyber posture) are tied together by only two node types—machines and exploits.

CyGraph (Server) compiles CyQL (DSL) queries to the native language for our graph database implementation (Cypher for Neo4j). The CyQL lexical analyzer and parser are generated via ANTLR (Parr, 2013). The lexical analysis groups input CyQL queries into tokens. The parser recognizes the tokens in terms of CyQL's grammar structure and maps them to a parse tree. CyGraph Server then iterates over the parse tree and generates corresponding Cypher query code.

ANTLR defines notation for specifying an input language's grammar, in Extended Backus–Naur Form (Information Technology, 1996). We thus define the grammar rules for CyQL. In CyQL, each function calls returns a

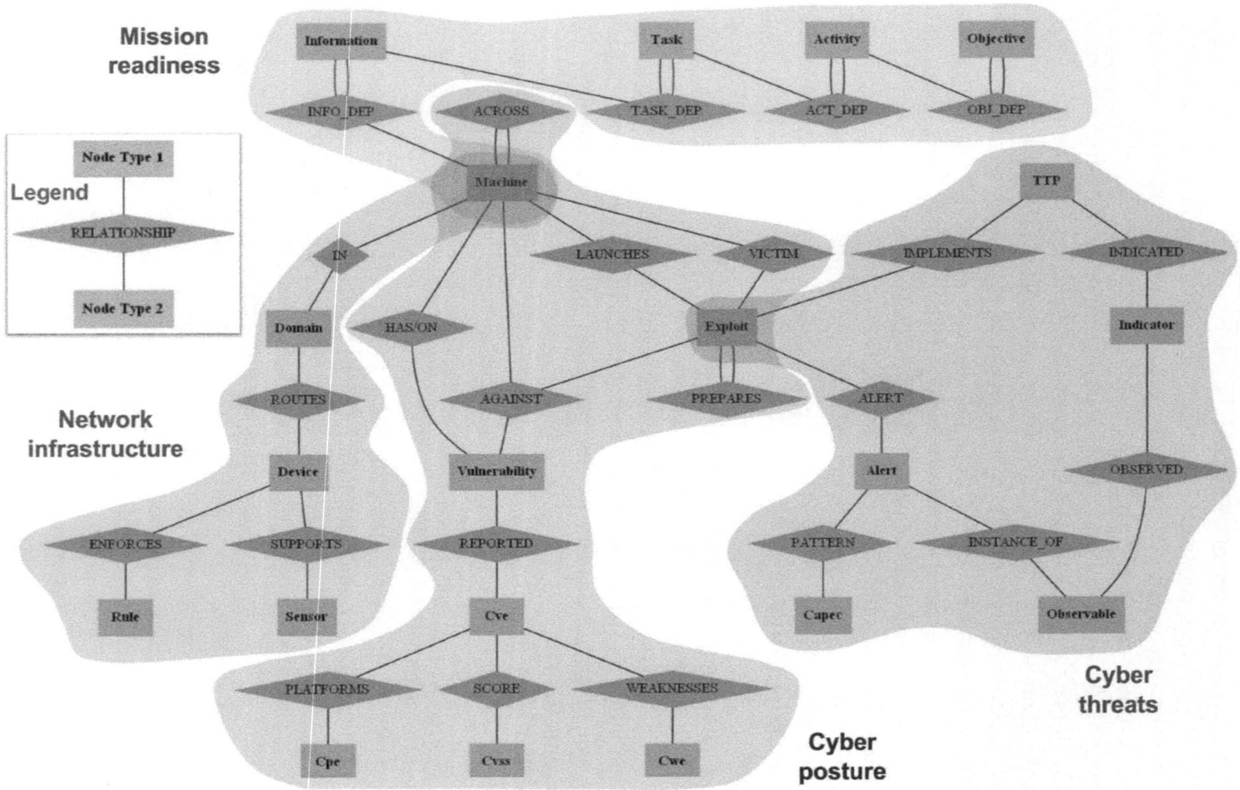

FIG. 6 Knowledge graph model for CyQL.

```
grammar CyQL;
@header {
    package org.mitre.cygraph.dsl;
}

query : subset (SETOP subset)* EOF ;
subset : call
              | '(' subset (SETOP subset)* ')' ;
call : FUNCTION '(' ')'
     | FUNCTION '(' paramList ')' ;

paramList : param (',' param)* ;
param : ID '=' disjunction
      | ID '=' value
      | ID '=' valueList ;

valueList : '[' value (',' value)* ']' ;
value : ID
      | INT
      | GLOB
      | STRING
      | IPV4
      | ipv4Range ;

disjunction : conjunction ('or' conjunction)* ;
conjunction : specifier ('and' specifier)* ;
specifier : '{' paramList '}'
          | '(' disjunction ')' ;

ipv4Range : IPV4 '-' IPV4
          | IPV4 '/' IPV4
          | IPV4 '/' INT ;

IPV4 : INT '.' INT '.' INT '.' INT ;
```

```
FUNCTION : 'network' | 'all' |
'exploitPaths' | 'correlateAlerts' |
'cveData' ;
SETOP : 'join' | 'except' | 'union' |
'intersect' ;
ID : [a-z][a-zA-Z0-9]* ;
GLOB : [a-zA-Z*][a-zA-Z0-9*_-]* ;
INT : [0-9]+ ;

STRING : '"'
          (   ESC
              |   ~('"'|'\\')
              )*
          '"' {
              setText(

org.antlr.v4.misc.CharSupport
.getStringFromGrammarStringLiteral(
   getText())
   );
}
;

fragment ESC :   '\\' (["\\/bfnrt] |
UNICODE) ;
fragment UNICODE : 'u' HEX HEX HEX HEX;
fragment HEX : [0-9a-fA-F];

COMMENT : '//' ~[\n]* '\n' -> skip;

BLOCK_COMMENT : '/*' .*? '*/' -> skip;
WS : [ \t\r\n]+ -> skip ;
```

FIG. 7 Grammar for prototype version of CyQL (CyGraph Query Language).

matching subgraph. Optionally, we can chain multiple function calls with the keyword "join." Each function (optionally) takes a list of parameters (name–value pairs). A value can be a single key-value pair with a primitive-type value, a list of literal values for a parameter (key), or a "specifier," which is a list of parameters and their values (surrounded by curly braces). Function calls cannot be nested, but specifiers can.

Fig. 7 shows the ANTLR grammar for CyQL. It defines a query as a collection of function calls (query types), such as mapping a network topology, correlating intrusion alerts, or traversing exploitable vulnerability paths. Function parameters can be used with Boolean operations (AND, OR, and NOT) to refine a query based on node properties and relationships. These results can then be expanded by JOINing other functions, to provide additional context. The CyQL grammar also defines rules for parsing cyber entities such as IP addresses, address ranges, and host names (including support for wildcard expressions).

3.5.2 Example CyQL Queries

As an example, consider a CyQL query for finding "exploit paths," i.e., sequences of vulnerabilities that an adversary could exploit for lateral movement through a network. These need not be simply linear chains. Rather, the query finds all possible paths (including branches/merges) between a specified set of machines. From a security analyst's perspective, this means "show me how an adversary can get from these machines to those machines."

While that kind of query answers the basic question about attacker reachability among network machines (exploitable paths), the analyst might want to refine the query further to provide more specific focus. For example, we might want to only include a certain group of machines (e.g., having a common hostname pattern) or include only those machines that have alerts for them.

On the other hand, the analyst might wish to expand a basic query by including additional information such as vulnerabilities associated with the machines in a set of exploitable paths. The query language should allow a rule as simple as *join vulnerabilities*() such that the appropriate vulnerability subgraphs are joined with the corresponding machine nodes of the exploit-paths subgraph.

The function *exploitPaths()* determines the structure and the types of edges in the matching subgraph. Its arguments get compiled to an SQL-like WHERE clause that constrains the node properties that match the query. For the *exploitPaths()* function, constraining arguments include starting and ending machines for an exploitable subgraph.

Consider this example CyQL query:

```
exploitPaths(start = ({subnet=1.1.3.0/24} or
            {ip=[1.1.4.32, 1.1.4.33]}) and
            {hostname=*-VM*}, end = {name = "DB Server"})
```

Here is the resulting query compiled from CyQL to Neo4j Cypher:

```
MATCH (start)-[r:AGAINST|VICTIM|ON|LAUNCHES|IN|ROUTES*]-(end)
WHERE ((start.subnet = "1.1.3.0/24" OR
            start.ip IN ["1.1.4.32", "1.1.4.33"])
    AND start.hostname =~ "^[a-zA-Z0-9_-]*-VM-[a-zA-Z0-9_-]*$")
AND (end.name = "DB Server")
RETURN start, r, end
```

The CyQL version of this query is much less verbose. It encapsulates the knowledge of allowed relationship types (AGAINST, VICTIM, ON, LAUNCHES, IN, and ROUTES) for subgraphs representing attack reachability between machines. This is based on the role of those particular relationship types in the CyGraph data model.

Fig. 8 is the resulting parse tree for this CyQL query. It shows the starting machines as a disjunction (OR) of the specified IP addresses, combined conjunctively (AND) with a wildcard expression for host names. The ending machine is simply matches a particular host name.

Fig. 9 shows how this query formulation, translation, and execution happens within the overall CyGraph architecture. The analyst formulates a query expressed in CyQL, which the client submits to the CyGraph Server. The service parses the CyQL query, verifies the types of function parameters, translates the query to equivalent Neo4j Cypher, and submits the resulting Cypher query to the database engine.

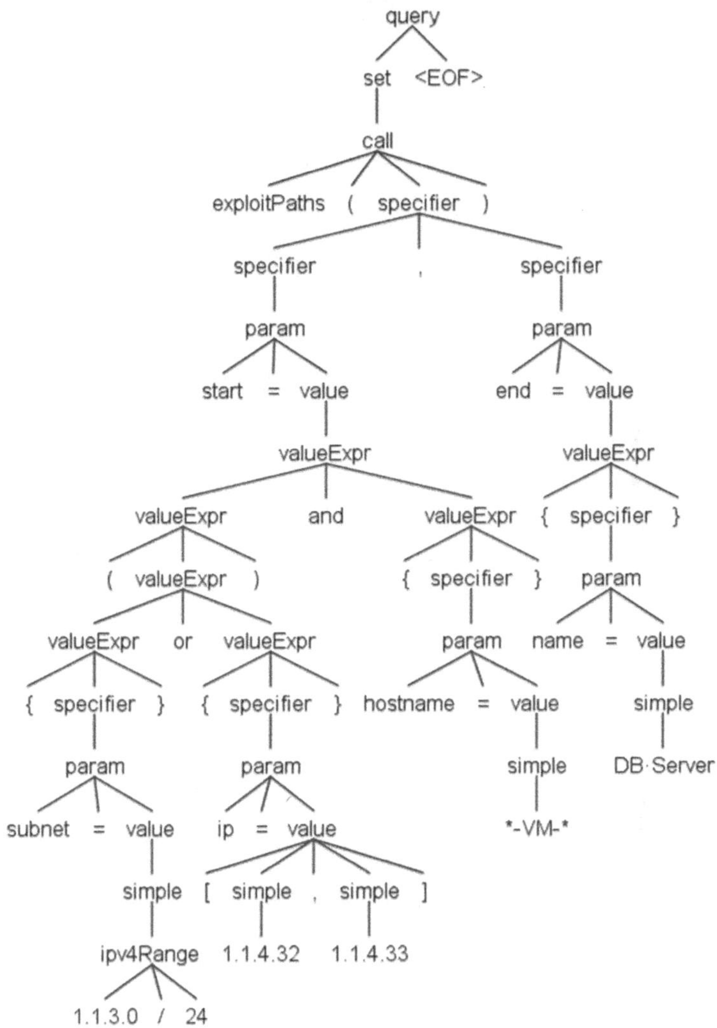

FIG. 8 Grammar parse tree, for example, CyQL query.

CyQL's "filter, then expand" model allows for flexible queries while maintaining a simple and readable syntax. As an example, consider a simple network topology query:

```
network()
```

Here is the result of compiling this CyQL query into Cypher:

```
MATCH (domain:Domain)-[r:ROUTES*]-(device:Device)
RETURN domain, r, device
```

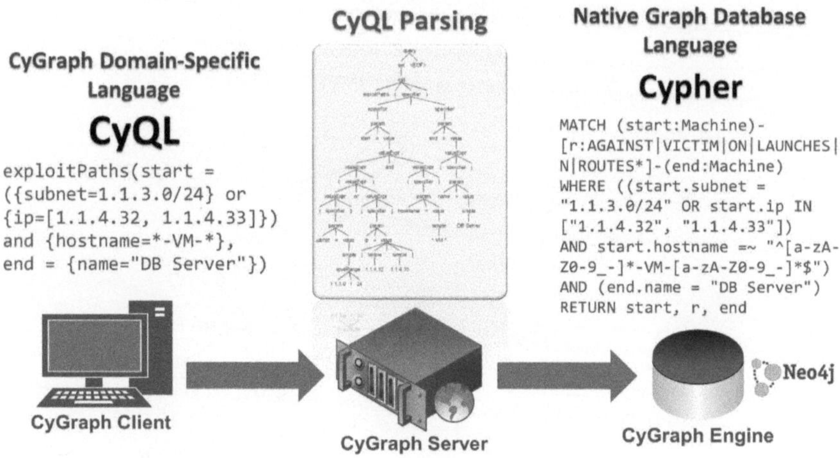

FIG. 9 CyQL domain-specific query language processing in CyGraph.

This query returns only the backbone of the network—the protection domains and devices such as routers and switches. Endpoint machines are not included in the resulting subgraph, but can easily be added:

```
network() join machines()
```

This compiles to:

```
MATCH (domain:Domain)-[r:ROUTES*]-(device:Device)
OPTIONAL MATCH (machine:Machine)
RETURN r, machine, domain, device
```

The *OPTIONAL MATCH* clause in Cypher functions similar to a *LEFT JOIN* in SQL. This query returns the network backbone and all machines on the network.

An analyst may want to consider only a subset of machines. This is done via parameters to the *machines()* function:

```
network() join machines(ip = 1.1.3.2)
```

```
MATCH (domain:Domain)-[r:ROUTES*]-(device:Device)
OPTIONAL MATCH (machine:Machine)
       WHERE machine.ip = "1.1.3.2"
RETURN r, machine, domain, device
```

Parameters can refer to properties on the machine nodes, or on related objects. For instance, we can include only the machines in certain domains:

```
network() join machines(domain = {name = DMZ}
       or {name = "Data Center"})
```

```
MATCH (domain:Domain)-[r:ROUTES*]-(device:Device)
OPTIONAL MATCH (machine:Machine), (machine)-[:IN]->(domain2)
       WHERE (domain2.name = "DMZ") OR (domain2.name = "Data Center")
RETURN r, machine, domain, device
```

Alternatively, we might be interested in machines with known vulnerabilities:

```
network() join machines(vulnerable = true)
```

```
MATCH (domain:Domain)-[r:ROUTES*]-(device:Device)
OPTIONAL MATCH (machine:Machine)
       WHERE (:Vulnerability)-[:ON]->(machine)
RETURN r, machine, domain, device
```

This query shows vulnerable machines on the network but does not include nodes for the vulnerabilities themselves. To add them, we join another function:

```
network() join machines(vulnerable = true) join vulnerabilities()
```

```
MATCH (domain:Domain)-[r:ROUTES*]-(device:Device)
OPTIONAL MATCH (machine:Machine)
       WHERE (:Vulnerability)-[:ON]->(machine)
OPTIONAL MATCH (vulnerability)-[:ON]->(machine)
RETURN r, machine, domain, device, vulnerability
```

Another way to expand a query is with the "!" operator. By default, CyQL functions return the smallest subgraph that makes sense. For instance, the *machines*() function returns only machine nodes, even if other types of nodes are used in the search parameters. Adding a "!" to the end of a function name makes it return all the nodes and relationships used in the query. As an example, consider this query (which has no "!" operator):

```
machines(domain = {name=DMZ}, vulnerabilities = {name=Heartbleed})
```

This returns machines in the "demilitarized zone" (DMZ) that are vulnerable to Heartbleed, but does not return the domain or vulnerability nodes. Alternatively, this version of the query includes the "!" operator:

```
machines!(domain = {name=DMZ}, vulnerabilities = {name=Heartbleed})
```

This returns the same machines as the first version, plus the DMZ domain node, the Heartbleed vulnerability node, and the relationships that connect them to the machines.

There is a difference between

```
machines!(vulnerabilities = {name = Heartbleed})
```

and

```
machines(vulnerabilities={name=Heartbleed}) join vulnerabilities()
```

in that the first query will return only one vulnerability node for Heartbleed, while the second will include any additional vulnerabilities on Heartbleed-vulnerable machines.

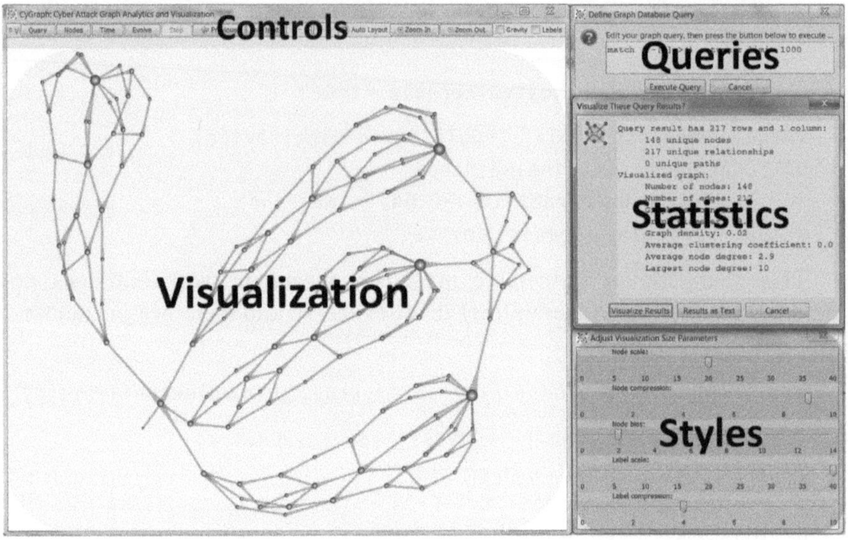

FIG. 10 CyGraph client-side user interface.

3.6 CyGraph Interactive Visualization

In the CyGraph architecture, client applications submit queries to the CyGraph service and process the query results. This decoupled architecture supports various clients with a variety of interactive modalities, including ad hoc queries, query expansion/pivoting/filtering, visual summarization, and dynamic graph evolution over time.

Fig. 10 shows one such CyGraph analysis/visualization client. This client allows an analyst to pose graph pattern-matching queries. The client then renders the query result (instances of a matched pattern) through interactive graph visualization. The client also has functions for graph statistics, styling, spatial layout, and evolution over time.

In general, adding constraints to queries (more specific patterns) yields smaller matched subgraphs. An analytic strategy is to begin with more general queries and then refine them as more is learned. This helps focus the analysis, manage complexity, and improve performance.

As an example, a query against a particular graph knowledge base yields a subgraph of 200,000 vertices and 400,000 edges, as shown on the *left side* of Fig. 11. The query itself completes in only a few seconds, but the vertex positioning needed for effective visualization (as shown in Fig. 12) takes about half an hour.[a] Using the same knowledge base, a more constrained graph

---

[a]This is for a Dell Latitude E6530 laptop running 64-bit Windows 7 Enterprise, with Intel Core i7-3720QM @ 2.60 GHz (8 logical cores, 4 physical and 2 logical per physical), and 8 GB of memory.

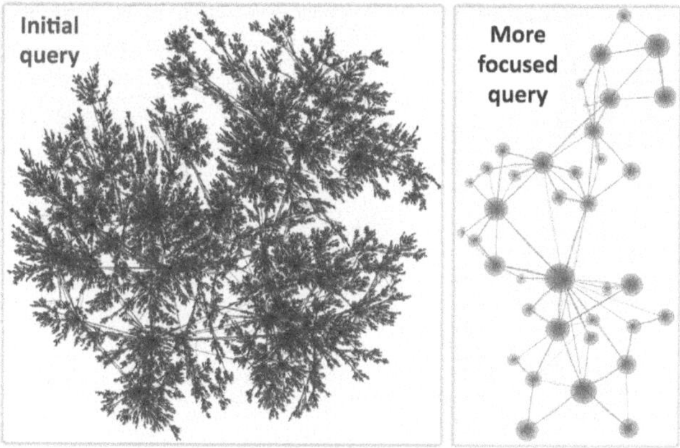

FIG. 11 Constraining CyGraph query results to subgraph of interest.

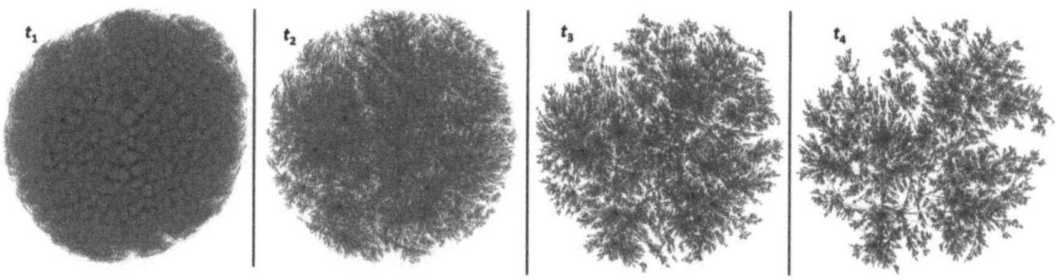

FIG. 12 Progression of graph visualization layout.

query yields a matching subgraph of 10,000 nodes and 20,000 edges, as shown in the right side of Fig. 11. In this case, graph visualization layout completes in only a few seconds.

CyGraph clients provide a variety of ways for interacting with graph query results. This includes filters on node and edge property values, to focus graph visualization (matched queries) on selected criteria. Property-dependent styling choices (e.g., node/edge size, color, transparency, edge routing, and arrow shapes) can also be defined. For example, Fig. 13 shows configuration dialogs for defining filter properties and background color.

CyGraph also supports visual clustering of graph nodes, based on property values or through manual selection. This is illustrated in Fig. 14.

4 EXAMPLE APPLICATIONS

Rather than relying on fixed analytics and visualizations, CyGraph gives the analyst the power and flexibility for crafting queries to solve the problem at hand. Complementary to queries that discover patterns of interest, CyGraph's interactive visualization conveys discovered patterns in ways that help induce faster learning and deeper understanding. This section describes a variety of applications that benefit from CyGraph analysis and visualization, for both real data sources as well as driven by simulations.

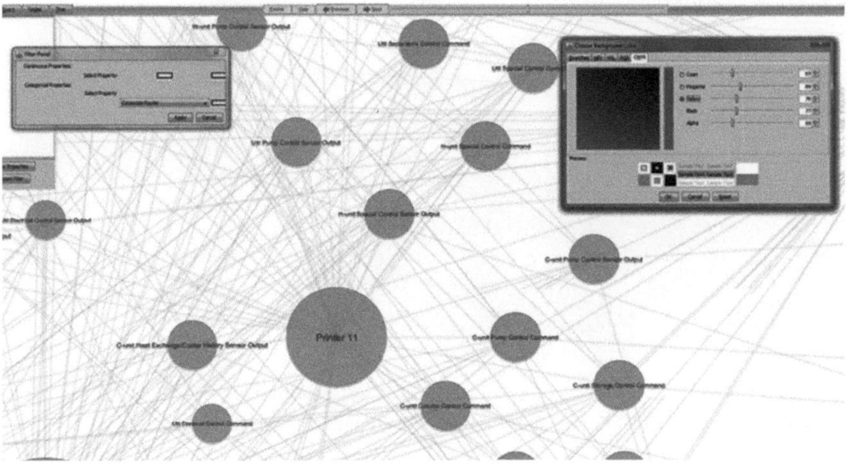

FIG. 13 Defining node filter properties and background color.

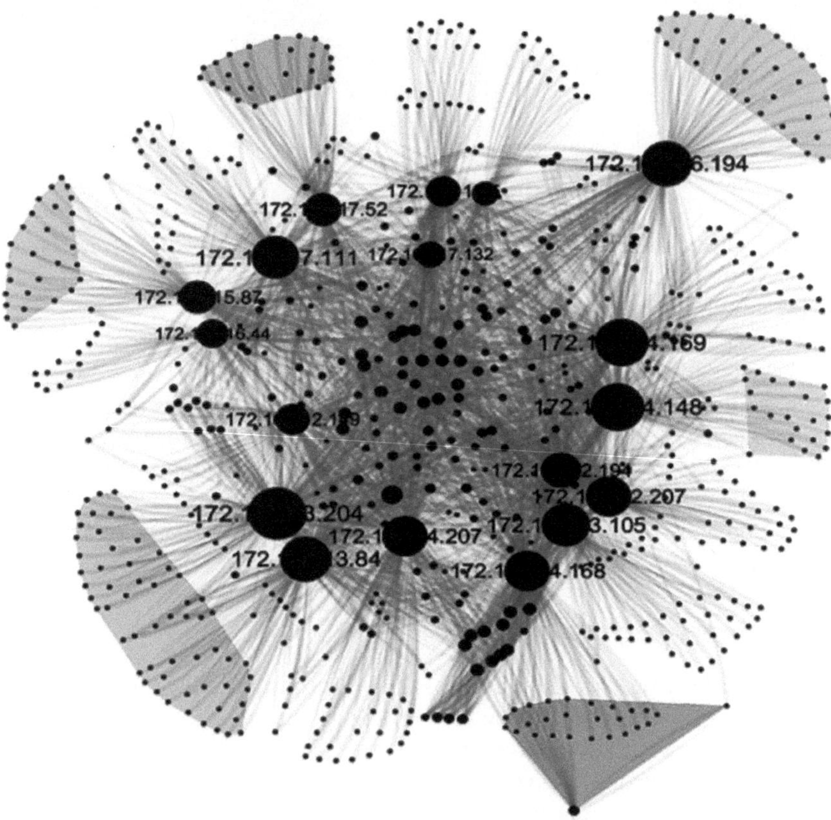

FIG. 14 Clusters of related nodes.

Section 4.1 describes query-driven analytics in CyGraph for extracting relevant portions of the knowledge base for solving particular cyber operational problems. Section 4.2 examines how CyGraph supports cybersecurity modeling and simulation tasks.

4.1 Cyber Analytics

CyGraph supports deep analytics within and across the layers of its knowledge graph stack. This includes network infrastructure that hosts cyber attacks and defenses, cyber posture gained by proactive network hardening in advance of attack, cyber threats (both potential and actual) against the network environment, and associated mission impacts.

Section 4.1.1 describes how CyGraph can leverage the TVA/Cauldron tool to map vulnerability exposures across network infrastructure to assess network security posture. Section 4.1.2 integrates such vulnerable attack paths with cyber threat information and actual alerts. Section 4.1.3 shows how CyGraph can analyze the impact of cyberattacks on an organization's business/mission functions. Section 4.1.4 presents a case study that illustrates a number of CyGraph analytic capabilities.

4.1.1 Network Infrastructure and Cyber Posture

A basic capability for assessing cyber posture is to map vulnerability exposures across network infrastructure. This helps prioritize host vulnerabilities, identify insecure access policy rules, and show how attackers can potentially leverage multiple vulnerabilities to incrementally penetrate a network. As shown in Fig. 15, this analysis requires an enumeration of host vulnerabilities (e.g., from a vulnerability scanning tool), a topology defining network segmentation and location of firewalls, and access rules for the firewalls. For this we can leverage the TVA/Cauldron tool, which analyzes the topology, vulnerabilities, and rules to map paths of exposed vulnerability across a network, known as an attack graph.

Fig. 16 shows the TVA/Cauldron attack graph for a particular network. This has machines grouped into subnets, with edges showing exposed vulnerabilities across subnets. Implicitly, machines within subnets (more generally known as protection domains) have full access to one another's vulnerabilities, i.e., a fully connected subgraph.

In the TVA tool, the attack graph is visualized in a predetermined way, with limited options for constraining the graph, e.g., attack start and goal (highlighted green (gray in the print version) and red (dark gray in the print version)). For managing attack graph complexity, it relies on visual aggregation, such as collapsing protection domains to single nodes to show attack reachability at a domain level and using a single edge to represent the full set of vulnerabilities exploitable from one host to another. In Fig. 16, all protection domains are expanded to show exploitation at the machine level. This visualization is somewhat cluttered, even after judicious manual positioning of machines after expanding subnet boxes, obscuring salient patterns.

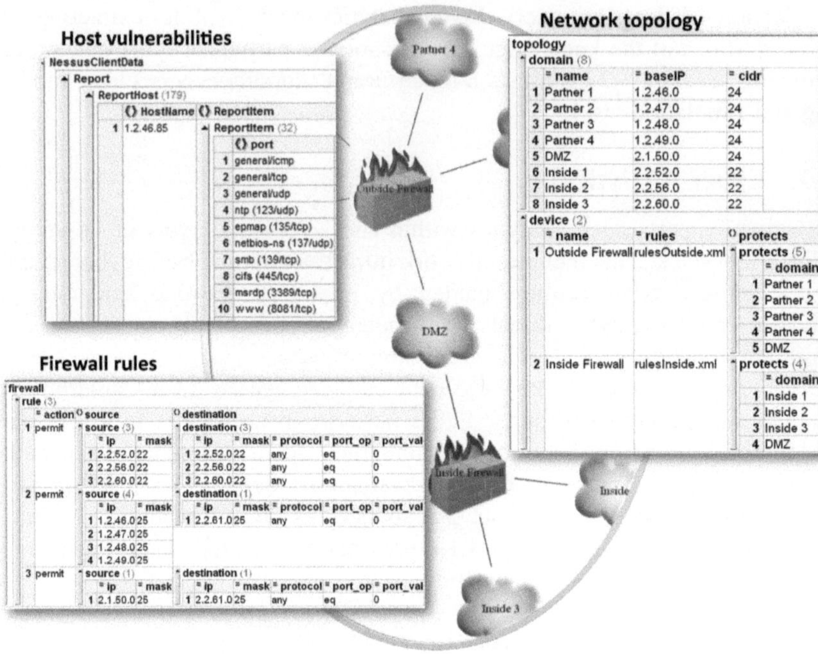

FIG. 15 Network topology, vulnerabilities, and firewall rules.

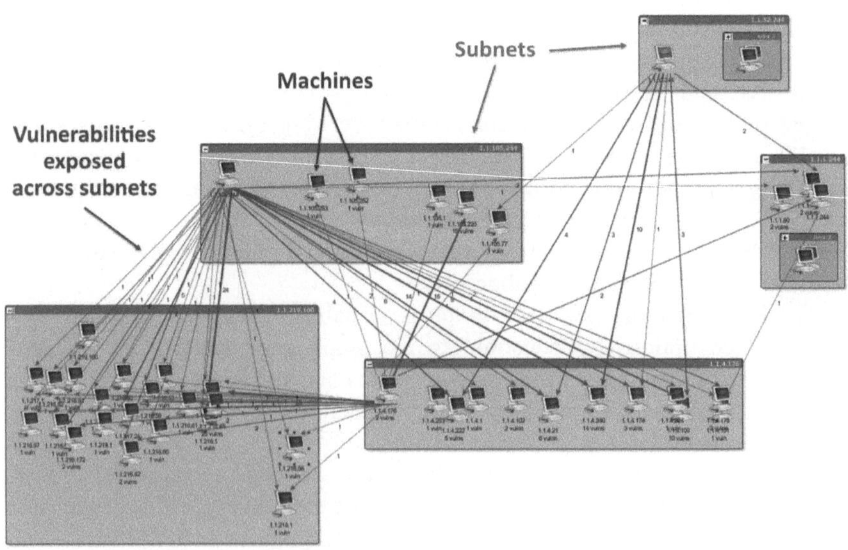

FIG. 16 TVA attack graph visualization.

We ingest this attack graph model into CyGraph, transforming it according to the knowledge graph model of Fig. 6. This has machine, vulnerability, and domain node types, and their corresponding relationship (edge) types. Then for analysis, a naïve initial query (requiring no knowledge of the underlying data model) could be simply "`MATCH ()-[r]->() RETURN r`." This pattern matches all relationships, yielding the full knowledge graph in Fig. 17. Note that this visualization shows all model relationships in full detail, including those that are aggregated or implicit in Fig. 16. Still, some key patterns are apparent, e.g., two machines (1.1.105.244 and 1.1.4.176), and to a lesser degree another one (1.1.52.244) that have the most extensive attack reachability (to exploitable vulnerabilities to other machines).

In CyGraph, queries can be successively refined (constrained) to focus the analytic results. For example, the details of the particular vulnerabilities might be less important than the attack relationships among the machines themselves. In this case, we can pose the query "`MATCH ()-[r:ACROSS|IN]->() RETURN r`." This pattern constrains relationships to only those linking machines across protection domains (ACROSS) and machine memberships within domains (IN). This yields the resulting matched subgraph in Fig. 18. This clearly shows the strong dominance of 1.1.105.244, 1.1.4.176, and 1.1.52.244, along with their reachable machines in common. The domain membership relationships show potential victim machines that are not directly exploitable across domains, but can be exploited with only one additional attack step.

For prioritizing vulnerabilities, an important strategy is to focus on those that are exposed across protection domains (Noel and Jajodia, 2014). In our model, that corresponds to the query "`MATCH ()-[r:ACROSS]->() RETURN r`." This yields the matched subgraph in Fig. 19, which just shows machines that can directly attack across protection domains. This clearly highlights the machines with heavy access to vulnerabilities, and the vulnerable machines that they can reach. In this figure (as for Figs. 17 and 18), edge directionality has arrow heads narrow and arrow tails wide.

This simple example illustrates some important properties of the CyGraph approach. In general, the cyber posture layer can be populated with any relationships capturing potential attacker advantages (at all layers of the network stack; Ritchey et al., 2002), e.g., host inventory agents mapped to reported vulnerabilities (Noel et al., 2009).

In more traditional cybersecurity tools, the analytic and visual behaviors are determined at design time and built into code. In CyGraph, the model structure is driven by the data sources and how they are transformed into a graph knowledge base. Then, given some understanding of the underlying knowledge model, the analyst can construct ad hoc queries to fine-tune analytic results. As we show in subsequent sections, we can enrich the graph knowledge base with additional elements (arbitrary nodes, relationships, and properties) relevant to cybersecurity and mission assurance in our environment. Richer knowledge supports more refined queries and specialized interactive visualizations.

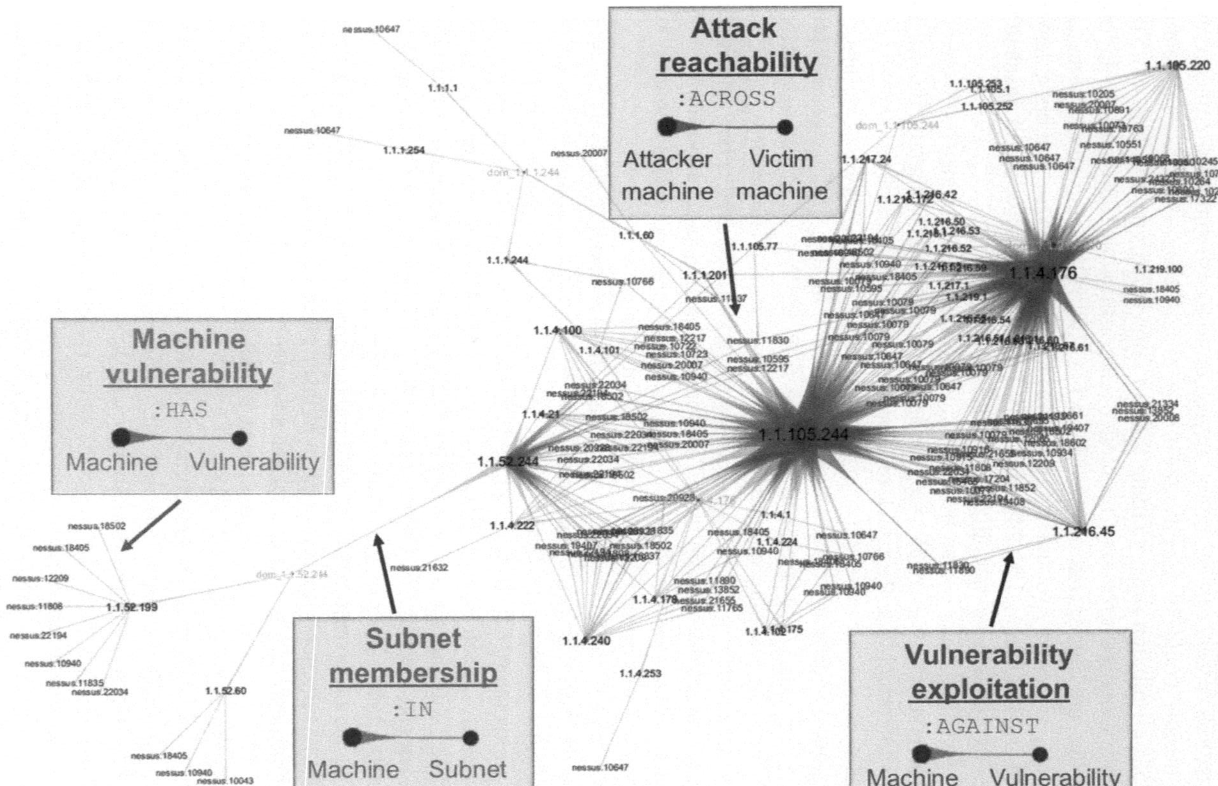

FIG. 17 Full graph of security posture relationships.

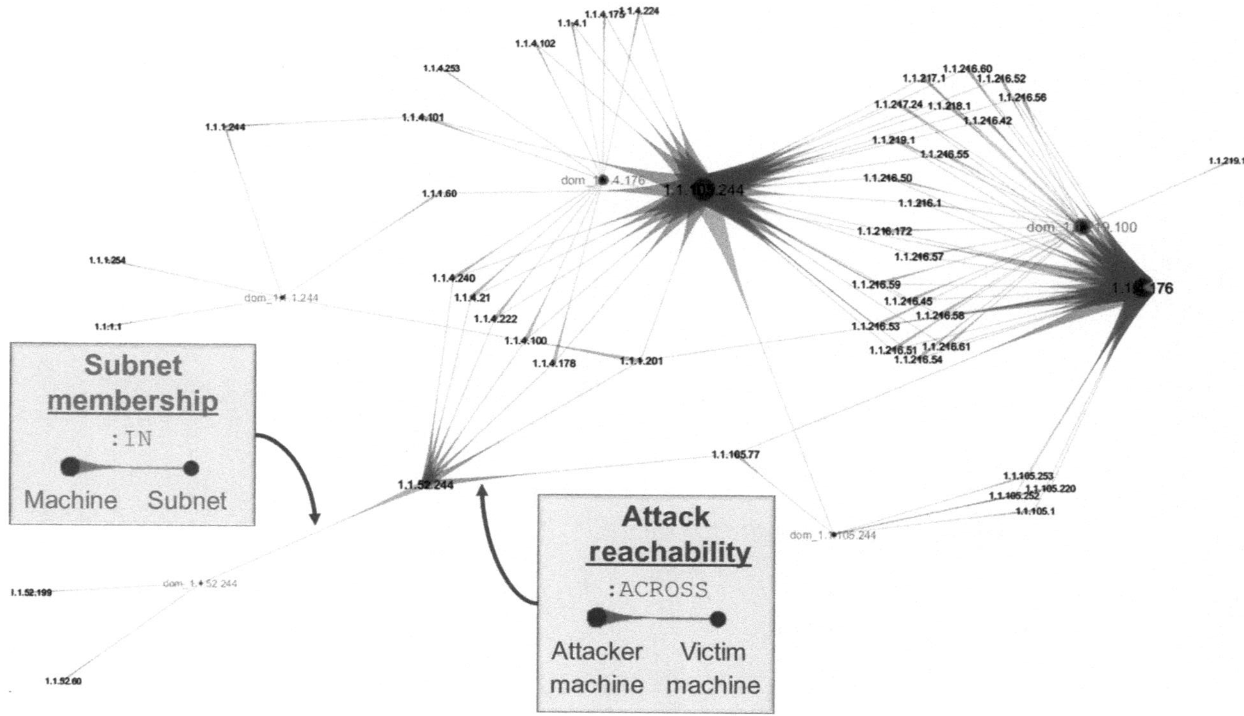

FIG. 18 Machines vulnerable across subnets and subnet memberships.

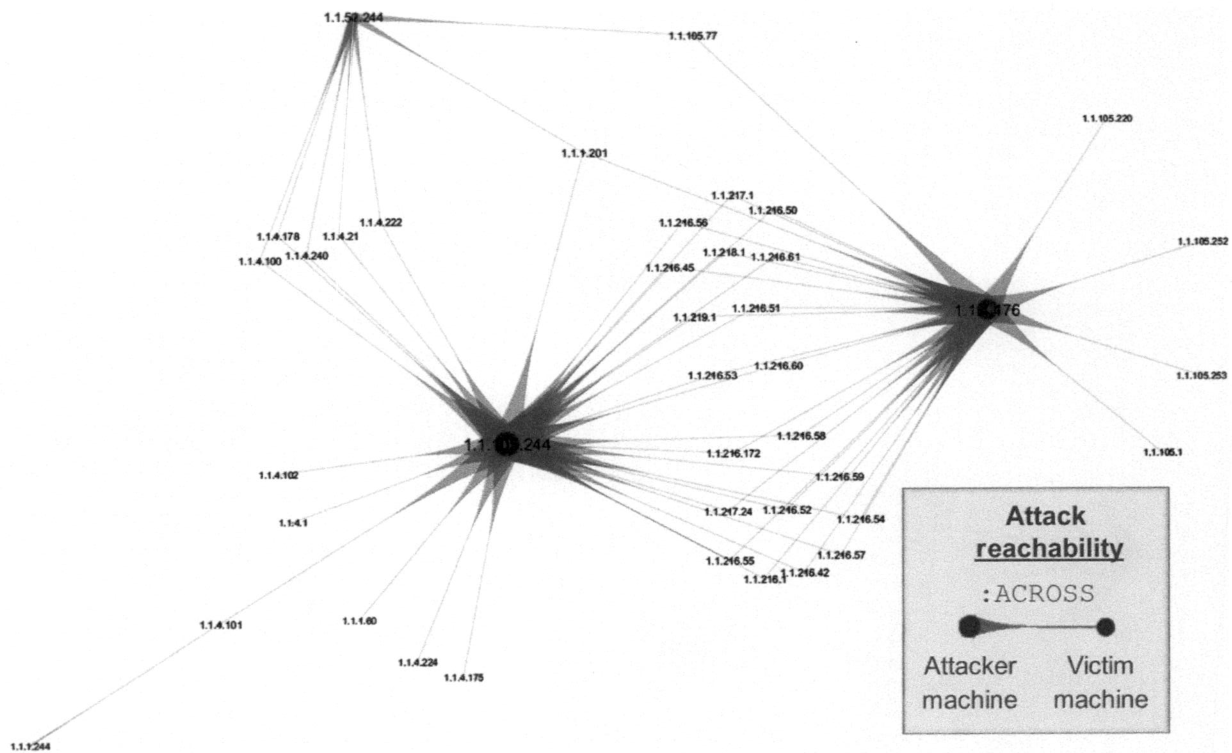

FIG. 19 Machines with vulnerabilities exposed across subnets.

4.1.2 Cyber Threats

In CyGraph, the cyber posture layer enumerates potentially exploitable vulnerabilities within the network infrastructure. This forms the context for understanding potential cyber threats and responding to actual ones. The population of tools such as CyGraph is greatly facilitated by standard languages for sharing threat intelligence, including attack patterns and associated observables.

An important resource for such threat intelligence sharing is the CAPEC™, a standardized catalog and taxonomy of attack patterns. CAPEC provides detailed characterizations of each attack pattern and organizes the patterns into a taxonomic hierarchy (general attack classes, their subclasses, and specific attacks). Navigating CAPEC on the web requires following parent–child hyperlinks embedded in the textual content. This does not lend well to understanding the overall hierarchical taxonomic structure. CyGraph tree visualization capabilities help in this regard, providing a variety of interactive visualization modalities for visualizing and navigating the CAPEC taxonomy (Noel, 2015).

The left side of Fig. 20 shows the CAPEC taxonomy. Here, nodes are attack pattern classes and edges are parent–child relationships (with replicated subtrees for the relatively few instances of classes with multiple parents). The right side of the figure shows one kind of interactive tree visualization employed by CyGraph, which has a Cartesian layout of nodes.

Other forms of tree visualization seek to maximize use of display space, e.g., sunburst and treemap visualizations. A type of treemap visualization in CyGraph with particularly desirable visual properties is the Voronoi treemap (Balzer and Deussen, 2005; Carrot Search, 2016). Fig. 21 shows a Voronoi treemap visualization for the CAPEC taxonomy. The left side of the figure is the initial view, which has the highest-level (most general) attack classes. The analyst can then drill down into successive levels (more specific attack classes). In this visualization, the area of each attack class is a function of the number of its child subclasses, so that more populated parts of the taxonomy are emphasized. Transparency and color muting provide some context of parent and child classes.

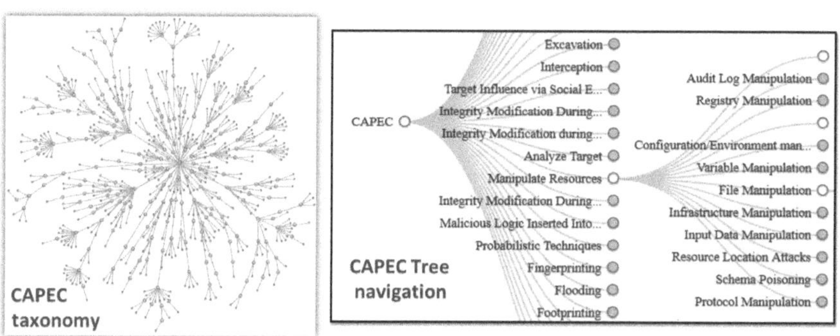

FIG. 20 CAPEC taxonomy of attack patterns.

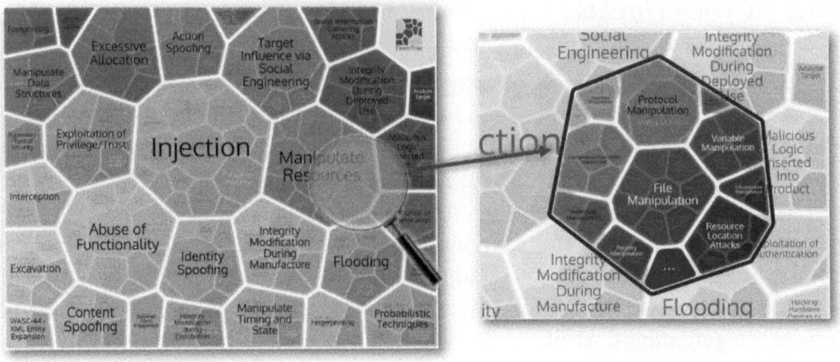

FIG. 21 Interactive Voronoi treemap of CAPEC taxonomy.

In CAPEC, each attack pattern includes the phases of attack (explore, experiment, and exploit) comprising it, including indicators for success/failure of each phase. This provides opportunities for understanding how each phase of a particular attack can be detected and thwarted.

Furthermore, certain attack patterns may yield outcomes that help enable other attack patterns, as in the example of Fig. 22. This shows interrelated attack patterns leading to exploitation of a database. Each main box is a CAPEC attack pattern, broken into attack phases. The CAPEC-170 pattern (web application fingerprinting) identifies details of the target database, which helps the attacker choose one of the three subsequent attacks leading to database compromise.

Another important resource for cybersecurity analysis is STIX™, a structured and extensible language for cyber threat intelligence. The unifying architecture of STIX encompasses cyber threat actors; their campaigns; their tactics, techniques, and procedures (TTPs); cyber incidents; attack indicators and observables; attack targets; and responsive courses of action. For example, Fig. 23 shows STIX content that describes a threat actor leveraging a CAPEC attack pattern (phishing) to deliver a particular kind of malware (The MITRE Corporation, 2016) as their TTPs. Another example is Fig. 24, in which STIX describes a particular indicator (intrusion detection rule) for data exfiltration. Particularly relevant to CyGraph models is STIX support for chaining of attacks (The MITRE Corporation, 2016).

For managing security information and events in their network environment, security teams often rely on SIEM products. These products collect data from a variety of sources (e.g., security devices, sensors, logs, and traffic) into a unified framework. ArcSight's Common Event Format (CEF) (ArcSight) is an interoperability standard for sharing such data. For example, it is possible to ingest STIX threat intelligence into ArcSight (via CEF) and correlate STIX fields with security events (Murdock and Pramanik, 2004).

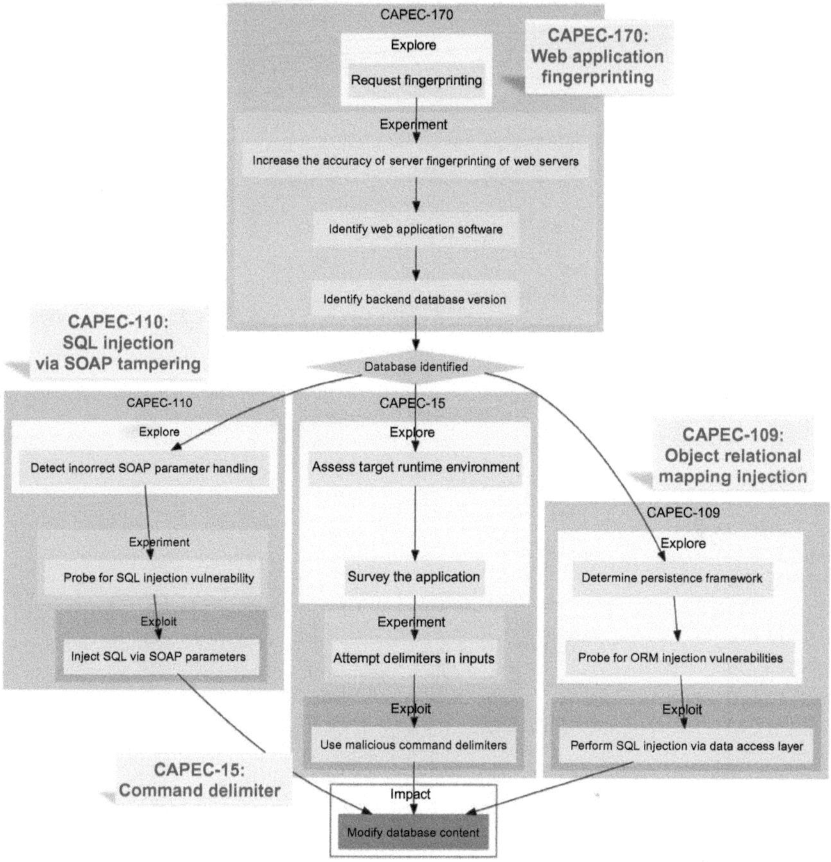

FIG. 22 Relationships within and between CAPEC attack patterns.

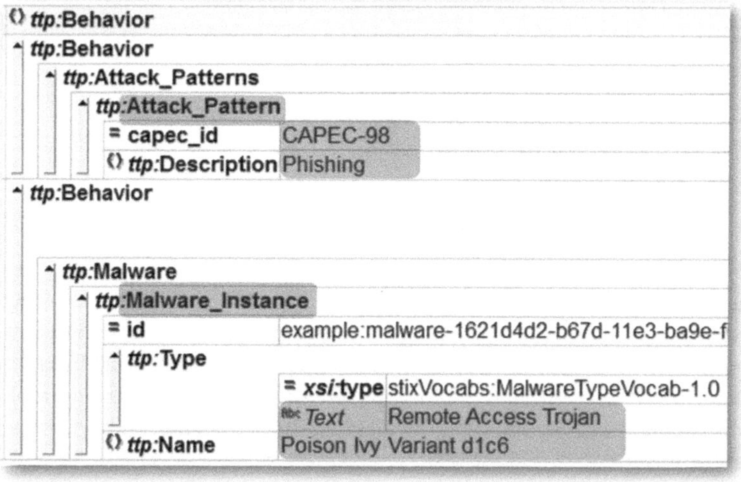

FIG. 23 STIX threat intelligence for attacker TTPs.

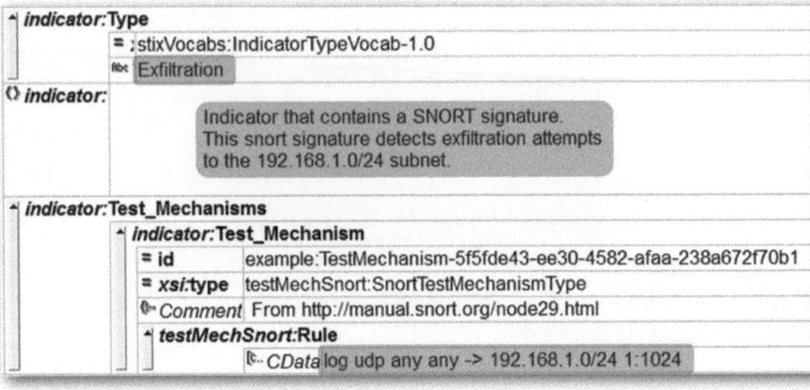

FIG. 24 STIX indicator for data exfiltration.

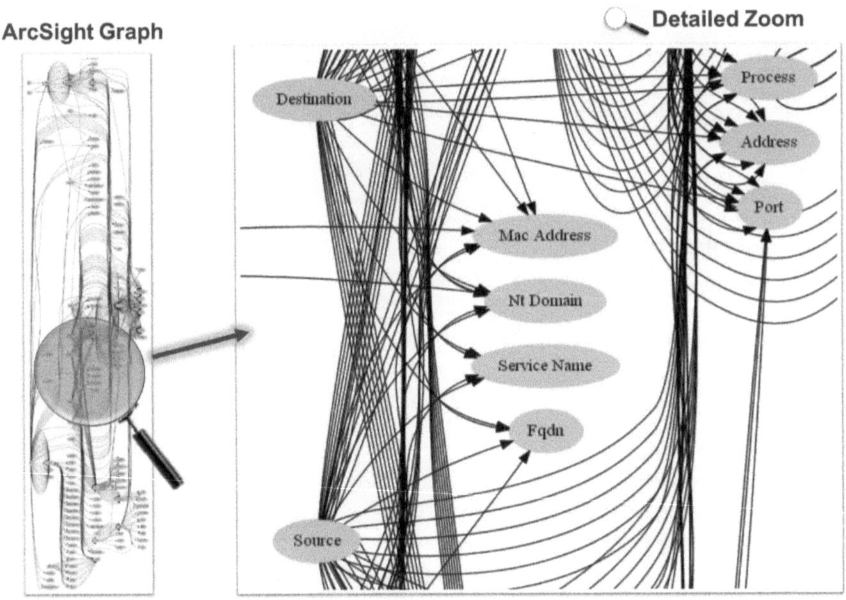

FIG. 25 Hierarchy of ArcSight field groupings.

We can leverage SIEMs for populating CyGraph models, for both cyber posture (e.g., vulnerabilities) and cyber threats (e.g., intrusion detection alerts). Fig. 25 shows the full range of ArcSight fields, which we cast as a directed acyclic graph for populating CyGraph. ArcSight organizes its data model as multilevel groupings of related fields. We capture this in graph relationships, where each edge denotes membership in a group (child as member of parent group).

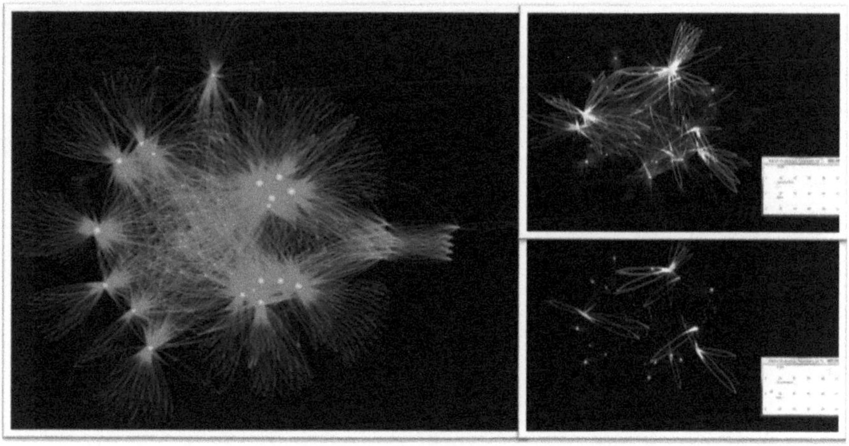

FIG. 26 Highlighting recent events in dynamically evolving graph model.

When the graph elements are labeled with time (network traffic, sensors, etc.), CyGraph can visualize graph evolution over time. When analyzing such time-dependent graph models, often it is important to maintain focus on recent events, so they do not become lost in the clutter of older ones. As shown in Fig. 26, CyGraph supports this through highlighting of recent events (edges). Here, network traffic is modeled with IP addresses as vertices, and edges for packets between them. More recent events (edges) are given brighter colors (nearly white), while older events are colored progressively darker (shades of blue (gray in the print version)) and more transparent.

4.1.3 Mission Dependencies

In CyGraph, the mission dependency layer analyzes the impact of cyberattacks on organizational functions (missions). It captures hierarchical dependencies among mission components, down to the cyber assets that support them. In practice, these dependencies are generally captured through manual modeling of mission dependencies, e.g., through CJA, CyCS, or CMIA. Fig. 27 is an example of such a mission dependency model ingested into CyGraph.

Through the mission dependencies layer of the knowledge graph, CyGraph shows transitive (*n*th order) mission effects of cyberattacks. For example, a CyGraph query can begin at the victim host of an attack and traverse the graph forward to enumerate the mission components that depend on it, showing impact on all effected levels of the mission dependency hierarchy. Such a query can also include potential next attack steps, following known vulnerability paths. A query could traverse in the opposite direction, to show the "cyber key terrain" supported by a given mission component. Overall, these kinds of analytic queries can tie together relationships among all layers of the CyGraph knowledge base.

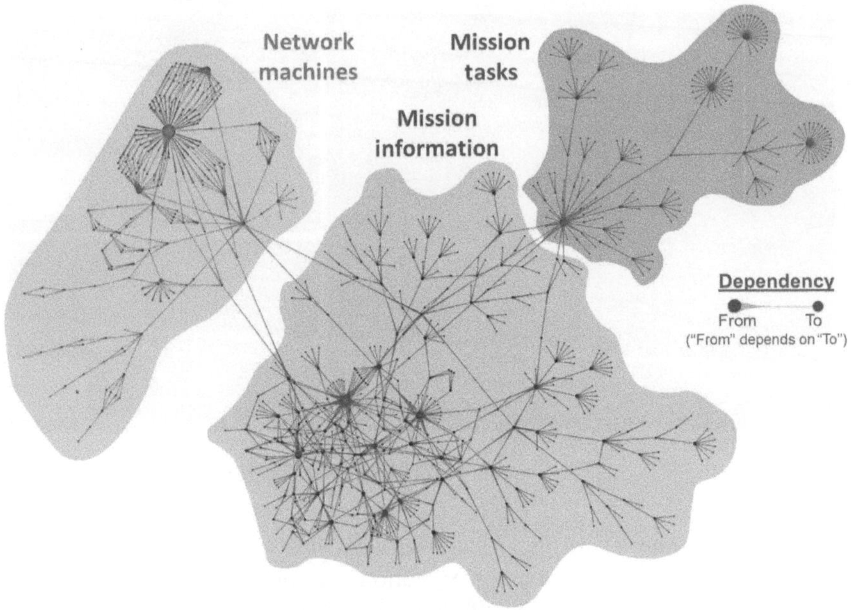

FIG. 27 Hierarchy of mission dependencies on cyber assets.

4.1.4 Case Study

This section presents a case study that illustrates various cybersecurity analytic capabilities in CyGraph. We compare a baseline model built by the TVA/Cauldron tool with a richer model built in CyGraph. CyGraph leverages the TVA/Cauldron output attack graph for the cyber posture layer of its knowledge graph. We then enrich this baseline model with intrusion detection alerts, vulnerability data from the NVD, and attack patterns from CAPEC™.

The left side of Fig. 28 shows the network architecture for this case study. In this architecture, the network is protected from the internet by an external firewall. Mission-critical servers are protected by a second (internal) firewall. There are workstations (represented by the vulnerable client machine in Fig. 28) that have two vulnerabilities—a remote buffer overflow in a web browser that allows arbitrary code execution, and stored credentials that can be stolen. The database backend server has an SQL injection vulnerability that is exploitable by an authenticated user. The Domain Name System (DNS) server in the network DMZ is vulnerable to a cache-poisoning attack that allows an attacker to redefine the internet address for a domain name, e.g., to an address under the attacker's control.

In this scenario, the vulnerable DNS service is exposed to the internet through the external firewall. The web client vulnerability is exposed to any malicious service to which the client browses, and the stored credential can be stolen by an attacker having sufficient control over this machine. The right side of Fig. 28 shows the resulting TVA/Cauldron attack graph, built from a

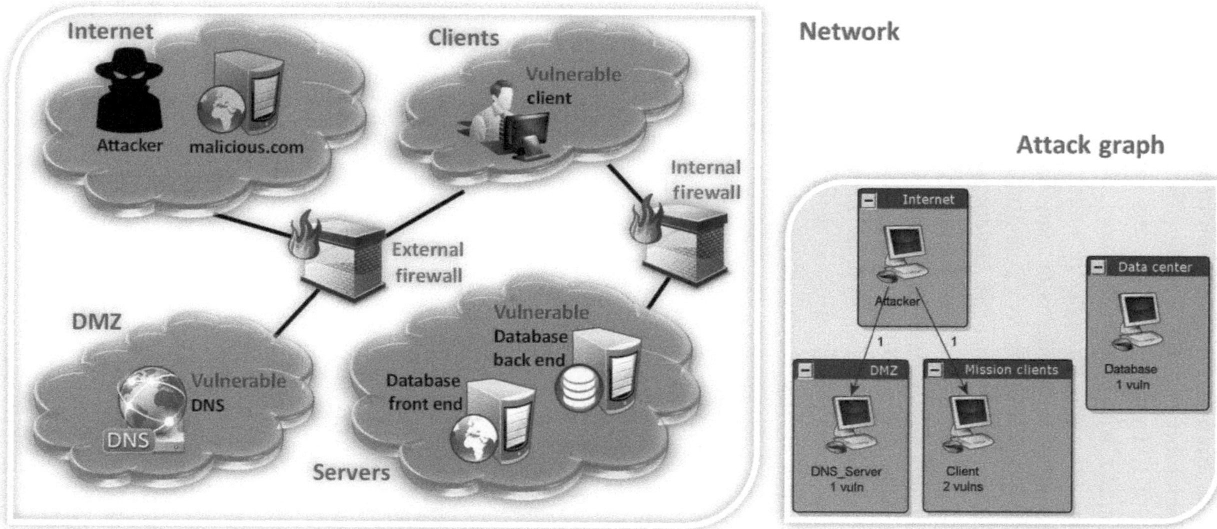

FIG. 28 Example network with TVA/Cauldron attack graph.

specification of the network topology, firewall rules, and vulnerability scan results. Of the four vulnerabilities reported in the scan, two are exposed to the internet, i.e., DNS server cache poisoning and web client remote buffer overflow. However, from its available information, the attack graph does not show any vulnerability exposures leading into the mission-critical servers.

Fig. 29 shows a CyGraph model for this scenario. It ingests the TVA/Cauldron attack graph, which gives machines, subnets (protection domains), and vulnerabilities (exposed through firewalls). We then add the network topology (connectivity among domains and firewalls), along with details about the vulnerabilities from the NVD and associated attack patterns from CAPEC. With the assumption that the servers are mission critical, this gives relationships in all four of the CyGraph knowledge graph layers.

This model of potentially exploitable attack paths provides context for responding to security events. When responding to events, queries anchored on events of interest allow the analyst to answer specific questions about the situation. In our scenario, an intrusion detection system (Snort) detects a buffer overflow attack against a network client machine (the cache-poisoning attack against the DNS server was not detected). To understand the context for this alert, the analyst submits this (Cypher) query:

```
MATCH paths = (:Machine)-[:SRC]->
   (:Alert {name:"Snort 33022"})-[:DETECTION]->
   (:Exploit)-[:AGAINST]->
   (:Vulnerability)-[:ON]->
   (:Machine)
RETURN paths
```

A literal translation of this query is “for this alert, show me the source (attacking) machine, and whether this alert is detection of exploitation against a vulnerability on a machine.”

Fig. 30 is the subgraph that matches this query. This shows that the victim of this alert does in fact have a vulnerability associated with an exploit (CAPEC attack pattern) that the intrusion signature (Snort 33022) detects. Here, there are additional relationships (not specified in the query) that Neo4j includes as associated with the returned nodes, i.e., DST for the alert’s destination machine and ENABLES representing the enabling of future possible exploitation.

Next, assume that a second alert is generated, which detects attempts at probing a web application for potential vulnerabilities. The analyst, already being suspicious about the first alarm (and the associated vulnerability), issues a query to analyze how the two alerts might be related. Here is the query:

```
MATCH paths = (:Alert {name:"Snort 33022"})-
   [:SRC|DST|DETECTION|ON|ENABLES|AGAINST|PREPARES*]->
   (:Alert {name:"Snort 1576"})
RETURN paths
```

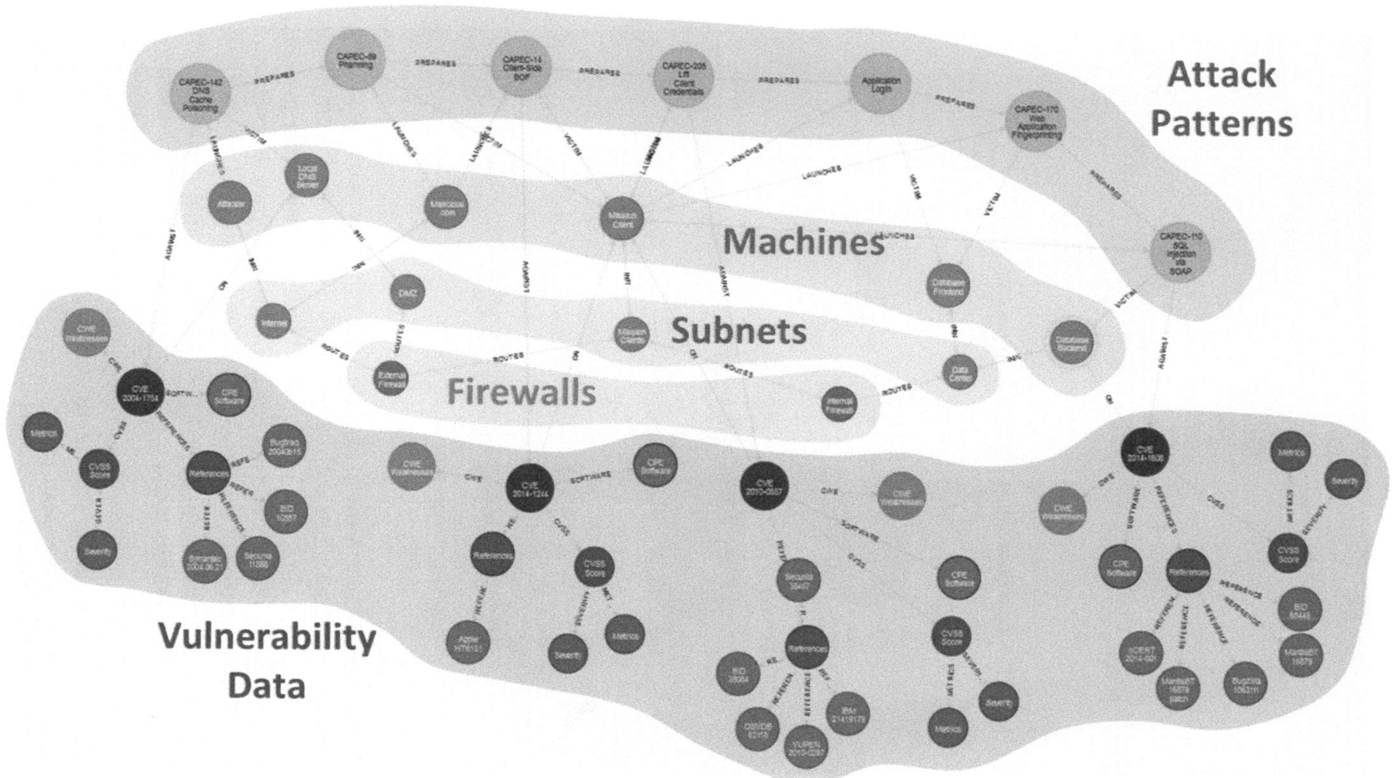

FIG. 29 Full cyber knowledge graph—ready for queries.

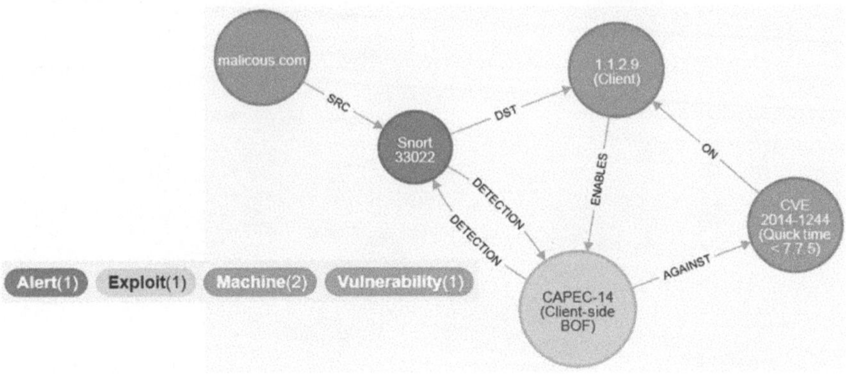

FIG. 30 Query showing context for an intrusion detection alert.

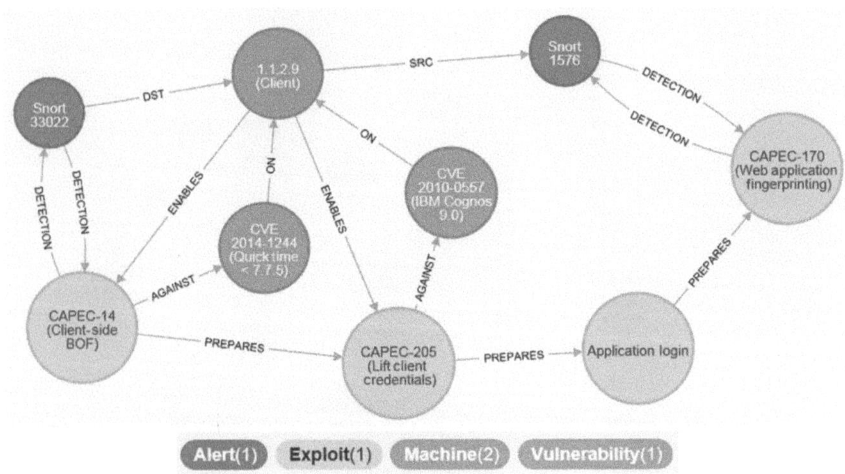

FIG. 31 Query showing relationships between two alerts.

This translates to "show me everything between these two alerts, using the relationship types about alerts and vulnerability exploitation." Here, the horizontal bars represent choice of relationship type, i.e., any of the given types will match. The asterisk at the end of the relationship types denotes traversal over an arbitrary number of relationships of these types, i.e., arbitrarily deep.

Fig. 31 is the resulting query match. This shows that the two alerts are indeed related, i.e., there is a chain of potential exploits linking them:

- Client-side buffer overflow against mission client.
- Lifting of database login credentials on client.
- Logging in to database from client.
- Fingerprinting to discover potential database vulnerabilities.

In this chain, two of the exploits are against known vulnerabilities. Just as for missed intrusion detections, vulnerability scanners do not always find existing vulnerabilities. In fact, in this case, the database login (after stealing the password) is essentially indistinguishable from a benign login and thus has no associated vulnerability or alert. Still, given this query result, the analyst might suspect that these are potentially multiple attack steps by the same threat actor.

Being concerned, the analyst might want to understand the next possible steps that the attacker could take. Here is a query to answer that:

```
MATCH paths = (:Alert {name:"Snort 1576"})-
   [:DST|DETECTION|ON|ENABLES|AGAINST|PREPARES*]->()
RETURN paths
```

This query translates to “show me everything that can happen after the second alert.” Again, this uses a choice of relationship types about alerts and vulnerability exploitation, with arbitrarily deep traversal.

Fig. 32 is the subgraph match for this query. This shows that the web application fingerprinting against the database frontend server (which was detected) prepares for a subsequent attack against the database backend. Exploitation of this vulnerability would let the attacker inject arbitrary SQL commands, e.g., to steal, corrupt, or destroy mission-critical information.

At this point, the defender suspects that a malicious attack is being launched from the client. To better understand potential response options, the analyst poses this query:

```
MATCH
   ()-[r:ROUTES*]->(),
   (:Machine {name:"1.1.2.9 (Client)"})-[i1:IN]->(),
   (:Machine {name:"1.1.3.4 (Database Frontend)"})-[i2:IN]->(),
   (:Machine {name:"1.1.3.5 (Database Backend)"})-[i3:IN]->()
RETURN r, i1, i2, i3
```

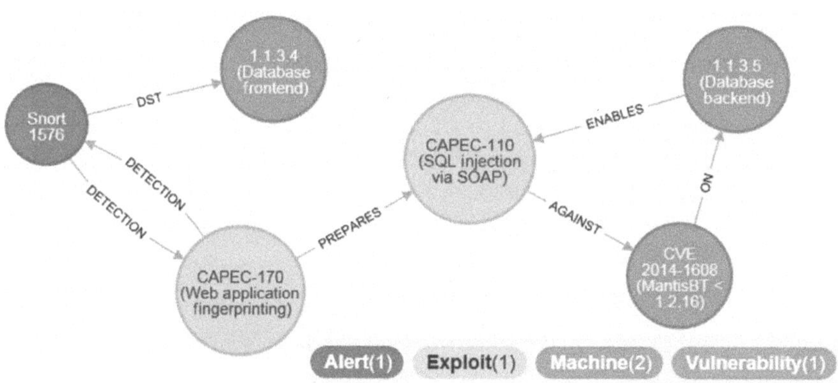

FIG. 32 Query showing potential mission impact.

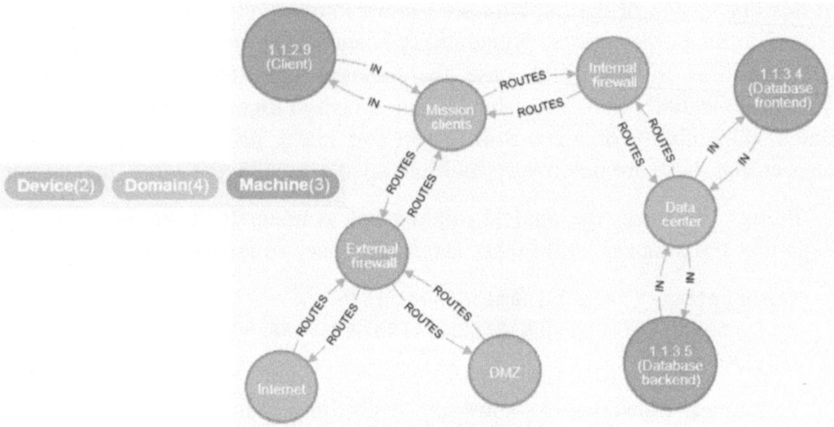

FIG. 33 Query showing potential firewall-blocking responses.

This translates to "show me the network topology, and how these three machines (client and two servers) connect to it."

Fig. 33 is the resulting query match. This shows that the internal firewall is in position to block traffic from the suspicious client to the mission-critical servers. Also, based on the correlation with the initial alert (client-side buffer overflow), the defender suspects that the network client is being controlled from the outside. For that, blocking via the external firewall is an option.

After the attack is thwarted, the defender wants to better understand how the attacker might have gained entry into the network. The defender poses this query:

```
MATCH paths = ()-[:PREPARES|ON|ENABLES|AGAINST*]->
    ()-[:PREPARES]->
    (:Exploit)-[:DETECTION]->
    (:Alert {name:"Snort 1576"})
RETURN paths
```

This translates to "show me all paths (of arbitrary depth) using relationship types about alerts and vulnerability exploitation (ignoring alert source/destination machines) that lead to an exploit detected by the initial alert."

Fig. 34 is the resulting query match. This suggests the cache-poisoning vulnerability as a likely precursor to the buffer overflow attack on the client, even though the cache-poisoning attack was not itself detected.

4.2 Cyber Modeling and Simulation

CyGraph can not only build a knowledge base from real network and threat data, but it can also be driven by simulations and perform model synthesis. Section 4.2.1 examines a CyGraph RESTful web service that visualizes

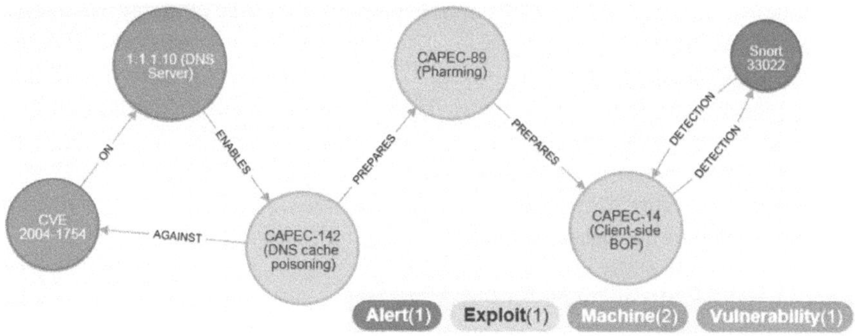

FIG. 34 Query discovering initial source of attack.

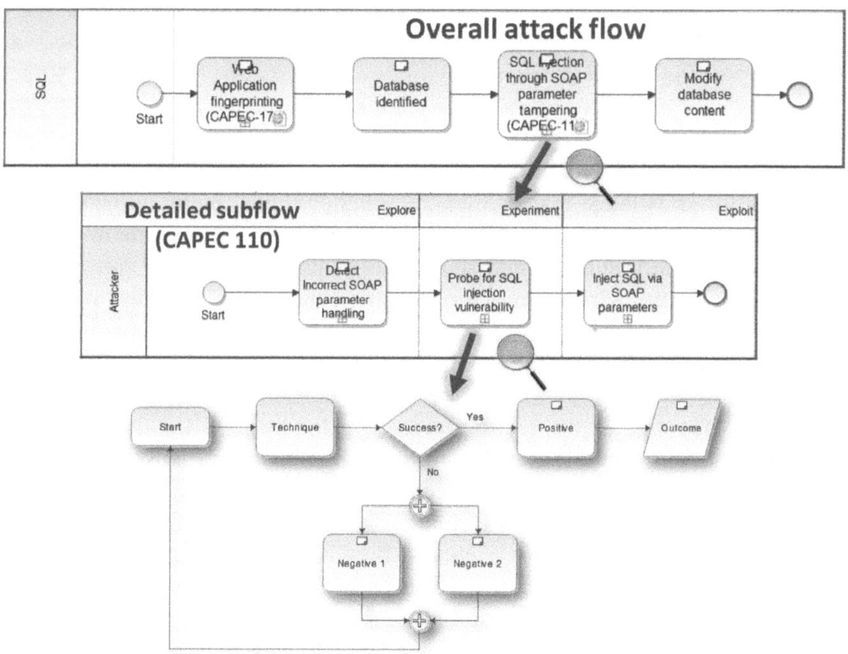

FIG. 35 Process flow model for simulation-driven analytics.

dynamic graph state changes. Section 4.2.2 applies CyGraph to the generation of node attributes for cyber graph models.

4.2.1 Simulation-Driven Visualizations

CyGraph includes a RESTful web service that visualizes dynamic state changes in graph models, e.g., driven by simulations. For example, Fig. 35 shows a process flow model for a multistep attack against a database, which is modeled in terms of CAPEC attack patterns. This is modeled in iGrafx (Object Management Group, 2011), a tool that combines Business Process Model and Notation (BPMN) (iGrafx, 2016) and discrete event simulation. Such models are hierarchical, with high-level processes decomposed into lower-level process flows. In Fig. 35, the highest-level flow is a sequence of CAPEC attack pattern executions. Each attack pattern is decomposed into the explore/experiment/exploit phases defined by CAPEC, which are in turn decomposed into the detailed steps for carrying out each attack phase.

For sufficient realism, process models of cyberattacks need to be coupled with models of the network environment and security posture. For example, particular attack classes are only successful against certain classes of vulnerabilities, and firewall policies control attacker access. These kinds of vulnerability/attack conditions differ for every network and are continually evolving. They should be captured and managed by automated tools such as CyGraph.

This kind of hybrid modeling for cyber and mission interactions was explored in the AMICA project (Analyzing Mission Impacts of Cyber Actions) (Noel et al., 2015). AMICA captures behavior (process models) for mission operations and cyber (attacker and defender) activities, for high-fidelity understanding and measurement of mission impact. AMICA leverages CyGraph for knowledge management, automated model building, and visualization of environmental constraints (network topology, attack graph, mission dependencies, etc.) and dynamic state changes under simulation runs.

In AMICA, the iGrafx simulation engine follows the time and resource-constrained process model for mission and cyber threads. It tests cyber environmental constraints as needed in the process flows, updating them whenever process tasks (i.e., for cyber attacker and defender) change environmental conditions. Throughout the entire process, CyGraph shows the dynamic state evolution of the network environment through animated visualization.

Fig. 36 is an example of CyGraph visualization driven by simulations. This shows visualizations for two simulation runs, with different firewall policy rules for each run. In this scenario, firewall rules enforce access for an internal network and an external partner network. Policy 1 is coarse grained, allowing only mission-required access across the partner network domains, but within each domain (internal and partner) allowing fully connected access (even among machines that do not need to communicate for the mission). Policy 2 is more fine grained. The internal and partner networks are each divided into multiple domains, and only mission-required access provided across each of the smaller domains. In the scenario, the attacker starts at a particular host machine in the partner network and follows a shortest path to reach a critical database server.

In Fig. 36, the attack graph layer (left column) shows potentially exploitable paths of vulnerability from machine to machine. The network topology layer (right column) shows the underlying connectivity among hosts, switches, routers, and firewalls. The attack graphs have edges highlighted

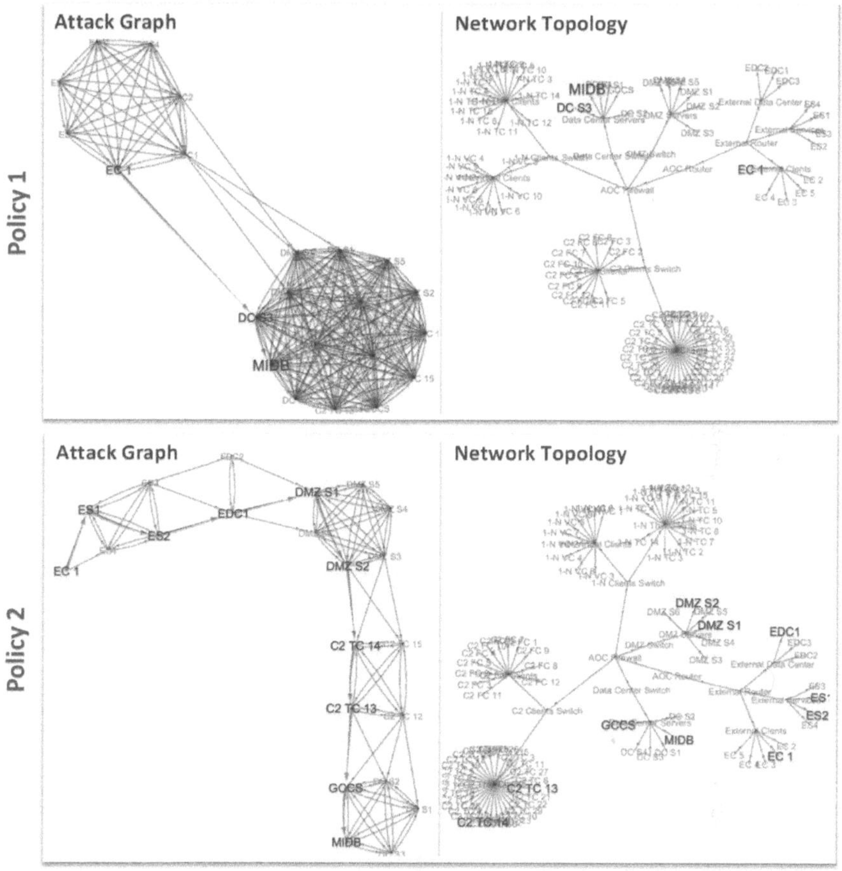

FIG. 36 CyGraph visualization driven by simulations.

representing paths of actual (simulated) exploitation, with the victim machines highlighted. The network topologies have these same victims highlighted. For the network topologies, each attack step (from one machine to another) involves multiple edges of the topology. Topology routes are generally shared over multiple attack steps, so that edge highlighting for multiple attack steps is not effective for topology graphs.

As shown in Fig. 36, the simplistic coarse granularity of Policy 1 allows the attacker to reach the target database service in only two attack steps. On the other hand, the more fine-grained (complex) and restrictive Policy 2 forces the adversary to attack nine hosts to compromise the target.

While not shown in this example, the state changes portrayed in CyGraph can be driven by both attacker and defender. For example, defender process flows can update the state of hosts and vulnerability exposures in response to attacks. In this way, the dynamic interplay of attackers and defenders under

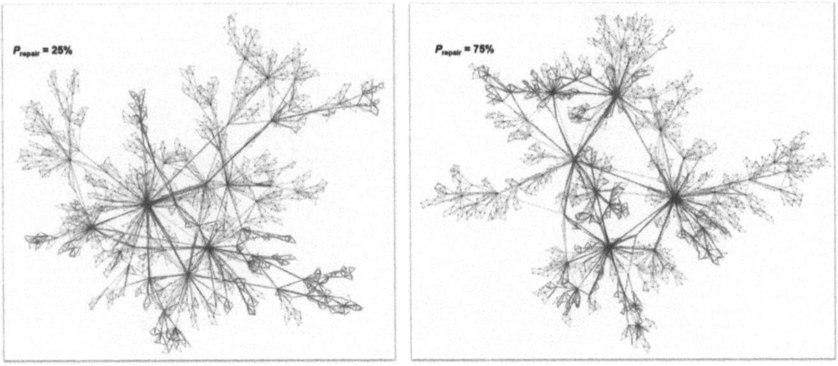

FIG. 37 Simulated interplay of cyber attacks and defenses.

simulation can be more deeply understood. Fig. 37 is an example of this. Here, the attacker is following a path of exploitation across the network. For each attack step, there is some probability that the attack is detected and thwarted by the defender. The figure shows simulation runs for two different defense-success probabilities, i.e., 25% and 75% (left and right sides of Fig. 37, respectively).

4.2.2 Cyber Model Synthesis

There are numerous practical reasons for the automatic generation of cyber graph models, e.g., for scalability testing or to avoid divulging sensitive real network data. There has been considerable progress in understanding the general laws and distinguishing characteristics of real-world graphs, for synthesizing graphs that are more realistic (Chakrabarti and Faloutsos, 2006). These generally focus on collective properties of graph nodes and edges, such as distribution of node degree, graph diameter, and community (clustering) structure. Thus, we can leverage established models for generating graphs (nodes and edges) themselves.

However, practical graph models also include various *attributes* for the nodes and edges in the model. For example, machines (nodes) in a network graph might include attributes such as numbers of vulnerabilities for each machine. We postulate that such attributes generally follow power-law distributions, in which the spectral density (data variance distributed over the frequency domain) varies inversely with frequency, known as $1/f$ models. Such distributions have been observed in many phenomenon (Wikipedia, 2016). One interpretation is a kind of preferential attachment in which some resource (e.g., vulnerabilities) are distributed among entities (e.g., machines) according to how many they already have—a kind of "like attracts like" argument (Merton, 1968). Another interpretation is a kind of stochastic noise process associated with a nonequilibrium system, subject to change driven by a flux of influences to/from other systems (Wikipedia, 2016).

In particular, we generate attribute values for nodes through pseudo-random noise, which has power-law distribution over a graph's spatial layout. This noise is two-dimensional (spatial), with a spectral density that varies inversely with spatial frequency. Such $1/f$ noise is also known as "pink noise" (Halley and Kunin, 1999), because of its color association with light spectra. It is intermediate between a purely random process having no memory of past values ("white noise") and random-walk processes that are dominated by their recent history ("brown noise").

We employ a typical approach for generating $1/f$ pink noise, which "reddens" white noise through a process of autoregression (smoothing). This is low-pass spatial filtering that reduces the higher-frequency components of the white noise (which itself contains equal amounts of all spatial frequencies). For white noise, the individual noise values are independent, i.e., uncorrelated over space. Pink noise varies more smoothly, so that nearby points are more correlated.

When we apply such $1/f$ (pink) noise distributions to generated graphs, nearby nodes are more likely to have similar attribute values, i.e., there is local correlation. For example, nearby nodes in a network structure might be more likely to be managed in a similar way, and thus have similar numbers of vulnerabilities. In rendering such discretized noise fields (two-dimensional arrays), we map noise values to display colors.

For generating pseudo-random $1/f$ spatial (two-dimensional) noise in CyGraph, we follow an established process (Vandevenne, 2004). We first create a two-dimensional array of random (white) noise values. We then apply linear interpolation as a smoothing process. This is done over multiple spatial scales (spatial frequencies), by successively varying the scale over which the smoothing is applied. This yields versions of the white noise smoothed at various scales, which we combine over all scales.

By controlling the relative contributions of smoothed noise at each scale, we obtain noise with different statistical distributions, as shown in Fig. 38. Noise signals with more smoothing (fewer high-frequency components) have stronger correlation among nearby values. For cyber graph models, this

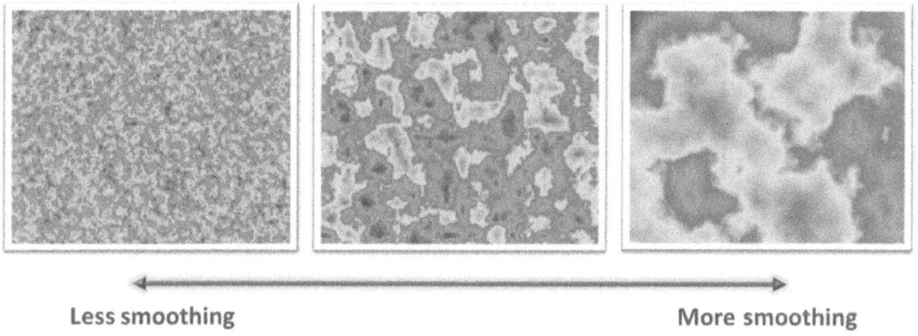

FIG. 38 Controlling variation in generated attribute values.

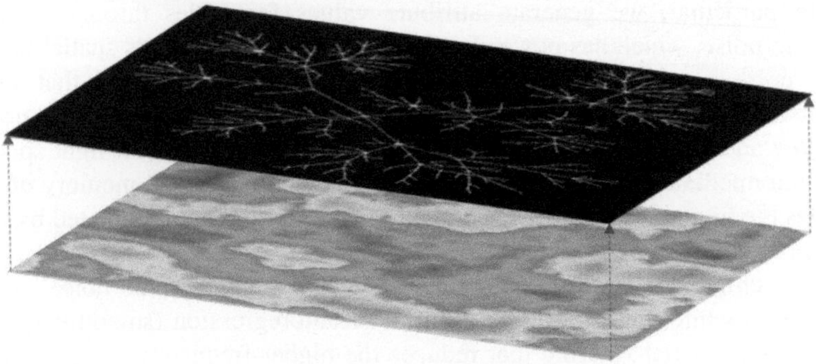

FIG. 39 Mapping correlated attribute values to graph nodes.

corresponds to environments with less variation in attribute values (e.g., vulnerabilities on machines).

Once we generate two-dimensional (spatial) $1/f$ noise with given statistical properties, we map the resulting noise array to a CyGraph model. In particular, we align the two-dimensional array of attribute values over a visualized graph, as shown in Fig. 39. The spatial coordinates of the graph nodes (e.g., network machines) are assigned attribute values (e.g., number of vulnerabilities) based on corresponding locations in the two-dimensional attribute (noise) array. The array values are not "noise" in the usual sense of unwanted contamination of a desired signal. Rather, they are the signal (node attribute) values themselves, as a model that is stochastic but still has locally correlated values (nearby nodes have similar attribute values).

In Fig. 39, the attribute values (two-dimensional $1/f$ noise array) are mapped to a color palette that helps distinguish noise values as bands of colors. The graph nodes are given the corresponding colors as projected from the noise array. The display also renders graph edges as the average of their endpoint node colors, to portray a visual transition.

5 SUMMARY

CyGraph is a flexible approach for dynamic cyber graph analytics and interactive visualization. It provides a graph knowledge base about attack vulnerability, threat indicators, and mission dependencies within a network environment. CyGraph builds a predictive model of possible attack paths and critical vulnerabilities and correlates network events to known vulnerability paths. It also includes dependencies among mission requirements and network assets, for analysis in the context of mission assurance. CyGraph has an open, extensible data model based on a layered property-graph formulation, which makes it easy to extend.

CyGraph brings together isolated data and events into an ongoing overall picture for decision support and situational awareness. It prioritizes exposed vulnerabilities, mapped to potential threats, in the context of mission-critical assets. In the face of actual attacks, it correlates intrusion alerts to known vulnerability paths, and suggests best courses of action for responding to attacks. For postattack forensics, it shows vulnerable paths that may warrant deeper inspection.

CyGraph incorporates an attack graph model that maps the potential attack paths through a network. This includes any network attributes that potentially contribute to attack success, such as network topology, firewall rules, host configurations, and vulnerabilities. The dynamically evolving attack graph provides context for reacting appropriately to attacks and protecting mission-critical assets. CyGraph then ingests network events such as intrusion detection alerts and other sensor outputs, including packet capture. It also incorporates mission dependencies, showing how mission objectives, tasks, and information depend on cyber assets.

For building its graph knowledge base, CyGraph leverages a number of standards under Making Security Measurable™, a suite of standardization initiatives being developed by MITRE and others in the cybersecurity community. These provide standard languages for a variety of input data sources for building CyGraph knowledge graphs, particularly STIX™ (Structured Threat Information eXpression), CAPEC™, and content in the NVD.

In the CyGraph architecture, the data model is based on a flexible property-graph formulation. This model is free to evolve with the available data sources and desired analytics, rather than being fixed at design time. The backend database is implemented in Neo4j, a NoSQL database optimized for graphs. This represents node adjacency via direct pointers, avoiding expensive join operations for graph traversal. It allows CyGraph to take advantage of database technology with graph traversal performance orders of magnitude better than relational databases. Query execution times depend only on the size of the traversed subgraph, independent of the size of the overall graph.

In CyGraph, RESTful web services provide interfaces for data ingest, analytics, and visualization. Data in the wild are mapped to the common CyGraph data model. The domain-specific CyQL supports flexible ad hoc queries against the CyGraph data model. The middle-tier CyGraph Service then compiles CyQL into lower-level native graph query language (Cypher for Neo4j). CyGraph provides a variety of clients for specialized analytic and visual capabilities of query results, including dynamic graph evolution, layering, grouping/filtering, and hierarchical views.

CyGraph has the potential to greatly reduce effort (e.g., within enterprise security operations centers) for prevention and response of cyberattacks, provide situational awareness, and assure missions. Existing tools for attack graph analysis employ specialized data structures and algorithms designed for solving specific problems, e.g., attack reachability or network hardening. Flexibility and extensibility in the face of dynamically evolving network environments and threats have not been first-class design criteria.

Overall, CyGraph introduces a novel unified data model that captures complex relationships among cyber security elements (network topology, firewalls, hosts, vulnerabilities, etc.), threat indicators (intrusion detection alerts, log file entries, suspicious traffic, etc.), and mission dependencies on cyber assets. This allows security operators to better understand the full scope of adversary activities, the relevance to known network vulnerability paths, and the potential impact to specific missions.

ACKNOWLEDGMENTS

This work was funded in part by the MITRE Innovation Program (project number EPF-14-00341), with Vipin Swarup as Cyber Security Innovation Area Leader. We wish to thank Bill Chan of MITRE for providing the architecture diagram for CAVE.

REFERENCES

Albanese, M., Jajodia, S., Noel, S., 2012. Time-efficient and cost-effective network hardening using attack graphs. In: Proceedings of the 42nd Annual IEEE/IFIP International Conference on Dependable Systems and Networks.

Amazon Web Services. AWS Management Portal for vCenter, 2016. https://aws.amazon.com/ec2/vcenter-portal/.

Andlinger, P., 2015. Graph DBMS Increased Their Popularity by 500% Within the Last 2 Years. http://db-engines.com/en/blog_post/43.

Angles, R., Gutierrez, C., 2008. Survey of graph database models. ACM Comput. Surv. 40 (1), 1–39.

ArcSight. ArcSight Publishes Open Standard Designed to Improve the Interoperability of Security and Compliance Systems. http://www.marketwired.com/press-release/arcsight-publishes-open-standard-designed-improve-interoperability-security-compliance-697462.htm.

Artz, M., 2002. NetSPA: A Network Security Planning Architecture. Master's thesis, Massachusetts Institute of Technology.

Baas, B., 2012. NoSQL spatial—Neo4j versus PostGIS. Master's thesis, Delft University of Technology.

Balzer, M., Deussen, O., 2005. Voronoi treemaps. In: Proceedings of the IEEE Symposium on Information Visualization.

Batra, S., Tyagi, C., 2012. Comparative analysis of relational and graph databases. Int. J. Softw. Eng. Soft Comput. 2, 509–512.

Bostock, M., Ogievetsky, V., Heer, J., 2011. D3: data-driven documents. IEEE Trans. Vis. Comput. Graph. 17 (12), 2301–2309.

Carrot Search. FoamTree: Interactive Voronoi Treemaps, 2016. http://carrotsearch.com/foamtree-overview.

Chakrabarti, D., Faloutsos, C., 2006. Graph mining: laws, generators, and algorithms. ACM Comput. Surv. 38, 1–69.

Cray. Urika-GD Product Brief, 2014. http://www.cray.com/sites/default/files/resources/Urika-GD%20Product%20Brief%20Online%205-page.pdf.

CyberAnalytix takes a 7-Year Path to $100 K, 2008. http://www.bizjournals.com/boston/blog/mass-high-tech/2008/05/cyberanalytix-takes-a-7-year-path-to-100k.html.

DB-Engines, 2015. DB-Engines Ranking—Trend of Graph DBMS Popularity. http://db-engines.com/en/ranking_trend/graph+dbms, March 2015.

Dutot, A., Guinand, F., Olivier, D., Pigné, Y., 2007. GraphStream: a tool for bridging the Gap between complex systems and dynamic graphs. In: Proceedings of the Emergent Properties in Natural and Artificial Complex Systems.

Graphviz—Graph Visualization Software. http://www.graphviz.org/.

Gudivada, V., Rao, D., Raghavan, V.V., 2016. Renaissance in database management: navigating the landscape of candidate systems. IEEE Comput. 49 (4), 31–42.

Gulabani, S., 2014. Developing RESTful Web Services with Jersey 2.0. Packt Publishing, Birmingham, UK.

Halley, J., Kunin, W., 1999. Extinction risk and the 1/f family of noise models. Theor. Popul. Biol. 56 (3), 215–230.

Harper, R., 2013. There Is Such a Thing as a Declarative Language, and It's the World's Best DSL. https://existentialtype.wordpress.com/2013/07/22/there-is-such-a-thing-as-a-declarative-language/.

Harris, E., Perlroth, N., 2014. Target missed signs of a data breach. The New York Times. http://www.nytimes.com/2014/03/14/business/target-missed-signs-of-a-data-breach.html.

Iannacone, M., Bohn, S., Nakamura, G., Gerth, J., Huffer, K., Bridges, R., Ferragut, E., Goodall, J., 2014. Developing an ontology for cyber security knowledge graphs. In: Proceedings of the 9th Annual Cyber and Information Security Research Conference.

iGrafx, 2016. http://www.igrafx.com/.

Information Technology—Syntactic Metalanguage—Extended BNF, 1996. International Standard ISO/IEC 14977:1996(E).

Ingols, K., Lippmann, R., Piwowarski, K., 2006. Practical attack graph generation for network defense. In: Proceedings of the 22nd Annual Computer Security Applications Conference.

Jajodia, S., Noel, S., O'Berry, B., 2005. Topological analysis of network attack vulnerability. In: Kumar, V., Srivastava, J., Lazarevic, A. (Eds.), Managing Cyber Threats: Issues, Approaches and Challenges. Springer, Berlin, Germany.

Jajodia, S., Noel, S., Kalapa, P., Albanese, M., Williams, J., 2011. Cauldron: mission-centric cyber situational awareness with defense in depth. In: Proceedings of the 30th Military Communications Conference.

Lippmann, R., Ingols, K., 2005. An annotated review of past papers on attack graphs. Technical report, MIT Lincoln Laboratory.

Martin, R., 2008. Making security measurable and manageable. In: MILCOM 2008—2008 IEEE Military Communications Conference, 16–19 Nov., pp. 1–9, San Diego, CA.

Merton, R., 1968. The Matthew effect in science. Science 159 (3810), 56–63.

Murdock, J., Pramanik, S., 2004. Correlating Advanced Threat Information Feeds. HP Protect.

Musman, S., Tanner, M., Temin, A., Elsaesser, E., Loren, L., 2011. Computing the impact of cyber attacks on complex missions. In: Proceedings of the IEEE International Systems Conference.

Nadkarni, A., Vesset, D., 2014. Worldwide Big Data Technology and Services 2014–2018 Forecast. International Data Corporation. IDC #250458.

National Vulnerability Database, 2016. https://nvd.nist.gov/.

Neo4j, 2016. http://neo4j.com/.

Ning, P., Xu, D., 2004. Hypothesizing and reasoning about attacks missed by intrusion detection systems. ACM Trans. Inform. Syst. Secur. 7, 591–627.

Noel, S., 2015. Interactive visualization and text mining for the CAPEC cyber attack catalog. In: Proceedings of the ACM Intelligent User Interfaces Workshop on Visual Text Analytics.

Noel, S., Heinbockel, W., 2015. An overview of MITRE cyber situational awareness solutions. In: Proceedings of the NATO Cyber Defence Situational Awareness Solutions Conference.

Noel, S., Jajodia, S., 2007. Attack graphs for sensor placement, alert prioritization, and attack response. In: Proceedings of the Air Force Cyberspace Symposium.

Noel, S., Jajodia, J., 2009a. Attack Graph Aggregation. US Patent 7,627,900.

Noel, S., Jajodia, S., 2009b. Advanced vulnerability analysis and intrusion detection through predictive attack graphs. In: Proceedings of the Armed Forces Communications and Electronics Association Critical Issues in C4I.

Noel, S., Jajodia, S., 2014. Metrics suite for network attack graph analytics. In: Proceedings of the 9th Annual Cyber and Information Security Research Conference.

Noel, S., O'Berry, B., Hutchinson, C., Jajodia, S., Keuthan, L., Nguyen, A., 2002. Combinatorial analysis of network security. In: Proceedings of the 16th Annual International Symposium on Aerospace/Defense Sensing, Simulation, and Controls.

Noel, S., Elder, M., Jajodia, S., Kalapa, P., O'Hare, S., Prole, K., 2009. Advances in topological vulnerability analysis. In: Proceedings of the Cybersecurity Applications & Technology Conference for Homeland Security.

Noel, S., Ludwig, J., Jain, P., Johnson, D., Thomas, R.K., McFarland, J., King, B., Webster, S., Tello, B., 2015. Analyzing mission impacts of cyber actions (AMICA). In: NATO Workshop on Cyber Attack Detection, Forensics and Attribution for Assessment of Mission Impact.

NSA-Funded 'Cauldron' Tool Goes Commercial, 2009. http://www.darkreading.com/nsa-funded-cauldron-tool-goes-commercial/d/d-id1131178.

O'Hare, S., Noel, S., Prole, K., 2008. A graph-theoretic visualization approach to network risk analysis. In: Proceedings of the Workshop on Visualization for Computer Security.

Object Management Group. Documents Associated with Business Process Model and Notation (BPMN) Version 2.0, 2011. http://www.omg.org/spec/BPMN/2.0/.

Ou, X., Govindavajhala, S., Appel, A., 2005. MulVAL: A Logic-Based Network Security Analyzer. In: Proceedings of the 14th USENIX Security Symposium.

Parr, T., 2013. The Definitive ANTLR 4 Reference. The Pragmatic Programmers.

RedSeal Networks, 2016. http://www.redsealnetworks.com/.

Ritchey, R., O'Berry, B., Noel, S., 2002. Representing TCP/IP connectivity for topological analysis of network security. In: Proceedings of the 18th Annual Computer Security Applications Conference.

Robinson, I., Webber, J., Eifrem, E., 2015. Graph Databases, second ed. O'Reilly, Sebastopol, CA.

Sanders, C., 2011. Practical Packet Analysis—Using Wireshark to Solve Real-World Problems, second ed. No Starch Press, San Francisco, CA.

Schweitzer, P., 2013. Attack–Defense Trees. Doctoral dissertation, University of Luxembourg.

Sheyner, O., Wing, J., 2004. Tools for generating and analyzing attack graphs. In: Proceedings of the Workshop on Formal Methods for Components and Objects.

Skybox. Risk Analytics for Cyber Security Management, 2016. http://www.skyboxsecurity.com/.

Structured Threat Information eXpression (STIX™)—A Structured Language for Cyber Threat Intelligence, 2016. https://stixproject.github.io/.

The MITRE Corporation. Making Security Measurable, 2013. http://makingsecuritymeasurable.mitre.org/.

The MITRE Corporation. Making Security Measurable—Directory of Efforts by Organization Name, 2014. http://makingsecuritymeasurable.mitre.org/directory/organizations/index.html.

The MITRE Corporation. The Common Attack Pattern Enumeration and Classification—A Community Resource for Identifying and Understanding Attacks, 2016. https://capec.mitre.org/.

The MITRE Corporation. Crown Jewels Analysis, 2009. http://www.mitre.org/publications/systems-engineering-guide/enterprise-engineering/systems-engineering-for-mission-assurance/crown-jewels-analysis.

The MITRE Corporation. Cyber Command System (CyCS), 2016. http://www.mitre.org/research/technology-transfer/technology-licensing/cyber-command-system-cycs.

The MITRE Corporation. Threat Actor Leveraging Attack Patterns and Malware, 2016. http://stixproject.github.io/documentation/idioms/leveraged-ttp/.

The MITRE Corporation. Kill Chains in STIX, 2016. http://stixproject.github.io/documentation/idioms/kill-chain/.

Vandevenne, L. Texture Generation using Random Noise, 2004. http://lodev.org/cgtutor/randomnoise.html.

vis.js—A Dynamic, Browser-Based Visualization Library, 2016. http://visjs.org/.

VMware. VMware Virtualization for Desktop & Server, Application, Public & Hybrid Clouds, 2016. http://www.vmware.com/.

Vukotic, A., Watt, N., Abedrabbo, T., Fox, D., Partner, J., 2015. Neo4j in Action. Manning Publications, Shelter Island, NY.

Wang, L., Yao, C., Singhal, A., Jajodia, S., 2006. Interactive analysis of attack graphs using relational queries. In: Damiani, E., Liu, P. (Eds.), Data and Applications Security XX. Lecture Notes in Computer Science. vol. 4127. Springer Berlin Heidelberg, Sophia Antipolis, France.

Wang, L., Jajodia, S., Singhal, A., Cheng, P., Noel, S., 2013. k-zero Day safety: a network security metric for measuring the risk of unknown vulnerabilities. IEEE Trans. Depend. Secure Comput. 11, 30–44.

Wikipedia. Power Law, 2016. https://en.wikipedia.org/wiki/Power_law.

Wikipedia. Non-Equilibrium Thermodynamics, 2016. https://en.wikipedia.org/wiki/Non-equilibrium_thermodynamics.

Wynn, J., Whitmore, J., Upton, G., Spriggs, L., McKinnon, D., McInnes, R., Graubart, R., Clausen, L., 2011. Threat assessment & remediation analysis (TARA): methodology description version 1.0. MITRE Technical report MTR110176.

Yates, C., Ladd, S., Devijver, S., 2006. Expert Spring MVC and Web Flow (Expert's Voice in Java). Apress, New York City.

Zadrozny, P., Kodali, R., 2013. Big Data Analytics Using Splunk. Apress, New York City.

Chapter 5

Cognitive Analytics: Going Beyond Big Data Analytics and Machine Learning

V.N. Gudivada*,[1], M.T. Irfan[†], E. Fathi* and D.L. Rao*
*East Carolina University, Greenville, NC, United States
[†]Bowdoin College, Brunswick, ME, United States
[1]Corresponding author: e-mail: gudivadav15@ecu.edu

ABSTRACT

This chapter defines analytics and traces its evolution from its origin in 1988 to its current stage—cognitive analytics. We discuss types of learning and describe classes of machine learning algorithms. Given this backdrop, we propose a reference architecture for cognitive analytics and indicate ways to implement the architecture. A few cognitive analytics applications are briefly described. The chapter concludes by indicating current trends and future research direction.

Keywords: Cognitive analytics, Text analytics, Learning analytics, Educational data mining, Cognitive systems, Cognitive computing, Personalized learning, Data science, Machine learning, Big data analytics, Business analytics

1 INTRODUCTION

Cognitive computing is a computational environment which is comprised of (1) a high-performance computing infrastructure powered by special processors such as multicore CPUs, GPUs, TPUs, and neuromorphic chips; (2) a software development environment with intrinsic support for parallel and distributed computing, and powered by the underlying computing infrastructure; (3) software libraries and machine learning algorithms for extracting information and knowledge from unstructured data sources; (4) a data analytics environment whose processes and algorithms mimic human cognitive processes; and (5) query languages and APIs for accessing the services of the cognitive computing environment. We have defined cognitive computing in terms of its functions, since it is not easy to define it precisely and completely by other methods. Cognitive analytics draws upon the cognitive computing

Handbook of Statistics, Vol. 35. http://dx.doi.org/10.1016/bs.host.2016.07.010

environment to generate actionable insights by analyzing diverse heterogeneous data sources using cognitive models that the human brain employs.

Classical symbolic and rule-based approaches to problems such as machine translation and speech-to-speech translation are being superseded by statistical learning approaches. For example, consider the problem of recognizing handwritten digits. Rule-based approaches entail developing a number of rules which aim to explicitly capture ways to write digits by different users. This results in too many rules. Furthermore, additional rules are needed to accommodate new users who might write a digit differently from the ways that were reflected in the current rule set. In contrast, artificial neural network (ANN) approaches use several small pieces of evidence in the form of features and combine them to produce higher-level features. ANN approaches are more robust as they perform better with data which is not seen in the training phase.

The ubiquity of big data (Gudivada et al., 2015a), abundant computing power, and the resurgence of neural network algorithms are providing *scalable solutions* to several difficult problems. The performance of newer approaches to problems that have been considered difficult for computers such as finding objects in images and classifying images rival human performance. For example, in the ImageNet Large-Scale Visual Recognition Challenge (Russakovsky et al., 2015), the error rate for some algorithms for detecting objects in video and scene classification is as low as 6%, whereas the error rate for humans is 5%. In another study with deep-learning algorithms (Goodfellow et al., 2014), Google reports 99.8% accuracy in recognizing CAPTCHA images on the hardest category of reCAPTCHA dataset. In another study at Facebook on the classification of images, Taigman et al. (2014) achieved an accuracy of 97.35% on the Labeled Faces in the Wild dataset using a nine-layer deep neural network. Finally, Lake et al. (2015) describe an approach called *Bayesian Program Learning*, which is used to recognize 1623 handwritten character sets from 50 languages with only a limited training. Though the above problems are diverse, deep neural network algorithms perform exceptionally well in all these domains.

The above approaches coupled with advances in information retrieval, natural language understanding (Gudivada et al., 2015b), artificial intelligence (AI), and machine learning are helping to usher in a new paradigm for strategic decision making. The term *data analytics* when used in a generic sense refers to any actionable information that results from computational analysis of data using mathematical and statistical methods. Data analytics is an interdisciplinary domain encompassing mathematics, statistics, and computer science. Implicitly, there is a domain associated with data analytics. The domain provides the data for analysis. The primary goal of data analytics is to gain insights into a process or problem so that the latter can be improved or solved. In other words, analytics is a data-driven approach to decision making and problem solving.

Though certain types of analytics are common across various application domains, they tend to vary significantly from domain to another. This has led to the proliferation of names such as business analytics, text analytics, image analytics, video analytics, graph analytics, spatial analytics, visual analytics, and cognitive analytics. However, irrespective of the domain, data analytics is comprised of three components: data acquisition and loading, methods and algorithms, and a computational platform that implicitly embodies workflows and best practices. The data acquisition and loading components enable the preparation of input data and loading it into the computational platform. Various algorithms and approaches for data analysis are provided by the methods and algorithms component. Lastly, the computational platform brings everything together as a system and provides interfaces for users and other applications to interact with it.

From a functional perspective, there are three categories of data analytics: descriptive, prescriptive, and predictive. Descriptive analytics provides a dashboard view of the current state of a system or process. It uses descriptive statistics and machine learning algorithms to provide insight into a system. The insight often reveals, in a process, for example, various steps in the process, how the steps are sequenced, what type of resources are consumed, and how much time is spent in each process. As another example, readability of English texts is determined by text analytics such as the Fry readability formula, Automated Readability Index, Flesch-Kincaid, Gunning-Fog, Coleman-Liau Index, and SMOG Index. Software metrics and measurements are analytics used to characterize properties of software. Such metrics include the number of classes, number of methods per class, depth of inheritance tree, number of interfaces, and total lines of code.

Prescriptive analytics is a natural outcome of descriptive analytics. It suggests ways to improve a current process or system using simulation and optimization algorithms. In the case of software metrics and measurements, prescriptive analytics specifies a range of values for each measurement such as bounds for number of methods in a class. Furthermore, it specifies refactoring techniques if a measurement is not within the specified range.

Predictive analytics enables answering "what-if" questions by building predictive models using inferential statistics and forecasting techniques. It enables organizations to make data-driven strategic decisions. Predictive models are built using the operational and historical data. They extract associations and other implicit relationships in the data to build the models. Various regression models such as linear, logistic, Lasso, ridge, Cox proportional hazards, and Bayesian are widely used. Logic regression, for example, is used in clinical trials and fraud detection to associate a probability with a binary outcome.

Like cognitive computing, cognitive analytics is pursued from two complementary perspectives. The first is driven by the computer science researchers in both industry and academia. Advances in big data, cloud computing, natural language understanding, and machine learning are enabling extraction of knowledge from vast repositories of unstructured data such as natural language text, images, video, and audio. From this group's perspective, the knowledge extracted from the unstructured data coupled with statistical inference and reasoning distinguishes cognitive analytics from business analytics. The second perspective is advanced by cognitive and neuroscience researchers. They employ theories of mind, functional areas of the brain, and cognitive models and processes. For example, an approach in this class might gather analytics about a cognitive process to validate the cognitive model as well as to improve the model (Chakraborty et al., 2014).

1.1 Chapter Organization

The overarching goal for this chapter is to present a unified approach to the emerging area of cognitive analytics. More specifically, in Section 2, we trace the evolution of data analytics and discuss central issues. Types of learning used in cognitive analytics are described at a conceptual level in Section 3. In Section 4, we discuss the following classes of machine learning algorithms: logistic regression, decision trees, support vector machines (SVMs), Bayesian networks (BNs), neural networks, and deep learning. This section also includes a discussion on machine learning frameworks and libraries.

We propose a reference architecture called Cognalytics for cognitive analytics in Section 5. This section also indicates how this architecture can be implemented using open source tools. Section 6 presents applications of cognitive analytics including learning analytics (LA), personalized learning, cognitive businesses, brain–computer interfaces (BCIs), and assistive technologies. Cognitive analytics trends and future research directions are described in Section 7. Section 8 concludes the chapter.

2 EVOLUTION OF ANALYTICS AND CORE THEMES

AI is a subfield of computer science and machine learning is a major area within AI. The recent emergence of big data and cloud computing created AI renaissance. The attendant media coverage of machine learning is making the latter a household name. This is also creating confusion and propagation of misinformation. In blogs and other self-published forums, some authors have declared AI and Computer Science as two distinct disciplines, likewise, AI and machine learning. The scope and the meaning of the term *analytics* are being reinvented.

You cannot manage what you do not measure is an old adage from the management world that is still true today in most organizations and academic disciplines. At the core of analytics are data, mathematical and statistical models built using this data. The types of data needed and the type of processing performed, and the variety of models built varies. The models are used for a broad range of purposes under the umbrella terms descriptive analytics, prescriptive analytics, and predictive analytics. AI, machine learning, distributed computing, and high-performance computing comprise the computational infrastructure to manage data and enable model building.

2.1 Multiple Perspectives

There exists multiple perspective on analytics. The Computer Science perspective is driven by technical considerations related to storing, managing, and querying data. In the early days, there was limited support for analysis. The business perspective views analytics from an organizational level and focuses on actionable insights into data. Visual analytics is a new area of analytics whose goal is analytical reasoning through interactive visual interfaces. Even more recently, other terms such as educational data mining (EDM), LA, and cognitive analytics have emerged.

Academia has responded to this unprecedented interest in analytics by creating new *interdisciplinary* degree programs primarily at the master's level. These programs fall into three categories: (1) programs that have the term analytics somewhere in their name—business analytics, health informatics, health care informatics, and nursing informatics. Other degree programs such as econometrics also fall under this category, though they do not explicitly use the term informatics in the name. These programs are typically administered or led by noncomputer science departments; (2) programs with names such as master of science in analytics and master of science in data science. These programs are typically led by computer science departments; and (3) numerous graduate certificates, tracks, and concentrations in analytics, data mining, knowledge discovery, machine learning, and big data.

2.2 Analytics Evolution

We trace the evolution of analytics from a Computer Science perspective as shown in Fig. 1. Basic analytics functions were part of the relational database management systems (RDBMS) from their early years. RDBMS served as *operational databases* for conducting day-to-day business transactions—online transaction processing (OLTP). Basic functions for descriptive statistics were provided. In subsequent years, more sophisticated functions were introduced under the name Statistics & SQL Analytics. They included functions for ranking of results, moving and cumulative aggregating values over a range of rows, lag and lead to access data from preceding and following rows, descriptive statistics, correlations, and linear regression. In the early days of RDBMS, analytic functions were implemented outside of the RDBMS system. Each analytic function was implemented by a standalone piece of code which made code optimization across RDBMS and analytic functions difficult. Recently, there have been efforts in implementing analytic functions within the database (Feng et al., 2012).

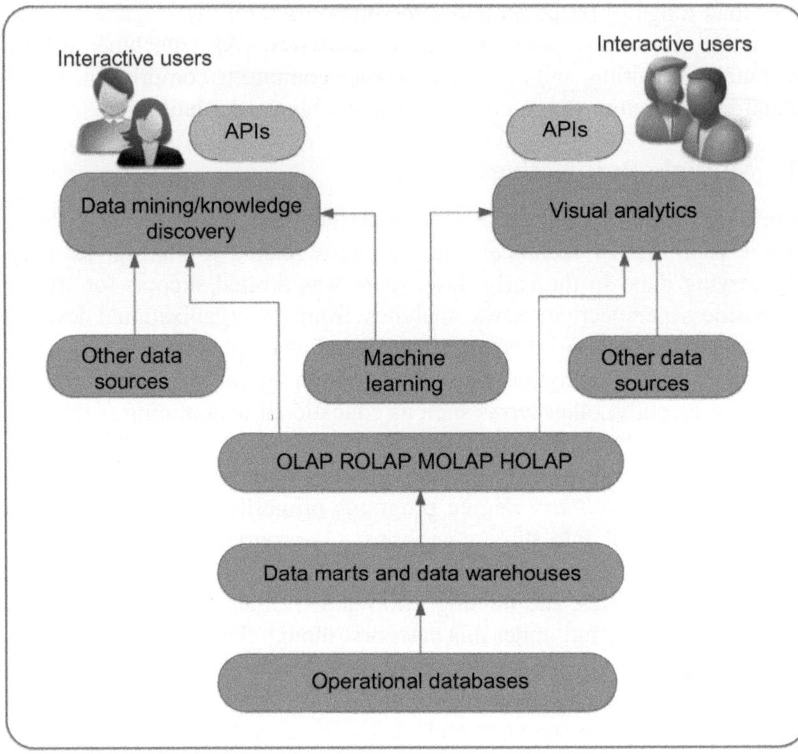

FIG. 1 Evolution of analytics.

2.3 Data Warehouses and Data Marts

The next stage in the evolution is the packaging of more advanced analytic functions into database systems for data marts and data warehouses. The latter are developed to help make data-driven strategic decisions—online analytical processing (OLAP). The terms *data mart* and *data warehouse* are sometimes used incorrectly as synonyms. A data warehouse is a consolidated and centralized repository of data extracted from various operational databases and other disparate data sources. A data mart, on the other hand, is a subset of a data warehouse, which targets the needs of a specific department of an organization. The data warehouse is like an enterprise *database schema*, whereas a data mart is akin to a *database view*. Data warehouses and data marts are used for generating customer and compliance reports, score cards, and dashboards. They are also used for planning, forecasting, and modeling. Extract, Transform, and Load (ETL) is a set of tools and processes that are used to design and implement data warehouses and data marts.

Both OLTP and some OLAP systems use SQL for performing analytic functions. SQL operators such as CUBE, ROLLUP, and GROUPING SET were specifically introduced for data analytics with data warehouses. An *OLAP cube* is a multidimensional data array, which is a generalization of a 2D or 3D spreadsheet. It can also be viewed as a logical structure that defines metadata. MDX (multidimensional expression) is a metadata-based query language to query OLAP cubes. Analytic operations on OLAP cube include slice (creating a new cube with fewer dimensions), dice (creating a new (smaller) cube by stating specific values for cube dimensions), drill down and drill up (navigating between the most detailed data level to the summarized data levels), roll-up (summarizing data along a specific dimension), and pivot (rotating the cube to view its various dimensions or faces).

2.4 ROLAP, MOLAP, and HOLAP

The third stage in the evolution is the emergence of ROLAP, MOLAP, and HOLAP. All three classes of cubes organize data in a way to enable efficient dimensional analysis. The first step in building a cube is to determine dimensions. For a sales department cube, for example, geographic region and industry classification are two dimensions. The next step is to determine the levels of data aggregation for each dimension. For the geographic region dimension, data aggregation levels include county, state, region, country, and continent. If the industry classification is energy utilities, data aggregation levels include natural gas, coal-powered electricity, wind, and solar.

ROLAP, MOLAP, and HOLAP are extensions of OLAP and are referred to as OLAP servers. A relational OLAP (ROLAP) server is an interface between an RDBMS warehouse and OLAP users. It implements navigational logic for the cube, sends SQL queries for execution to the underlying warehouse, and provides additional tools and services. ROLAP servers tend to suffer from performance since data needs to be fetched from the warehouse on the fly.

In contrast with ROLAP, MOLAP cubes extract data a priori from a warehouse and store the data in the cube itself. All the calculations are precomputed at the time of the cube creation. This contributes to superior performance but limits the amount of data handled by the MOLAP cube. Also, MOLAP consumes additional storage space. HOLAP is a hybrid server which combines best of both ROLAP and MOLAP. HOLAP is scalable like ROLAP and provides superior performance like MOLAP.

2.5 Data Mining/Knowledge Discovery

Independent of analytics evolution, machine learning (ML) emerged in parallel as a subdiscipline of AI. The majority of machine learning algorithms fall into the following broad categories: decision trees, associative rule learning, genetic algorithms, refinement learning, random forest, SVMs, BNs, neural networks, and deep learning.

The next stage in the analytics evolution is the emergence of data mining (aka knowledge discovery). Data mining is a synergistic confluence of databases, statistics, AI, and ML. Its goal is to find anomalies and discover hidden patterns and correlations in data to enable the generation of actionable intelligence. Such intelligence has been used to increase revenues, improve customer relationship, reduce operating costs, and make strategic decisions. A significant task in data mining is locating the relevant data and preparing the data for ingestion into ML algorithms.

2.6 Visual Analytics

Visual analytics is a relatively new field and developed independently of data mining. Like data mining, it draws data from several sources including RDBMS, OLAP cubes, and other sources such as social media. Visual analytics combines automatic and visual analysis methods with human interactive exploration. It is based on the premise that combining the quantitative capabilities of computers and the cognitive capabilities of humans leads to powerful ways to create new knowledge. Interactive exploration and visual manipulation play a central role in visual analytics. Both data mining and visual analytics systems are available as cloud services. Their functionality is accessed through APIs.

2.7 Cognitive Analytics

Cognitive analytics is the natural evolution of both data mining and visual analytics. Cognitive analytics removes humans from the loop and is completely automated. It is in a formative stage now and there is tremendous interest from both industry and academia. However, industry is primarily driving both research and development. Cognitive analytics draws upon advances in several areas and combines the computing and cognitive science approaches. Fig. 2 shows a functional view of cognitive analytics. Data for cognitive analytics comes from several sources and includes structured, semistructured, and unstructured data. Furthermore, it employs knowledge structures such as taxonomies and ontologies to enable reasoning and inference. Extraction of both low-level features and high-level information is crucial to cognitive analytics.

Shown in the wide rectangle in Fig. 2 are internal components of cognitive analytics engine. Various knowledge representation structures are needed to represent and reason with knowledge. An assortment of machine learning algorithms and inference engines are also needed. The domain cognitive models capture domain-specific cognitive processes to enable cognitive style problem solving. The learning and adaptation component improves system performance by learning from previous interactions with the users.

In contrast with all other analytics, cognitive analytics generates multiple answers for a question and assigns a degree of confidence to each answer.

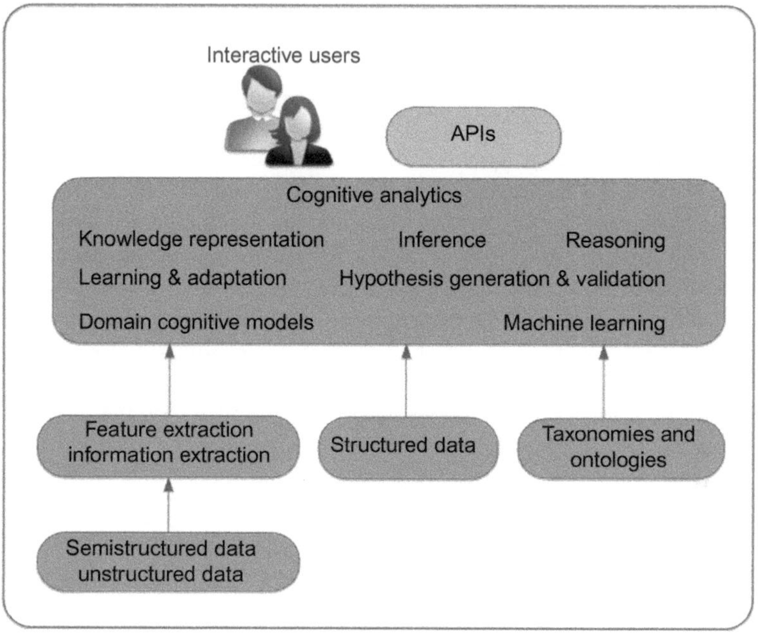

FIG. 2 Conceptual view of cognitive analytics.

In other words, cognitive analytics uses probabilistic algorithms to come up with multiple answers with varying degrees of relevance. Noncognitive analytics, in contrast, employ deterministic algorithms and compute only one answer for any question. Computing multiple answers requires another component, which is labeled Hypothesis Generation & Validation. This component is pioneered by IBM and is responsible for generating multiple hypotheses for a question, gathers evidence for each hypothesis, and using the evidence scores the relevance of a hypothesis as an answer to the question.

In summary, analytics comes in many forms with varying functional capabilities. Each form reflects the underlying technologies and the characteristics of the domain which propels the form. Regardless of these differences, we can forge a generic architecture for cognitive computing. Implementation of such an architecture requires a platform with the following characteristics: infrastructure for data cleaning, transformations, and fusion; a set of both deterministic and probabilistic algorithms for computing analytics; a learning component powered by a domain cognitive model; an array of machine learning frameworks for hypothesis generation, evidence gathering, and scoring hypotheses; and a high-performance computing system with scalability, performance, and elasticity. In Section 5, we propose a reference architecture for cognitive analytics and discuss ways to implement the architecture.

3 TYPES OF LEARNING

There are two broad classes of learning: supervised and unsupervised. Supervised learning involves learning from examples, which is a set of associations between inputs and outputs. This is akin to how children learn to read and write—a teacher presents letters of the alphabet and utters the corresponding sounds. Repeating the process with the same examples will gradually train students' biological neural networks to associate symbols with sounds.

Training data is comprised of two parts: input and expected output. Let (i, o) be an item in the training dataset, which specifies that when the system is presented with input i, it should output o. The training data is a set of n such pairs: $\{(i_1, o_1), (i_2, o_n), \ldots, (i_n, o_n)\}$. A trained model should work correctly if the examples in the training set are given to the model again. For example, given i as input to the trained model, it should produce o as the output.

A reasonable criterion should be defined to measure the error between the correct output and the outcome generated by the model. The main task of supervised learning is to minimize the error function. This is similar to how the teacher corrects students in their initial attempts to read or write, which gradually minimizes the error function of their biological neural network model. In addition to the error function, other characteristics of the model include the number of model parameters and modeling flexibility (Battiti et al., 2008). Decision trees, neural networks, regression, and Bayesian classification are examples of supervised learning algorithms.

Unsupervised learning algorithms draw inferences from datasets consisting of just input data without labeled responses. Unsupervised learning infers a function which describes the hidden structure from unlabeled data. Since the examples given to the learner are unlabeled, there is no error or reward signal to evaluate a potential solution. This distinguishes unsupervised learning from supervised learning and reinforcement learning (see Section 3.2). K-means clustering, genetic algorithms, and simulated annealing are examples of unsupervised learning algorithms.

For cognitive analytics, unsupervised algorithms have an edge over the supervised ones. In the big data context, we do not know the patterns in the data *a priori*. Furthermore, training data may not be available. Unsupervised learning algorithms are better suited for this scenario. Unsupervised learning algorithms are also used to automatically generate test data. The latter is employed to train supervised learning algorithms. For complex question-answering (Q/A) environments such as the Jeopardy! game, several hypotheses are generated as candidate answers; evidence is gathered and used to rank hypotheses. In such Q/A environments, it is advantageous to use supervised learning to generate some hypotheses and unsupervised learning to generate additional hypotheses. This approach benefits from both types of learning and the resulting system performs more robustly. There are many applications such as real-time fraud detection, continuous security vulnerability assessment, computer vision, and natural language understanding for which unsupervised learning suits well.

3.1 Active Learning

Active learning is a special case of semisupervised learning. The key hypothesis of active learning is to allow the learning algorithm to choose the data from which to learn. In other words, the learning algorithm can interactively query the user (or another information source) to obtain the desired output(s) for the input instance on hand. This is expected to result in better performance as well as less training. The advantage of active learning over supervised learning is that the former obviates the need for thousands of labeled instances for training (Settles, 2009). This is especially important for cognitive analytics where unstructured data is abundant, but has no labels.

Active learning is also called query learning and optimal experimental design. Algorithms are needed to determine which data points should be labeled with the expected result. These algorithms are also known as query strategies and include uncertainty sampling—label only those points for which the current model is least certain about the correct output; query by committee–label those points for which the *committee* disagrees the most, where the committee is comprised of a variety of models which are trained on the current labeled data; expected model change—label those points which would change the current model the most; expected error reduction—label those points which would reduce the model's generalization error the most; and many more.

3.2 Reinforcement Learning

Humans and theoretical systems have various types of learning in common. Learning by imitating a teacher is the most standard but not the only way of transferring knowledge. In real life, we notice the outstanding tendency of children to try dangerous things such as placing fingers in an electrical receptacle without a guiding teacher. Depending on the outcome or experience, a child may repeat this activity again or may never repeat it. This type of learning is called *reinforcement learning*. The latter is a form of supervised learning.

Reinforcement learning is inspired by behaviorist psychology. The latter is concerned with how to take actions by an agent in an unknown environment to maximize some notion of cumulative rewards. In the example of bicycle riding, positive rewards can be in the form of admiring friends and negative ones can be injuries to biological tissues. But after some trials with the goal of maximizing the positive rewards, learning occurs (i.e., can ride the bicycle now). Initially, the system is not explicitly trained and receives feedback on its performance when it is in production. In a sense, reinforcement learning is trial-and-error learning.

A reinforcement learning environment is formulated as a Markov decision process (MDP). Many reinforcement algorithms use dynamic programming techniques. These algorithms do not need the knowledge about the MDP. When exact methods become infeasible, these algorithms target large MDPs. The basic reinforcement learning model is comprised of (a) a set of environment states S, (b) a set of actions A, (c) stochastic rules that govern state transitions, (d) rules to compute the immediate reward of a transition, and (e) rules to describe an agent's observations. Reinforcement learning is especially well suited to problems which include a long- vs short-term reward trade-off as in elevator scheduling, and robot control.

Reinforcement learning serves as a theoretical tool for investigating the principles of autonomous agents learning to act in their environment. The agents strive to improve their behavior through their interactions and experiences with other agents in the environment. Reinforcement learning has been used as a practical computational tool for designing and constructing autonomous agents for domains such as robotics, combinatorial search problems, and industrial manufacturing.

3.3 Ensemble Learning

Ensemble learning is based on multiple learning models, which are strategically generated, and optimally combined for solving problems such as classification (Polikar, 2009). The idea behind this is that two minds are better than one. Also, to make strategic decisions, we solicit input from multiple sources and combine or rank the sources. An ensemble itself is a supervised learning algorithm. Ensemble learning systems are also called multiple classifier systems.

Ensemble algorithms yield better results if there are significant differences or diversity among the models. For example, more random decision trees lead to a stronger ensemble than entropy-reducing decision trees. However, selecting a range of strong learning algorithms, though they may significantly diverge, is the key for good performance. Common types of ensembles include (a) Bayes optimal classifier—is an ensemble of all the hypotheses in the hypothesis space, (b) Bayesian parameter averaging—approximate the Bayes Optimal Classifier by sampling hypotheses from the hypothesis space and combining them using Bayes' rule, (c) Bootstrap aggregating (bagging)—building multiple models, typically of the same kind, from different subsamples of the training dataset, (d) Boosting—building multiple models, typically of the same type, each model learns to fix the prediction errors of a prior model in the chain, (e) Stacking—building multiple models, typically of differing types, and having a supervisor model that learns how to best combine the predictions of the primary models, and (f) Bucket of models—a model selection algorithm is used to choose the best model for each problem, and the model selection is based on cross-validation.

4 MACHINE LEARNING ALGORITHMS

In this section, we review, compare, and contrast several standard machine learning (ML) algorithms that have been widely used in cognitive analytics. Before we delve into the specifics of the algorithms, we discuss considerations that are common to machine learning algorithms in general.

Input, output, and model: The input dataset to a machine learning algorithm is usually composed of many rows, where each row is called an *example*. Each example represents one *data point* and is composed of multiple *feature values* and optionally some *target values*. All feature values of an example collectively form a *feature vector*. All the examples in the dataset usually have the same number of elements in their feature vectors and also the same number of target values. The feature vector gives a quantitative representation of the example in terms of its *features*. Finding "good" features is extremely important and is more of an art than science. The target values, if present, ascribe a *labeling* to the example. The two mainstream classes of machine learning algorithms—supervised and unsupervised (see Section 3)—differ due to the presence or absence of the target values in an input dataset. The output of a machine learning algorithm is concerned with predicting the target values for a new feature vector.

The most challenging part of a machine learning algorithm is choosing an underlying model for mapping feature vectors to target values. These models are usually predictive and very rarely are explanatory. The models also have parameters that must be instantiated using the input dataset, and this process is called learning. The difficulty in choosing a model stems from the fact that often an infinite number of candidate models exist, even if the class of models is restricted. Choosing a model from the candidate set involves a balancing act which we discuss next.

Classification vs regression: We use the terms *target values* and *labels* interchangeably. A classification problem involves assigning a class label for a given feature vector. For example, an email message is classified as *spam* or *not spam*. The terms *spam* and *not spam* are the class labels. In other words, the output is a (class) label. In other examples such as weather prediction, the target value is a scalar value—the probability of a weather event. Target values can also be vectors.

Based on the type of target values desired, a machine learning algorithm deals with either a *classification* problem or a *regression* problem. The main difference between the two lies in a discrete vs continuous range of target values. In a classification problem, there are two or more discrete classes. Each example belongs to one of these classes. In a large majority of classification problems, the class labeling of the input examples is specified. The main objective of a classification algorithm is to accurately predict class labels for previously unseen examples. Classification problems of this type where the input examples are labeled fall into the supervised learning category (see Section 3). In contrast, unsupervised classification problems start with unlabeled input examples. The job of a classification algorithm in this case is to predict which examples belong to the same class. The interpretation of the classes in the unsupervised classification scenario is provided by human domain experts. Regression problems share the same problem structure as classification problems, with the key difference that the target values are no longer discrete labels.

Measuring predictive performance: Prediction is the prime goal of machine learning algorithms in general. Therefore, the algorithms are evaluated in terms of their predictive performance. Several technical issues arise while measuring the predictive performance, which are referred to as *overfitting* and *underfitting*.

Consider a supervised classification problem, where a set of labeled examples are available in the input dataset. Following is a standard mechanism for applying a machine learning algorithm to the dataset. First, the input dataset is partitioned into three nonoverlapping subsets—training, validation, and test. The size of these three sets is a design choice. The *training set* is used during the *training phase* to help instantiate the parameters of the model so that the model predicts well. How well the model predicts is measured using the *validation set*. The latter provides a set of examples that the algorithm has not seen during the training phase and is used to select a model that may not perform the best with respect to the training set but performs very well with respect to previously unseen examples. There is an important reason for measuring the accuracy of the model using the validation set instead of the training set. Imagine that the model does a perfect job predicting target values given feature vectors of examples in the training set.

On the surface, it may seem like perfect learning. However, it could very well be the case that the model simply "memorized" the training examples. In that case, given a previously unseen example, the model will most likely perform poorly. This is known as *overfitting*. Therefore, the main goal of the validation set is to prevent overfitting by choosing a model that may not be perfect with respect to the training examples, but does the best job when faced with unseen examples. In the last phase, the *test set* is used to measure the accuracy of the algorithm, which can differ from training and validation accuracies. Note that the test set is used only for unbiased evaluation of the model.

Underfitting is the opposite of overfitting. Underfitting means that the model is not sophisticated enough to capture the richness of the data. Large error margins for both the training and validation sets indicate underfitting, while very low error margin for the training set and very high error margin for the validation set indicate overfitting. Both overfitting and underfitting are undesirable and one of the challenges of machine learning lies in finding the sweet spot between them.

The subsequent discussion in this section will primarily focus on supervised classification problems, which are most prevalent in the real world.

The reader may consult Murphy's book (2012) for a complete coverage of machine learning algorithms.

4.1 Logistic Regression

Logistic regression is essentially a classification algorithm. The word "regression" in its name comes from its close sister in the regression domain known as *linear regression*. Given that the classes are discrete in supervised classification problems, the goal for the algorithms is to find the *decision boundaries* among the classes. Decision boundaries separate examples of one class from another. Depending on the problem instance, decision boundaries may be complex and nonlinear in geometric shape. In general, different machine learning algorithms have different assumptions regarding the shape of decision boundaries. In the case of logistic regression, the assumption is that decision boundaries are linear. That is, they are hyperplanes in the high-dimensional feature space, where the dimension of the feature space is simply determined by the number of elements in the feature vector of a training example.

The logistic regression model parameters are roughly the weights for the features. Each weighted feature vector is mapped to a value between 0 and 1 via the S-shaped logistic function. This value is interpreted as the probability of an example belonging to a particular class. The learning algorithm tunes the weights in order to correctly classify the training examples. The issue of avoiding overfitting inevitably arises here. The gradient descent method and several variants of it are popular for tuning the weights. Once the weights are chosen, the logistic function is applied to any unseen example to obtain the probability of it belonging to a class.

Due to the simplistic assumption of linear decision boundaries, logistic regression is often times the *first go-to* algorithm for classification problems. Also, because of the linear, noncomplex decision boundaries, logistic regression is known to be less prone to overfitting. Intuitively, overfitting occurs when we try to correctly classify every single training example by arbitrarily wiggling the decision boundary. Additionally, gradient descent typically works very fast and thus makes the training phase of logistic regression quick. All of these advantages justify the popular application of logistic regression to a variety of classification problems. On the down side, however, the simplistic modeling assumptions may lead to underfitting for rich and complex datasets.

Logistic regression has been used in a variety of application areas. Honorio and Ortiz (2015) has used it to learn the structure and parameters of a social network model that captures the strategic behavior of individuals. The model has been used to find the most influential individuals in a network (Irfan and Ortiz, 2011, 2014). Logistic regression has also been used in GIS (Ayalew and Yamagishi, 2005; Lee, 2005), email spam filtering (Chang et al., 2008), and other problems within natural language processing

(Jurafsky and Martin, 2009; Nadkarni et al., 2011), speech recognition (Jurafsky and Martin, 2009), finance (Laitinen and Laitinen, 2001; Maher and Sen, 1997), and the broader domain of pattern recognition (Bishop, 2006).

4.2 Decision Trees

The Classification and Regression Tree (CART) method was originally presented by Breiman et al. (1984) during the 1980s. This has led to tremendous interest in decision tree learning. In the supervised classification setting, the objective of decision tree learning is to compute a special type of tree that can classify examples to classes. The notions of training, validation, and test sets as well as overfitting vs underfitting issues apply for decision trees too. The underlying model in decision tree learning is a tree in graph-theoretic sense. However, we must also recognize a stylized control flow that is superimposed on the tree structure. Each internal node of the tree, including the root, asks a decision-type question. Based on the answer for an example, we next traverse one of the children of that internal node. Once we reach a leaf node, we are certain to know the classification of the example according to the decision tree, since each leaf node is annotated with a class label.

In addition to CART, there are many other learning algorithms for finding the "best" tree for a classification problem. Most modern algorithms like Iterative Dichotomiser 3 (ID3) (Quinlan, 1986) and its successors C4.5 (Quinlan, 2014) and C5 (Quinlan, 2016) use information theoretic measures, such as entropy, to learn a tree. Entropy can be thought of as a measure of uncertainty. Initially, the whole training set, consisting of examples of different classes, will have a very high entropy measure. ID3 and its successors repeatedly partition the training set in order to reduce the sum of the entropy measures of the partitions. Usually, a greedy strategy is employed for this purpose. The algorithm chooses a feature and partitions the training set based on that feature. The feature is chosen with the goal of minimizing the sum of the entropy measures across the resulting partitions. The same procedure is recursed on each partition, unless all the examples in that partition belong to the same class.

One big advantage of decision tree learning over other learning methods such as logistic regression is that it can capture more complex decision boundaries. Decision tree learning is suitable for datasets that are not linearly separable—there exists no hyperplane that separates out examples of two different classes. The ability of decision trees to capture complex decision boundaries is sometimes its own pitfall, since this can lead to overfitting unless certain other techniques like "pruning the tree" is applied.

A few other advantages have made decision trees popular. First, they often lead to a clear visualization of how the machine learning algorithm performs classification. Second, the training phase is usually fast and scalable to large-scale data. Lastly, decision trees have been widely used in various ensemble learning methods, such as AdaBoost (Freund and Schapire, 1995;

Freund et al., 1999) and random forests (Breiman, 2001; Ho, 1995). Random forests belong to a broader umbrella of machine learning techniques known as *bagging*. Bagging techniques are especially geared for tackling overfitting. In random forests, multiple decision trees are learned, which collectively build a graph-theoretic forest. A new feature vector is classified differently by different decision trees in the forest. These individual classifications are aggregated to output the final classification.

4.3 Support Vector Machine

SVM is one of the most widely applicable machine learning algorithms (Bell, 2014; Shalev-Shwartz and Ben-David, 2014). Since Vapnik and Chervonenkis presented SVM during the 1960s, there has been a tremendous amount of work extending it in multiple directions. We will present the key idea behind SVM and its advantages. Schölkopf's book (1999) is a comprehensive reference on this topic.

Consider a restricted classification setting where the training set consists of only examples that belong to two classes, and the examples are linearly separable. Because of linear separability assumption, there exist hyperplanes that separate out the examples of the two different classes. In fact, there exist an infinite number of such hyperplanes. The central idea in SVM is to choose that particular hyperplane which sits "right in the middle" in between the examples of the two classes. Mathematically speaking, SVM chooses the hyperplane that maximizes the minimum distance between that hyperplane and all the examples. In other words, the hyperplane is equidistant from the examples of the two classes that are closest to it. In SVM terminology, two times the distance between the hyperplane and the points closest to it is known as the *margin*. As a result, SVM is also known as a maximum margin classifier. Maximizing the margin, or equivalently selecting that particular hyperplane in between the examples of the two different classes, is extremely significant. This leads to a good generalization for classifying previously unseen examples.

One of the reasons SVM is so widely applicable is that it can be easily extended to complex instances that are not linearly separable. This is done by mapping the training examples to a higher-dimensional space where they become linearly separable in conjunction with the *kernel trick* (Aizerman et al., 1964; Boser et al., 1992) to keep computation manageable.

Another reason in favor of its applicability is a subtle issue that accounts for the name "support vector." Not all training examples are equally important. In fact, since the decision boundary only depends on the training examples closest to it, it suffices to define the underlying model of SVM in terms of only those training examples. Those examples are called support vectors. Although the original dataset may contain a very large number of examples, the number of support vectors is usually very small. This makes SVM amenable to large-scale data, including streaming data. SVM is also memory efficient for many applications.

SVM algorithms have been successfully used for classifying images in extremely large repositories such as Instagram. They have also been used to analyze natural language text and web documents (Tong and Koller, 2001). In the medical domain, SVMs have been used to classify proteins into their functional family (Cai et al., 2003).

4.4 Artificial Neural Networks and Deep Learning

ANNs, or simply neural networks, belong to the broader class of biologically inspired computation. ANNs are modeled after how the neurons in the brain "fire" and how one neuron's firing affects other neurons connected to it. One of the earliest and most influential models of a neuron is attributed to McCulloch and Pitts (1943), who combined biology and mathematics to model the firing of a neuron as a threshold function. Subsequently, Rosenblatt (1958) presented the first algorithm, known as the *perceptron* algorithm, to learn the parameters of the simplest type of neural networks that can successfully deal with linearly separable classification problems.

Advances in high-performance computing and algorithmic progress advanced ANNs to address problems where the class boundaries are not linearly separable. This led to a further explosion of growth and interest in neural networks during the 1980s. In fact, many considered ANNs as a "one size fits all" framework, which ultimately created its own downfall. In particular, Geman et al. (1992) showed that neural networks are prone to overfitting and underfitting problems. Furthermore, they demonstrated that for a neural network to be effective for various problem instances, it needs to be complex and enough data is needed to effect good learning.

There are many variants of neural networks, but we will discuss the most prevalent one known as *feed-forward neural network*. A feed-forward network consists of neurons placed in multiple layers. The first layer is known as the *input layer*, and the last layer the *output layer*. All the layers in between input and output layers are *hidden layers*. The outputs of neurons in one layer are fed as inputs to the neurons of the next layer. The parameters of the model are the weights of the connections between neurons of two consecutive layers and a threshold value for each neuron. The weights indicate connection strengths and the threshold value determines whether or not a neuron fires.

Given a training dataset, the neural network is designed in such a way that the number of neurons in the input layer is equal to the number of features. Also, the number of neurons in the output layer is equal to the number of target values. Apart from these restrictions, there is no hard and fast rule regarding the number of hidden layers and the number of neurons in a hidden layer. Often times, these are determined by experimenting with several different network architectures and choosing one based on cross-validation

(Murphy, 2012). There are many algorithms for learning the parameters of a neural network (Murphy, 2012), but the most influential one is known as the *backpropagation algorithm* (Werbos, 1974).

Neural networks have many success stories like handwriting recognition and stock market prediction. However, due to issues of network complexity and amount of training data required, enthusiasm about neural networks somewhat subsided during the 1990s. With incredible advances in high-performance parallel computing and the emergence of big data (Gudivada et al., 2015a), neural networks reemerged under a new name—deep learning (LeCun et al., 2015). The power of deep learning comes from scalability—the number of hidden layers, and not from new or complex algorithms. Deep-learning algorithms have been making one breakthrough after another in several different areas, including image classification (Krizhevsky et al., 2012), speech recognition (Hinton et al., 2012), and machine translation (Sutskever et al., 2014).

Perhaps the primary reason for the popularity of deep learning is automatic feature extraction. Traditionally, features are carefully hand-crafted by humans. However, it has been shown that for image recognition tasks, deep learning automatically extracts image features in a hierarchical fashion—beginning from the edges in the images to higher-level features progressively (LeCun et al., 2015). This automatic extraction and representation of features has hugely outperformed many standard features like the well-known *sift* features. The latter have been used by the computer vision community for years. As a result, deep learning has initiated a paradigm shift in computer vision.

A major drawback of deep neural networks is that they cannot explain their decisions. From a user perspective, it is an oracle and a blackbox. Designing critical systems with the blind faith that deep learning will pick the "right" features is not a sound engineering design principle.

4.5 Bayesian Networks

Probabilistic approaches to real-world problems are omnipresent today. One of the key challenges in these approaches is the representation of the joint probability of a set of random variables, whose size is exponential in the number of random variables. However, most problems show some type of probabilistic structure in the sense that not every random variable is *conditionally dependent* on every other random variable. In such cases, we can succinctly represent the probability structure. Probabilistic graphical models (Koller and Friedman, 2009) deal with problems where there is a graphical structure among the random variables in terms of the conditional dependencies. BNs are probabilistic graphical models where the graph (or network) among the random variables is a directed acyclic graph (DAG). Each node in the DAG is a random variable and each directed edge from a node A to a node B

represents A's direct influence on B. The directed edges may not necessarily encode causality—in most cases they do not.

In addition to being a data structure for a compact representation of *joint probabilities*, a BN also represents *conditional independence* among its random variables. Interestingly, these two representational aspects of BNs are equivalent. The conditional independence property states: given the nodes that have a directed edge to a node A, the node A is conditionally independent of all the nodes that cannot be reached from A via a directed path. A more technical concept known as *d-separation* deals with the issue of whether two nodes are conditionally independent given a third node, based on the graph structure and irrespective of the actual probability distribution. D-separation is algorithmically well understood, although in the worst case it takes exponential time in the size of the graph (Koller and Friedman, 2009).

The key machine learning problems in the BN setting are learning the parameters (i.e., conditional probabilities) given the graph structure and learning the structure of the graph and the parameters given a probability distribution. For the former, several well-known techniques such as Maximum Likelihood and expectation maximization are widely used. The latter problem is more involved and often times requires searching for a graph structure in the huge space of all possible graphs. Various optimization techniques are used for this task.

Today, BNs boast a wide range of practical applications in diverse fields, such as bioinformatics (Zou and Conzen, 2005), image processing (Mittal, 2007), risk analysis (Weber et al., 2012), and engineering (Heckerman et al., 1995), to name just a few.

4.6 Libraries and Frameworks

Many libraries and frameworks are available for developing cognitive analytics applications. TensorFlow is an open source software library from Google for numerical computation using data flow graphs (Abadi et al., 2016). The library is optimized for execution on clusters and GPU processors. Among many other applications, TensorFlow is a deep-learning platform for computational biologists (Rampasek and Goldenberg, 2016).

Apache Singa is a general purpose, distributed neural platform for training deep-learning models over large datasets. The supported neural models include convolutional neural networks, restricted Boltzmann machines, and recurrent neural networks.

Torch7, Theano, and Caffe are the other deep-learning frameworks which are widely used. The Torch is a GPU-based scientific computing framework with wide support for machine learning algorithms. It provides an easy to use and fast scripting language called LuaJIT, which is implemented using the C language and CUDA. It comes with a large number of community-developed packages for computer vision, signal processing, and machine learning.

Theano is a Python library which is highly suited for large-scale, computationally intensive scientific investigations. Mathematical expressions on large multidimensional arrays can be efficiently evaluated. It tightly integrates with Numpy. Access to the underlying GPU hardware is transparent. Also, it performs efficient symbolic differentiation. Lastly, extensive unit-testing and self-verification functions are integrated into Theano, which enables diagnosing several types of errors in code.

Caffe is particularly suitable for convolutional neural networks and provides options for switching between CPUs and GPUs through configuration parameters. It has been stated that Caffe can process over 60 million images per day with a single Nvidia K40 GPU.

Massive Online Analysis (MOA) is a popular framework for data stream mining. The machine learning algorithms provided by the framework are suitable for tasks such as classification, regression, clustering, outlier detection, concept drift detection, and recommender systems.

MLlib is Apache Spark's machine learning library. Tasks that can be performed using the MLlib include classification, regression, clustering, collaborative filtering, and dimensionality reduction. mlpack is a C++-based machine learning library, which can be used through command line as well as C++ classes.

Pattern is a web mining module for the Python programming language. It features tools for data mining, natural language processing, clustering, network analysis, and visualization. *Scikit-learn* is another Python framework for machine learning, which is implemented using NumPy, SciPy, and matplotlib. Using the included machine learning algorithms, tasks such as clustering, classification, and regression can be accomplished.

Shogun is one of the oldest machine learning libraries, which is written in C++. However, it provides bindings for other languages such as Java, Python, C#, Ruby, R, Lua, Octave, and Matlab. *Veles* is a C++, distributed platform for developing deep-learning applications. Trained models can be exposed through REST API. Using Vales, widely recognized neural topologies such as fully connected, convolutional, and recurrent networks can be trained. Deeplearning4J, neon, and H2O are other libraries for deep learning.

Mahout is an Apache machine learning project. Mahout library is especially suited for execution on cluster computers and GPUs. Also, it tightly integrates with Hadoop Map/Reduce distributed processing framework. Logistic regression classifier, random forest decision trees, K-means clustering, and naive Bayes classifier algorithms are available in Mahout. Apache R project is a sophisticated platform for statistical computing. It features a comprehensive set of machine learning and visualization algorithms.

Amazon Machine Learning is a cloud-hosted service for creating machine learning models without knowing the internal details of machine learning algorithms. This service provides easy access to the data stored in Amazon S3, Redshift, and RDS. Azure ML Studio is a similar service from Microsoft.

5 COGNITIVE ANALYTICS: A COVETED GOAL

The term *cognition* refers to how humans acquire and use knowledge through their senses, learn from interactions and experiences in their environment, and acquire and improve their ability to perform functions such as walking, talking, driving a car, and problem solving. It is hypothesized that cognition is enabled by the higher-level functions of the brain. A *cognitive process* refers to the specific steps the brain uses for accomplishing tasks such as perception, planning, language acquisition, and thinking. Cognitive processes are different from deterministic algorithms. They elegantly cope with data which is ambiguous, uncertain, incomplete, and inconsistent using probabilistic algorithms.

A *cognitive model* is a blueprint for a cognitive process. In other words, a cognitive model *explains* a cognitive process. A set of cognitive processes endows humans' intelligent behavior. Machine cognition is similar to human cognition. Machine cognition targets computers to perform tasks at a performance level which rivals humans. Cognitive analytics is an emerging area and is currently going through a formation stage. It is expected to evolve rapidly and make its way into many commercial software applications.

A *software architecture* defines the overall structure of a software application and specifies its components and their functional characteristics, as well as communication among the components. Some architectures are generic and are used to build a class of software applications, whereas others are specific to just one application. A *cognitive architecture* is a supposition about *fixed* structures of the mind and interactions among them to endow intelligent behavior to humans and machines. The means used to realize cognitive architecture in humans and computers are different. The underlying infrastructure for human cognition is the mind and brain, whereas algorithms and computers encompass the infrastructure for machine cognition. Some cognitive architectures are generic enough to serve as a blueprint for multiple cognitive models. In this section, we describe *Cognalytics*—a proposed architecture for cognitive analytics—and also discuss ways to implement the architecture.

5.1 Cognalytics: A Reference Architecture for Cognitive Analytics

Shown in Fig. 3 is Cognalytics, a high-level reference architecture for implementing cognitive analytics. It is a layered architecture and the circled numbers on the right denote layer numbers. We use the terms system and architecture synonymously, and the context should help to elucidate the intended meaning.

Layer 1 is the *physical data layer* which stores unstructured, semistructured, and structured data. It also stores open source taxonomies and ontologies such as DBpedia and WordNet. Some data in the physical data layer is static or changes rarely, and other data is dynamic and changes with time.

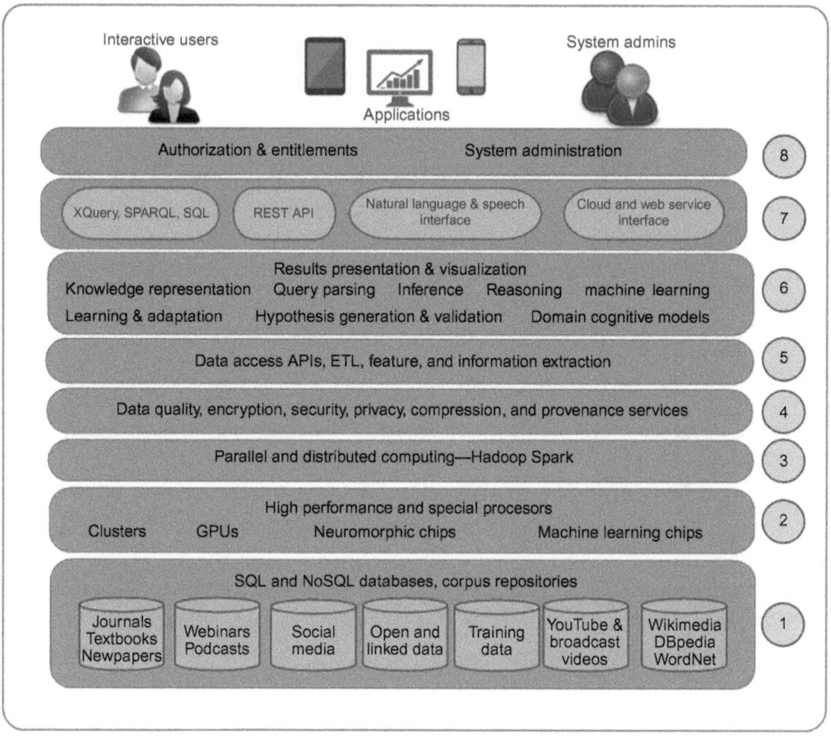

FIG. 3 Cognalytics: a reference architecture for cognitive analytics.

This suggests that static data should be physically stored, whereas the dynamic data should be stored logically. In the latter case, the system knows the requisite information about how to fetch this data as needed from the sources. Even the physically stored static data needs to be kept in sync with their sources. These are optimization issues and are not part of the high-level architecture. Since the data is large and heterogeneous, suitable database management systems (DBMS) should be used. They include both relational and NoSQL databases (Gudivada et al., 2016). Natural language texts are stored as text corpora.

Extremely large data volumes and attendant compute-intensive processing require high-performance computing and distributed processing techniques to meet stringent query latency requirements. Layer 2 addresses this need and is referred to as the *physical hardware layer*. Layer 3 provides a virtual machine and abstractions over layer 2 so that cognitive analytics applications can effectively leverage the computing power of layer 2. Layer 3 is referred to as the *hardware abstractions layer*.

Layer 4 provides data services which are implemented using the abstractions provided by layer 3. Functions featured by the data services layer range a broad spectrum from data cleaning, data quality assessment, compression, and encryption to ensuring privacy and maintaining data provenance. Not every cognitive analytics application may need all these data services. However, the services are generic, relatively low level, and are relevant across a broad range of cognitive analytics applications. Layer 4 is referred to as the *low-level data services layer.*

Layer 5 provides high-level data services. Application developers can specify workflows using the low-level data services of layer 4 and execute them. This layer also provides ETL tools for data integration and creates data warehouses and data marts. Lastly, it provides software tools and libraries for extracting both features and information from semistructured and unstructured data. This layer is referred to as the *high-level data services layer.*

Layer 6 is the core of the Cognalytics reference architecture. It features an assortment of machine learning algorithms, domain cognitive models, and inference and reasoning mechanisms including spatial and temporal reasoning. To facilitate inference and reasoning, several knowledge representation schemes are provided. The Learning & Adaptation subsystem is responsible for storing episodic and other types of knowledge and enables learning, adaptation, and evolution. The Query Parsing subsystem is responsible for parsing queries and identifying subqueries of a query. The Hypothesis Generation & Validation subsystem is responsible for providing several answers to a problem and assigning a degree of confidence to each answer. The Results Presentation & Visualization subsystem provides multimodal interfaces for presenting results. It also features functionality for interactive exploration of results through visualization. This layer is referred to as the *cognitive analytics layer.*

Layer 7 provides access to both interactive users and external systems through declarative query languages and APIs. Queries can be specified using natural language text as well as spoken language. This layer also exposes Cognalytics functions as cloud and web services. These services enable developing cognitive analytics applications without having to deal with the internal complexities of the Cognalytics architecture and its implementation. This layer is referred to as the *API layer.*

Layer 8 provides two major functions. The System Administration subsystem provides functions for creating users and associating them with roles. A role specifies a set of predetermined system functions that the role bearer can execute. The Authorization & Entitlement subsystem is responsible for authenticating users and ensuring that the users execute only functions for which they have authorizations. This layer is referred to as the *administration layer.*

5.2 Implementing Cognalytics

Implementing Cognalytics architecture requires substantial effort. Numerous open source libraries and tools are available to ease the effort. Furthermore, one can select the best library or framework from among the choices for each subsystem. We describe implementation layer-wise starting with the bottom layer. The tools and frameworks we indicate in this section are open source unless specified otherwise.

5.2.1 Physical Data Layer

PostgreSQL is an open source RDBMS which provides high availability, horizontal scalability, and performance. Replication and auto-sharding features are also available. It is an ideal choice for storing structured data. As of this writing, there are over 300 DBMS available for data management and most of these systems are open source (Solid IT, 2016). A number of NoSQL databases are available for storing text corpora and other unstructured data. Virtuoso, Sedna, BaseX, and eXist-db are native XML databases. Database systems for time series data include InfluxDB, RRDtool, Graphite, and OpenTSDB. Jena, Virtuoso, and Sesame are database systems for RDF data. For graph data management, Neo4j, OrientDB, Titan, Virtuoso, and ArangoDB are popular choices. The reader should consult (Solid IT, 2016) to explore unprecedented choices for data management.

5.2.2 Physical Hardware Layer

Though one can develop the computing infrastructure for Cognalytics in-house, it is often economical to use a cloud platform such as Amazon Web Services. On the other hand, developing an in-house infrastructure has its advantages. Special compute processors such as neuromorphic chips and neural network accelerators are available for developing the infrastructure. For example, True North (Merolla et al., 2014) is a brain-inspired neuromorphic chip. It is a self-contained chip with 5.4 billion transistors. True North features 1 million programmable neurons, 256 million programmable synapses on the chip, 4096 parallel and distributed cores which are interconnected via an on-chip mesh network, and 400 million bits of local on-chip memory. How True North has been used to implement convolutional networks for classification problems is described in Esser et al. (2016).

A class of microprocessors, called AI accelerators, are emerging to accelerate machine learning algorithms. For example, tensor processing units are application-specific processors developed for Google's TensorFlow framework (TensorFlow, 2016). As of this writing, Nvidia released Tesla P100 GPU chip. The chip specifically targets machine learning algorithms that employ deep learning. Tesla P100 features 150 billion transistors on a single chip. DGX-1, Nvidia's newest supercomputer, is powered by 8 Tesla P100 GPUs and ships with deep-learning software preinstalled. Zeroth is a cognitive computing platform developed by Qualcomm. The platform runs on a neural processing unit AI accelerator chip and deep-learning algorithms are available through an API. The latter is specifically designed for mobile devices to process image and speech data. Other neurocomputing engines include Chen et al. (2015), Du et al. (2015), Kim et al. (2015), and Liu et al. (2013).

5.2.3 Hardware Abstractions Layer

This layer provides libraries and frameworks to ease the application development process using specialized processors such as neuromorphic chips. The libraries and frameworks enable application developers to write code without concerns for the underlying special hardware. The application code is automatically transformed to enable efficient execution. Currently, Hadoop and Spark are popular choices for realizing this layer. Typically, neuromorphic and other chip manufactures provide APIs using which applications development can be accelerated. As neuromorphic processors' use becomes more widespread, we expect more advanced libraries and frameworks.

5.2.4 Low-level Data Services Layer

Data ingestion into a cognitive analytics system is a major task given that the volume of data is generally petabyte scale and, in some cases, even exabytes. Sqoop and Flume are two tools in the Hadoop ecosystem for extracting data from different sources and loading it into the Hadoop Distributed File System. Sqoop is used for extracting and loading structured data, whereas Flume does the same for unstructured data.

Many cognitive analytics applications acquire data from diverse data vendors to complement internally generated data. Algorithms and workflows for data cleaning are required for detecting and eliminating duplicates, resolving conflicting and inconsistent data, inferring missing data, detecting integrity constraint violations, and detecting and resolving outliers. Ganti and Sarma (2013) discuss a few popular approaches for developing data cleaning solutions. Other works in this direction include Osborne (2012) and McCallum (2012).

Protecting privacy rights is a tremendous challenge. Differential privacy restricts users' access to data based on their job roles. Data encryption simplifies data security and privacy protection. Especially in the medical and health care domains, the notion of *personally identifiable information* is central. Some techniques such as *data perturbation* enable data analytics without compromising privacy requirements. Data perturbation is a more effective approach for privacy preservation of electronic health records than deidentification and reidentification procedures.

Provenance involves maintaining a history of processing that has been applied to a data item. The history is maintained in the form of metadata graphs, which grow very rapidly. Analyzing these graphs is computationally expensive (Cheah, 2014). Tracking provenance may not be a concern in some cognitive analytics applications. The Open Provenance Model is a collection of specifications for implementing provenance. Pentaho Kettle, eBioFlow, PLIER, and SPADE are tools for implementing provenance.

Given the data volumes, data compression is an important consideration. Generally, text compression requires lossless algorithms—original data and data recovered from the compressed data are identical. Image and video data may tolerate some data loss when decompressed. RainStor/Teradata, a database specifically developed for big data, seems to provide a compression ratio of 40:1, and in some cases, the ratio is as high as 100:1.

5.2.5 High-Level Data Services Layer

Tools are required for integrating data from multiple sources. This data fusion requires normalizing data so that it conforms to a canonical form, identifying related data about an entity from different sources, specifying transformation rules, and resolving any conflicts. ETL are a set of tools which originated from the data warehousing area are used for this purpose. Scriptella, KETL, Pentaho Data Integrator—Kettle, Talend Open Source Data Integrator, Jaspersoft ETL, GeoKettle, Jedox, Apatar, CloverETL, and HPCC Systems are excellent ETL tools.

Pivotal Greenplum (originally Greenplum Database) is a massively parallel data warehouse. Greenplum branched off from PostgreSQL and added several data warehousing features. Pivotal Greenplum is uniquely suitable for big data analytics. Apache MADlib is a library for scalable in-database analytics (Hellerstein et al., 2012). MADlib provides parallel implementations of machine learning and mathematical and statistical functions. MADlib currently supports Pivotal Greenplum, PostgreSQL and Apache HAWQ (Hadoop Native SQL platform) databases, and data warehouses. Many NoSQL databases compute analytics efficiently in batch mode using MapReduce frameworks (Gudivada et al., 2016).

Several tools are available for extracting features and information from unstructured data, primarily natural language text. Apache UIMA project provides frameworks, tools, and annotators for facilitating the analysis of unstructured content such as text, audio, and video. Tools from the Stanford NLP group for solving major computational linguistics problems include statistical NLP, deep-learning NLP, and rule-based NLP. Other tools for solving natural language problems include openNLP and GATE. Apache Lucene Core is a full-featured text search engine Java library. GPText from Greenplum is a statistical text analysis framework optimized for execution on parallel computing platforms. GPText is also available as a cloud service (Li et al., 2013).

SyntaxNet is an open source neural network framework for developing natural language understanding systems. Parsey McParseface is a pretrained SyntaxNet model for parsing the standard English language. TensorFlow is another software library for machine intelligence. NuPIC is a platform for cognitive computing, which is based on a theory of neocortex called Hierarchical Temporal Memory (HTM).

Weka 3 is a Java software library for data mining. The R project provides a platform for statistical computing and visualization. OpenCV and ImageJ are libraries for computer vision tasks. Praat is a tool for speech manipulation, analysis, and synthesis. openSMILE is another tool for extracting audio features in real time.

5.2.6 Cognitive Analytics Layer

This layer brings all components and subsystems together by serving as an integrator and coordinator. Some of the libraries and tools we indicated in Section 5.2.5 are also useful for implementing this layer. This is because the distinction between low- and high-level features is subjective and fluid. And so is the distinction between data and information, as well as information and knowledge.

There are several tools for implementing this layer. Their functions are often complementary and multiple tools are needed. FRED is a machine reader for the Semantic Web (Presutti et al., 2012). It parses natural language text in 48 languages and transforms it to linked data. It is available as both a REST service and a Python library suite. Apache Stanbol is a software stack and reusable set of components for semantic content management. Federated knOwledge eXtraction Framework (FOX) is a tool for RDF extraction from text using ensemble learning (Speck and Ngonga Ngomo, 2014). Named Entity Recognition and Disambiguation (NERD) is another framework which unifies 10 popular named entity extractors and compares their performance (Rizzo and Troncy, 2012). Accurate Online Disambiguation of Named Entities in Text and Tables (AIDA) is another tool for extracting named entities from natural language texts (Yosef, 2016). AlchemyAPI provides 12 semantic text analysis APIs for natural language understanding (Feyisetan et al., 2014). Machine learning libraries for data mining include PyML, Apache Mahout, MLib, dlibml, WEKA, and scikit-learn.

There are several options for implementing the Results Presentation & Visualization subsystem. Results presentation is tied to web application development frameworks used for implementing Cognalytics. User interface development frameworks such as Bootstrap, Foundation, GroundworkCSS, Gumby, HTML KickStart, IVORY, and Kube provide rich functionality for results presentation and navigating the application. D3, chart, dygraphs, FusionCharts, and Highcharts are libraries for visualization which run in a web browser.

5.2.7 API Layer

Cognalytics provides several APIs for interaction with the outside world. SQL, SPARQL, and XQuery are standard languages for querying RDBMS, RDF, and native XML databases declaratively. Representational State Transfer (REST) is a minimal overhead Hypertext Transfer Protocol (HTTP) API for interacting with Cognalytics system. REST uses four HTTP methods GET (reading data), POST (writing data), PUT (updating data), and DELETE (removing data). Natural language and query interfaces provide a natural means for interacting with the system. The first two classes of interfaces primarily serve the needs of interactive users who pose structured queries, whereas the last class enables a more powerful and flexible way to submit queries.

5.2.8 Administration Layer

System administration functions include user management, system monitoring, backup and recovery, and access control. System monitoring, backup, and recovery functions are typically integrated into an organization-wide application. User management functions include user creation and assigning roles to users. Single sign-on (SSO) is a user authentication service that permits the same login ID and password to access multiple systems across an organization. Often software libraries combine authentication and authorization functions into one component.

Shibboleth is an open source software which provides federated identity solution. It enables users to connect to applications within and outside an organization using SSO. Apache Shiro is a Java security framework for integrating authentication, authorization, cryptography, and session management functions into applications. Other solutions include OpenDJ, OpenIDM, OpenAM, and DACS.

6 COGNITIVE ANALYTICS APPLICATIONS

Though data warehouses-driven analytics has been in existence for over 28 years (Devlin and Murphy, 1988), only recently has there been tremendous thrust on incorporating unstructured data into data analytics. The power of cognitive analytics stems from the complementary and synergistic value heterogeneous data sources bring. Cognitive analytics applications range from improving student engagement and developing intervention measures, developing more effective Intelligent Tutoring Systems (ITS) to developing cognitive assistants and personalized learning environments.

6.1 Learning Analytics

EDM and LA are two areas in the education and learning domain that draw upon data analytics. EDM can be viewed as descriptive analytics. The current EDM systems are tied to course management systems (CMS) such as Blackboard and Moodle, which provide structured data for analytics. Such data includes the number of CMS logins, time spent on each learning activity, and test scores. Based on this data, students are classified into various groups and appropriate intervention measures are designed for each group. There is no human involvement in this process.

LA takes EDM a step further. LA combines EDM with human judgment (Siemens, 2012). It is best viewed as prescriptive analytics. It uses machine learning algorithm techniques to reveal hidden patterns and generate actionable intelligence. The latter is used to design personalized intervention measures. In addition to structured data, LA includes semistructured data such as emails and discussion board postings into analytics. Recent efforts in LA aim to propel both EDM and LA into the realm of predictive analytics and beyond into cognitive analytics.

6.2 Personalized Learning

Personalized learning can be defined from multiple perspectives. One approach is to allow learners to proceed at their own pace. The order in which topics are learned by one user may be different from the order of topics for another learner. In other words, learners are not bound to a lock-step synchronization scheme. They are free to explore topics on a subject in any order, only constrained by the prerequisite dependencies. Another aspect is automated generation of assessments, providing contextualized and incremental scaffolding, and supplying immediate feedback on assessments. Descriptive analytics will help to suggest next topics to pursue to the learner. A personalized learning system, called ISPeL, which is based on the above principles is described in Gudivada (2016). How ISPeL can be extended to incorporate cognitive analytics is also described.

6.3 Cognitive Businesses

This is perhaps the single domain that has been deeply impacted by cognitive analytics already. Cognitive businesses are those that use cognitive analytics for both operational management and strategic decision making. The primary thrust is on extracting information from natural language texts and combining it with structured data. Cognitive analytics uses are vast and varied. It has been used to improve workflow processes, detecting fraud before it happens, ensuring regulatory compliance, repurposing content, and knowledge management. Technology companies such as IBM, Nvidia, Google, Microsoft, LinkedIn, Facebook, and Netflix have already incorporated cognitive analytics into their software products.

Multiple Sclerosis Association of America uses cognitive analytics and natural language understanding to return evidence-based answers to clinicians' complex questions. To find an answer, their system parses a corpus of 1500 question-and-answer pairs and also incorporates content from medical resources. Baylor College of Medicine used IBM Watson to develop Baylor Knowledge Integration Toolkit (KnIT). The latter's goal is to help researchers by discovering patterns in the research literature. KnIT helped researchers identify proteins that modify p53, a protein related to many cancers. The system analyzed 70,000 scientific articles on p53 to predict other proteins that turn on/off of p53's activity. This finding was accomplished in a matter of weeks, which would have taken researchers years without IBM Watson.

6.4 BCI and Assistive Technologies

The human brain is perhaps the most complex system in terms of its structure and function. Functional magnetic resonance imaging and electroencephalogram are two functional brain imaging techniques that help to establish an association between brain and behavior. BCI is a new technology that provides a direct communication pathway between a wired brain and an external device such as a robot or wheel chair. Cognitive analytics offers an exciting opportunity to develop new assistive technologies using BCI. The study reported in Harnarinesingh and Syan (2013) discusses how a three-axis industrial robot was used to generate writing. The next logical step is to investigate connecting the brain and the robot using BCI. This is just one example, and cognitive analytics and BCI have the potential to help develop many assistive technologies to help physically impaired people.

7 CURRENT TRENDS AND RESEARCH ISSUES

Cognitive analytics will be increasingly driven by special computing processors that mimic the neural computations of the brain. Advances in neuroscience and cognitive science are critical for propelling neuromorphic computing further. It is ironic that computing discipline itself will be enabling new discoveries in these sciences. Rapid advances in big data will exacerbate the need to move more and more processing into hardware to meet performance at scale requirements.

There is a mismatch between the current programming languages and software development environments with respect to neuromorphic architectures powered by neurosynaptic cores. IBM has already begun designing simulators and programming environments including a new programming language and associated libraries for the True North processor. Nvidia, Google, and Facebook have similar projects in the pipeline.

Cognitive computing and cognitive analytics will play a transformational role in the Internet of Things (IoT) domain. *Embedded analytics* in general and cognitive IoT in particular will endow wireless sensors and cameras to perform intelligent processing at the source. This has multiple benefits including improved data quality, adaptive sampling to reduce the volume of streaming sensor data, and increased opportunities for a community of sensors to work as collaborative agents. Another use for embedded analytics is to integrate discovered actionable insights into products that would benefit from such insights. More and more future applications will have embedded analytics, which will enable them to deliver additional value. For example, wearable medical devices will not only be able to generate timely alerts but also provide contextualized information about how to react to the alerts.

Current research in both feature and information extraction from unstructured data is primarily focused on natural language text. Recently, revived interest in neural computing and in particular convolutional networks has begun to yield new algorithms and approaches to image classification and object recognition problems. Similar emphasis is needed for speech and video data.

It is hypothesized that the human brain uses statistical learning. Creating neural models to simulate the brain is not easy given the current computing processors. The emergence of neuromorphic chips offers hope and excitement.

Currently, neuromorphic chips can simulate neurons in the order of millions and synaptic connections in the order of billions. To propel cognitive analytics to the next stage, we need neuromorphic processors that can simulate neurons in the order of billions and synaptic connections in the order of trillions.

It is widely believed that our brain uses statistical learning. Creating neural models to simulate a brain is not easy. Current artificial neural models: nodes in the order of millions and connections in the order of billions. What we want is nodes in the order of billions and connections in the order of trillions.

Cognitive analytics will play an increasingly critical role in smart cities. Insights will help in planning evacuation routes, prioritize resource allocations in implementing disaster relief, optimize energy usage, promote public safety, and prevent maintenance on city infrastructure. Personalized learning will be another beneficiary of cognitive analytics. However, significant research is needed in these areas to reap the benefits.

8 CONCLUSIONS

In this chapter, we defined analytics and traced its evolution. Cognitive analytics is pursued from two complementary perspectives: computer science, and cognitive and neurosciences. This chapter focused primarily on the computer science perspective. We introduced learning types and discussed several classes of machine learning algorithms. We proposed a reference architecture for cognitive analytics and indicated ways to implement it. We described a few cognitive analytics applications and indicated current trends and future research in cognitive analytics. Cognitive computing and analytics have immense potential to contribute to a new generation of applications for which learning is intrinsic and communication is through spoken and written natural language. It is a proven technology which is searching for applications.

Cognitive computing and analytics are more than just the AI. For example, AI is 1 of the 28 APIs provided by the IBM Watson. Cognitive computing in general and cognitive analytics in particular exacerbate data security, privacy, and provenance issues. There are also practical concerns with cognitive analytics. Would this technology lead to significant unemployment? Would it only enable people to do their jobs better or totally replace them? On a philosophical level, how far can we advance cognitive technologies? Will they advance to a level at which they surpass human intelligence? If so, what are its implications to individuals as well as the society at large?

The computing industry has never been before invested this level of effort and resources into machine learning research. The availability of inexpensive, cloud-based computing power, and the ubiquity of big data are the catalysts for the transformational advances we are witnessing in machine learning and cognitive computing. The synergistic confluence of computing, neuroscience, and cognitive science is poised for groundbreaking discoveries and compelling cognitive applications in the years ahead.

REFERENCES

Abadi, M., Agarwal, A., Barham, P., Brevdo, E., Chen, Z., Citro, C., Corrado, G.S., Davis, A., Dean, J., Devin, M., et al., 2016. TensorFlow: large-scale machine learning on heterogeneous distributed systems. ArXiv preprint arXiv:1603.04467.

Aizerman, A., Braverman, E.M., Rozoner, L., 1964. Theoretical foundations of the potential function method in pattern recognition learning. Autom. Remote Control. 25, 821–837.

Ayalew, L., Yamagishi, H., 2005. The application of GIS-based logistic regression for landslide susceptibility mapping in the Kakuda-Yahiko mountains, central Japan. Geomorphology 65 (1), 15–31.

Battiti, R., Brunato, M., Mascia, F., 2008. Reactive Search and Intelligent Optimization. Operations Research/Computer Science Interfaces, vol. 45. Springer Science & Business Media, Berlin, Germany.

Bell, J., 2014. Machine Learning: Hands-on for Developers and Technical Professionals. John Wiley & Sons, Hoboken, NJ.

Bishop, C.M., 2006. Pattern recognition and machine learning. Information Science and Statistics. Springer, New York, NY.

Boser, B.E., Guyon, I.M., Vapnik, V.N., 1992. A training algorithm for optimal margin classifiers. In: Proceedings of the Fifth Annual Workshop on Computational Learning Theory. ACM, New York, NY, pp. 144–152.

Breiman, L., 2001. Random forests. Mach. Learn. 45 (1), 5–32.

Breiman, L., Friedman, J., Stone, C.J., Olshen, R.A., 1984. Classification and Regression Trees. CRC Press, Boca Raton, FL.

Cai, C.Z., Han, L.Y., Ji, Z.L., Chen, X., Chen, Y.Z., 2003. SVM-Prot: web-based support vector machine software for functional classification of a protein from its primary sequence. Nucleic Acids Res. 31 (13), 3692–3697.

Chakraborty, A., Harrison, B., Yang, P., Roberts, D., St. Amant, R., 2014. Exploring key-level analytics for computational modeling of typing behavior. In: Proceedings of the 2014 Symposium and Bootcamp on the Science of Security, Raleigh, North Carolina, USA, HotSoS'14. ACM, New York, NY, pp. 34:1–34:2. http://doi.acm.org/10.1145/2600176.2600210.

Chang, M.-W., Yih, W.-T., Meek, C., 2008. Partitioned logistic regression for spam filtering. In: Proceedings of the 14th ACM SIGKDD International Conference on Knowledge Discovery and Data Mining. ACM, New York, NY, pp. 97–105.

Cheah, Y.-W., 2014. Quality, retrieval and analysis of provenance in large-scale data. Ph.D. thesis, Indiana University, Indianapolis, IN. Plale, Beth.

Chen, P.-Y., Kadetotad, D., Xu, Z., Mohanty, A., Lin, B., Ye, J., Vrudhula, S., Seo, J.-S., Cao, Y., Yu, S., 2015. Technology-design co-optimization of resistive cross-point array for accelerating learning algorithms on chip. In: Proceedings of the 2015 Design, Automation & Test in Europe Conference & Exhibition, DATE'15. EDA Consortium, San Jose, CA, pp. 854–859.

Devlin, B.A., Murphy, P.T., 1988. An architecture for a business and information system. IBM Syst. J. 27 (1), 60–80.

Du, Z., Ben-Dayan Rubin, D.D., Chen, Y., He, L., Chen, T., Zhang, L., Wu, C., Temam, O., 2015. Neuromorphic accelerators: a comparison between neuroscience and machine-learning approaches. In: Proceedings of the 48th International Symposium on Microarchitecture, MICRO-48. ACM, New York, NY, pp. 494–507.

Esser, S.K., Merolla, P.A., Arthur, J.V., Cassidy, A.S., Appuswamy, R., Andreopoulos, A., Berg, D.J., McKinstry, J.L., Melano, T., Barch, D.R., Nolfo, C.D., Datta, P., Amir, A., Taba, B., Flickner, M.D., Modha, D.S., 2016. Convolutional networks for fast, energy-efficient neuromorphic computing. Comput. Res. Reposit. abs/1603.08270. http://arxiv.org/abs/1603.08270.

Feng, X., Kumar, A., Recht, B., Ré, C., 2012. Towards a unified architecture for in-RDBMS analytics. In: Proceedings of the 2012 ACM SIGMOD International Conference on Management of Data, SIGMOD'12. ACM, New York, NY, pp. 325–336. http://doi.acm.org/10.1145/2213836.2213874.

Feyisetan, O., Simperl, E., Tinati, R., Luczak-Roesch, M., Shadbolt, N., 2014. Quick-and-clean extraction of linked data entities from microblogs. In: Proceedings of the 10th International Conference on Semantic Systems, SEM '14. ACM, New York, NY, pp. 5–12.

Freund, Y., Schapire, R.E., 1995. A decision-theoretic generalization of on-line learning and an application to boosting. In: European Conference on Computational Learning Theory. Springer, Berlin, Germany, pp. 23–37.

Freund, Y., Schapire, R., Abe, N., 1999. A short introduction to boosting. J. Jpn. Soc. Artif. Intell. 14 (771–780), 1612.

Ganti, V., Sarma, A.D., 2013. Data Cleaning: A Practical Perspective. Synthesis Lectures on Data Management, Morgan & Claypool Publishers, Williston, VT.

Geman, S., Bienenstock, E., Doursat, R., 1992. Neural networks and the bias/variance dilemma. Neural Comput. 4 (1), 1–58.

Goodfellow, I.J., Bulatov, Y., Ibarz, J., Arnoud, S., Shet, V., 2014. Multi-digit number recognition from street view imagery using deep convolutional neural networks. https://arxiv.org/pdf/1312.6082.pdf. arXiv.org.

Gudivada, V.N., 2016. Cognitive analytics driven personalized learning. Educ. Technol. Mag. Special Issue on Big Data in E-Learning, Educational Technology Publications, Englewood Cliffs, NJ, in press.

Gudivada, V., Baeza-Yates, R., Raghavan, V., 2015a. Big data: promises and problems. IEEE Comput. 48 (3), 20–23.

Gudivada, V., Rao, D., Raghavan, V., 2015b. Big data driven natural language processing research and applications. In: Govindaraju, V., Raghavan, V., Rao, C.R. (Eds.), Big Data Analytics, Handbook of Statistics. vol. 33. Elsevier, Amsterdam, The Netherlands, pp. 203–238.

Gudivada, V., Rao, D., Raghavan, V., 2016. Renaissance in database management: navigating the landscape of candidate systems. IEEE Comput. 49 (4), 31–42.

Harnarinesingh, R.E.S., Syan, C.S., 2013. Investigating the feasibility of a robot-based writing agent. In: Proceedings of the Second International Conference on Innovative Computing and Cloud Computing, ICCC'13. ACM, New York, NY, p. 60:60:65. 60.

Heckerman, D., Mamdani, A., Wellman, M.P., 1995. Real-world applications of Bayesian networks. Commun. ACM 38 (3), 24–26.

Hellerstein, J.M., Ré, C., Schoppmann, F., Wang, D.Z., Fratkin, E., Gorajek, A., Ng, K.S., Welton, C., Feng, X., Li, K., Kumar, A., 2012. The MADlib analytics library: or MAD skills, the SQL. Proc. VLDB Endow. 5 (12), 1700–1711. http://dx.doi.org/10.14778/2367502. 2367510.

Hinton, G., Deng, L., Yu, D., Dahl, G.E., Mohamed, A.-R., Jaitly, N., Senior, A., Vanhoucke, V., Nguyen, P., Sainath, T.N., et al., 2012. Deep neural networks for acoustic modeling in speech recognition: the shared views of four research groups. IEEE Signal Process. Mag. 29 (6), 82–97.

Ho, T.K., 1995. Random decision forests. Proceedings of the Third International Conference on Document Analysis and Recognition, vol. 1, pp. 278–282.

Honorio, J., Ortiz, L., 2015. Learning the structure and parameters of large-population graphical games from behavioral data. J. Mach. Learn. Res. 16, 1157–1210.

Irfan, M.T., Ortiz, L.E., 2011. A game-theoretic approach to influence in networks. In: Proceedings of the Twenty-Fifth AAAI Conference on Artificial Intelligence, AAAI'11. AAAI Press, San Francisco, CA, pp. 688–694.

Irfan, M.T., Ortiz, L.E., 2014. On influence, stable behavior, and the most influential individuals in networks: a game-theoretic approach. Artif. Intell. 215, 79–119.

Jurafsky, D., Martin, J.H., 2009. Speech and Language Processing: An Introduction to Natural Language Processing, Computational Linguistics, and Speech Recognition, second ed., Pearson Prentice Hall, Upper Saddle River, NJ.

Kim, Y., Zhang, Y., Li, P., 2015. A reconfigurable digital neuromorphic processor with memristive synaptic crossbar for cognitive computing. ACM J. Emerg. Technol. Comput. Syst. 11 (4), 38:1–38:25. http://dx.doi.org/10.1145/2700234.

Koller, D., Friedman, N., 2009. Probabilistic Graphical Models: Principles and Techniques. MIT Press, Cambridge, MA.

Krizhevsky, A., Sutskever, I., Hinton, G.E., 2012. ImageNet classification with deep convolutional neural networks. In: Advances in Neural Information Processing Systems. Neural Information Processing Society (NIPS), Inc., La Jolla, CA, pp. 1097–1105.

Laitinen, E.K., Laitinen, T., 2001. Bankruptcy prediction: application of the Taylor's expansion in logistic regression. Int. Rev. Financ. Anal. 9 (4), 327–349.

Lake, B.M., Salakhutdinov, R., Tenenbaum, J.B., 2015. Human-level concept learning through probabilistic program induction. Science 350 (6266), 1332–1338. http://dx.doi.org/10.1126/science.aab3050. http://science.sciencemag.org/content/350/6266/1332.

LeCun, Y., Bengio, Y., Hinton, G., 2015. Deep learning. Nature 521 (7553), 436–444. http://dx.doi.org/10.1038/nature14539.

Lee, S., 2005. Application of logistic regression model and its validation for landslide susceptibility mapping using GIS and remote sensing data. Int. J. Remote Sens. 26 (7), 1477–1491.

Li, K., Grant, C., Wang, D.Z., Khatri, S., Chitouras, G., 2013. GPText: greenplum parallel statistical text analysis framework. In: Proceedings of the Second Workshop on Data Analytics in the Cloud, DanaC'13. ACM, New York, NY, pp. 31–35. http://doi.acm.org/10.1145/2486767.2486774.

Liu, B., Hu, M., Li, H., Chen, Y., Xue, C.J., 2013. Bio-inspired ultra lower-power neuromorphic computing engine for embedded systems. In: Proceedings of the Ninth IEEE/ACM/IFIP International Conference on Hardware/Software Codesign and System Synthesis, CODES + ISSS'13. IEEE Press, Piscataway, NJ, pp. 23:1–23:1.

Maher, J.J., Sen, T.K., 1997. Predicting bond ratings using neural networks: a comparison with logistic regression. Intell. Syst. Acc. Finan. Manag. 6 (1), 59–72.

McCallum, Q.E., 2012. Bad Data Handbook: Cleaning up the Data so You Can Get Back to Work. O'Reilly Media, Sebastopol, CA.

McCulloch, W.S., Pitts, W., 1943. A logical calculus of the ideas immanent in nervous activity. Bull. Math. Biophys. 5 (4), 115–133.

Merolla, P.A., Arthur, J.V., Alvarez-Icaza, R., Cassidy, A.S., Sawada, J., Akopyan, F., Jackson, B.L., Imam, N., Guo, C., Nakamura, Y., Brezzo, B., Vo, I., Esser, S.K., Appuswamy, R., Taba, B., Amir, A., Flickner, M.D., Risk, W.P., Manohar, R., Modha, D.S., 2014. A million spiking-neuron integrated circuit with a scalable communication network and interface. Science 345 (6197), 668–673. http://dx.doi.org/10.1126/science.1254642.

Mittal, A., 2007. Bayesian Network Technologies: Applications and Graphical Models: Applications and Graphical Models. IGI Global, Hershey, PA.

Murphy, K.P., 2012. Machine Learning: A Probabilistic Perspective. MIT Press, Cambridge, MA.

Nadkarni, P.M., Ohno-Machado, L., Chapman, W.W., 2011. Natural language processing: an introduction. J. Am. Med. Inform. Assoc. 18 (5), 544–551.

Osborne, J.W., 2012. Best Practices in Data Cleaning: A Complete Guide to Everything You Need to Do Before and After Collecting Your Data. SAGE Publications, Thousand Oaks, CA.

Polikar, R., 2009. Ensemble learning. Scholarpedia 4 (1), 2776. http://www.scholarpedia.org/article/Ensemble_learning.

Presutti, V., Draicchio, F., Gangemi, A., 2012. Knowledge extraction based on discourse representation theory and linguistic frames. In: Proceedings of the 18th International Conference on Knowledge Engineering and Knowledge Management, Galway City, Ireland, EKAW'12. Scite Press, Setúbal, Portugal, pp. 114–129.

Quinlan, J.R., 1986. Induction of decision trees. Mach. learn. 1 (1), 81–106.

Quinlan, J.R., 2014. C4.5: Programs for Machine Learning. Elsevier, Amsterdam, The Netherlands.

Quinlan, J.R., 2016. C5.0: Programs for Machine Learning. RuleQuest Research, Empire Bay. https://www.rulequest.com.

Rampasek, L., Goldenberg, A., 2016. TensorFlow: biology's gateway to deep learning? Cell Syst. 2 (1), 12–14.

Rizzo, G., Troncy, R., 2012. NERD: a framework for unifying named entity recognition and disambiguation extraction tools. In: Proceedings of the Demonstrations at the 13th Conference of the European Chapter of the Association for Computational Linguistics, EACL'12. Association for Computational Linguistics, Stroudsburg, PA, pp. 73–76. http://dl.acm.org/citation.cfm?id=2380921.2380936.

Rosenblatt, F., 1958. The perceptron: a probabilistic model for information storage and organization in the brain. Psychol. Rev. 65 (6), 386.

Russakovsky, O., Deng, J., Su, H., Krause, J., Satheesh, S., Ma, S., Huang, Z., Karpathy, A., Khosla, A., Bernstein, M., Berg, A.C., Fei-Fei, L., 2015. ImageNet large scale visual recognition challenge. Int. J. Comput. Vis. 115 (3), 211–252. http://dx.doi.org/10.1007/ s11263-015-0816-y.

Schölkopf, B., Burges, C.J., 1999. Advances in Kernel Methods: Support Vector Learning. MIT Press, Cambridge, MA.

Settles, B., 2009. Active learning literature survey. University of Wisconsin–Madison. Computer Sciences Technical Report 1648.

Shalev-Shwartz, S., Ben-David, S., 2014. Understanding Machine Learning: From Theory to Algorithms. Cambridge University Press, Cambridge.

Siemens, G., 2012. Learning analytics: envisioning a research discipline and a domain of practice. In: Proceedings of the 2nd International Conference on Learning Analytics and Knowledge, Vancouver, British Columbia, Canada. ACM, New York, NY, pp. 4–8.

Solid, I.T., 2016. Knowledge base of relational and NoSQL database management systems. http://db-engines.com/en/ranking. Retrieved: July 2016.

Speck, R., Ngonga Ngomo, A.-C., 2014. Ensemble learning for named entity recognition. In: Mika, P., Tudorache, T., Bernstein, A., Welty, C., Knoblock, C., Vrandečić, D., Groth, P., Noy, N., Janowicz, K., Goble, C. (Eds.), The Semantic Web—ISWC 2014. Lecture Notes in Computer Science, vol. 8796. Springer, Berlin, Germany, pp. 519–534. http://svn.aksw.org/papers/2014/ISWC_EL4NER/public.pdf.

Sutskever, I., Vinyals, O., Le, Q.V., 2014. Sequence to sequence learning with neural networks. In: Advances in Neural Information Processing Systems. Neural Information Processing Systems Foundation, Inc., La Jolla, CA, pp. 3104–3112.

Taigman, Y., Yang, M., Ranzato, M., Wolf, L., 2014. DeepFace: closing the gap to human-level performance in face verification. In: Proceedings of the 2014 IEEE Conference on Computer Vision and Pattern Recognition, CVPR'14. IEEE Computer Society, Washington, DC, pp. 1701–1708. http://dx.doi.org/10.1109/CVPR.2014.220.

TensorFlow, 2016. An open source software library for numerical computation using data flow graphs. https://www.tensorflow.org/.

Tong, S., Koller, D., 2001. Support vector machine active learning with applications to text classification. J. Mach. Learn. Res. 2, 45–66.

Weber, P., Medina-Oliva, G., Simon, C., Iung, B., 2012. Overview on Bayesian networks applications for dependability, risk analysis and maintenance areas. Eng. Appl. Artif. Intell. 25 (4), 671–682.

Werbos, P., 1974. Beyond regression: new tools for prediction and analysis in the behavioral sciences. Ph.D. thesis, Harvard University, Cambridge, MA.

Yosef, M.A., 2016. U-AIDA: a customizable system for named entity recognition, classification, and disambiguation. Ph.D. thesis, Saarland University.

Zou, M., Conzen, S.D., 2005. A new dynamic Bayesian network (DBN) approach for identifying gene regulatory networks from time course microarray data. Bioinformatics 21 (1), 71–79.

Chapter 6

A Cognitive Random Forest: An Intra- and Intercognitive Computing for Big Data Classification Under Cune Condition

S. Suthaharan[1]
The University of North Carolina at Greensboro, Greensboro, NC, United States
[1]Corresponding author: e-mail: s_suthah@uncg.edu

ABSTRACT

In this chapter, a cognitive computing model, a cognitive computing architecture, and a cognitive computing system are proposed and tested to address a big data classification problem. The proposed cognitive computing model is called the STE-M model and it adopts the standard components, senses (S), thoughts (T), experiences (E), and memory (M), of human cognition to describe the processes involved in cognitive computing for big data classification. Similarly, the proposed cognitive computing architecture is called the cognitive random forest and it amalgamates the STE-M model and a set of random forest classifiers to enhance continuous learning. It also includes intra- and intercognitive computing models to connect STE-M and random forest models and improve classification accuracy with spatial and temporal reasoning. A cognitive computing system is also proposed and it is used to validate the proposed cognitive computing architecture. Experiments with a robotic navigation scenario under different environmental conditions show that the proposed cognitive random forest is capable of handling the environmental conditions what we called the cune conditions for big data.

Keywords: Big data, Noisy data, Big data classification, Cognitive computing, Machine learning, Random forest, Intracognitive computing, Intercognitive computing

1 INTRODUCTION

The particular practical problem that is addressed in this chapter is the robotic navigation under uncertain, unpredictable, and unstructured environmental

Handbook of Statistics, Vol. 35. http://dx.doi.org/10.1016/bs.host.2016.07.006

conditions. The combined constraint that resulted from these three conditions is typical to a big data environment; hence, we have named it a cune condition for simplicity and future references. The cune condition is defined as more complex perturbation than a simple noise degradation. It is clearly defined with other terminologies in the next section.

Human activities in an environment have become more complex than ever before due to advances in current technologies, and global mobility. While these advancements bring advantages to human society, they also create threats to safety and security of the environment and the human society. Therefore, the environmental events must be observed and studied to alleviate the environmental problems. The use of robots might be a solution to facilitate environmental monitoring. The robotic navigation generally involves computer vision and cognitive computing; therefore, it is important to study natural images and understand the issues related to cognitive computing models.

The environmental events generated by the human activities form big data which may not be easily processed by robots dynamically to extract knowledge without proper interaction between human and computing devices. The big data is originally defined by three characteristics (called 3Vs) of the environmental events: volume, velocity, and variety. Subsequently these Vs have been increased to 4Vs, 5Vs, and now 7Vs. By understanding the basic characteristics of these Vs, they can be mapped to three characteristics (Suthaharan, 2014): cardinality, continuity, and complexity. This 3Cs definition is sufficient in general to define and study big data classification problems.

The extraction of knowledge from a big data environment will be more efficient if big data classification approaches are integrated. In contrast to traditional data collection systems, a big data system can bring a massive amount of dynamically changing data to the classification problem domain (Suthaharan, 2015). Additionally the big data environment may be noisy; hence, the complexity of the problem can be increased tremendously with the environmental noise combined with other big data characteristics. This problem may be defined as the *classification under noisy environment* and it is called *cune* condition in this paper. According to urban dictionary, "cune" means the ability of a mind to reshape reality, and this type of reshaping of a mind is highly possible in a big data environment. In cune condition, the data is reshaped by the environmental perturbation and it can affect the performance of the machine learning models and algorithms. For example, the cune condition can create label noise (Biggio et al., 2011; Frenay and Verleysen, 2014), with incomplete data (Gustavo et al., 2003; Ramoni and Sebastiani, 2001), inaccurate data (Suthaharan, 2015), and imbalanced data (Akbani et al., 2004; Kotipalli and Suthaharan, 2014). Another form of reshaping, especially on natural scene image data, may easily occur in a big data environment due to restructuring of edge details of the objects in a scene (Suthaharan, 2000).

Machine learning models are trained using learning algorithms and data sets (Fig. 1 depicts a simple architecture). It simply shows a system that

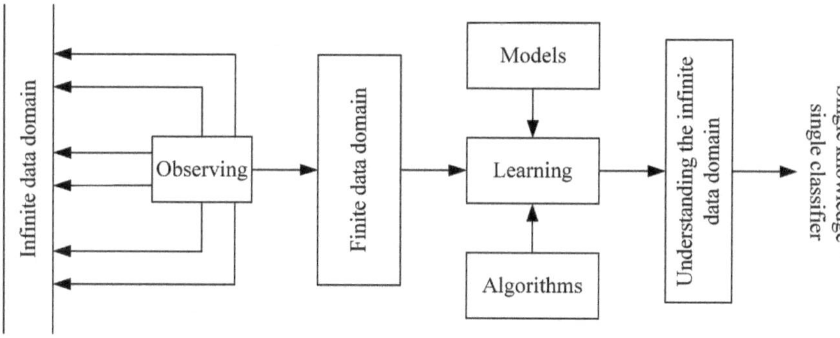

FIG. 1 An illustration of the standard machine learning system.

observes an infinite domain (e.g., natural environment) and generates data on a finite domain. The system then uses the finite domain to learn and understand the environment using machine learning models and algorithms. It is expected that the models are tested on similar data (which is called the seen data) to perform their operations effectively and work as expected on the similar data. However, with a mobile environment (e.g., robotic navigation), the data is dynamic. Therefore, if the learned model is applied to such distinct data sets (which is called the unseen data), then high likely they will not perform effectively as required. This is called the unseen data problem and this problem can be significantly increased under cune condition in a big data environment. The current machine learning techniques seek for an optimal solution, which in general leads to a single solution or a classifier.

Most commonly used machine learning techniques are support vector machine (Hearst et al., 1998; Jeyakumar et al., 2012), decision tree (Breiman et al., 1984; Murthy, 1998), random forest (Breiman, 1996, 2001), and deep learning (Hinton et al., 2012; Wan et al., 2013). The single classifier techniques may not be suitable for big data classification because of the 3Cs characteristics of the data and the cune condition of the environment. Therefore, it is now recommended or preferable to explore multiple classifiers for big data analytics and classification problems. Cognitive computing has recently been proposed and it is capable of finding such solutions. Therefore, the goal of the proposed approach is to combine both the machine learning techniques (Bishop, 2006; Hastie et al., 2009) and cognitive computing models (Baranyi and Csapo, 2012; Hurwitz et al., 2015) to address big data classification under cune conditions.

In this paper, a novel cognitive computing model, called the STE-M, is proposed and integrated with a set of standard random forest classifiers. This cognitive computing architecture is called *cognitive random forest* where the symbols S, T, E, and M represent senses, thoughts, experiences, and memory components of human cognition, respectively (Weseley and McEntarffer, 2010). It also integrates intra- and intercognitive computing models to connect

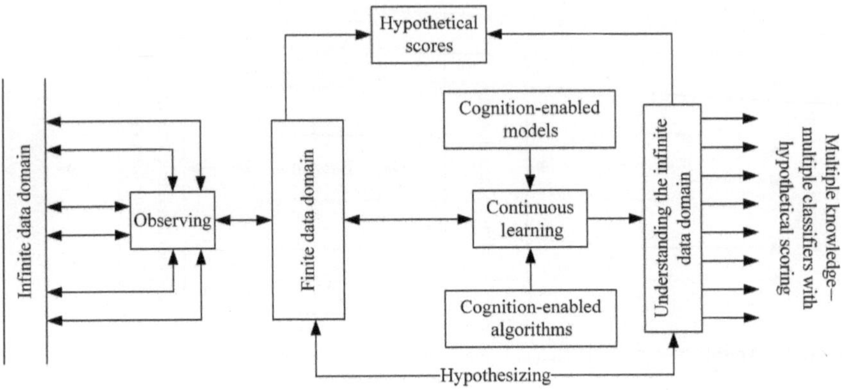

FIG. 2 A typical cognition-enabled machine learning system for big data analytics.

the STE-M and random forest models. The random forest has been selected because of its capability to generate multiple classifiers with bootstrap sampling and highly suitable for big data classification and continuous learning; however, it is not completely capable of handling big data problems under cune conditions. A simple cognition-enabled computing model is illustrated in Fig. 2. It reflects the processes shown in Fig. 1 with cognitive computing components added to the models, learning, and algorithm modules. In addition, hypothesizing and hypothesis scoring are integrated into the model. As stated in Hurwitz et al. (2015), hypothesizing and hypothesis scoring are two important tasks for cognitive computing.

The rest of the paper is organized as follows: Section 2 provides details of terminologies used in this chapter to improve readability and dependability. It defines several terminologies including cognitive computing models, cognitive computing architectures, cognitive computing systems, and cune conditions. A brief discussion on the random forest technique is presented in Section 3. The proposed STE-M model is presented in Section 4 in detail. Section 5 is completely dedicated to the proposed cognitive computing architecture. Section 6 presents the proposed cognitive computing system that is used to test the cognitive random forest. In Section 7, experimental procedures, results, and validation are presented. The software packages used in the experiment are Matlab (http://www.mathworks.com/products/matlab/) and R (https://cran.r-project.org/). Finally, Section 8 presents conclusions.

2 TERMINOLOGIES

This section is dedicated to defining various terminologies presented in this chapter. They are grouped into three sections based on cognitive computing models, cognitive computing architectures, and cognitive computing systems. These components are the major contributors to cognitive computing approaches adopted in computer science.

2.1 Cognitive Computing Models

The *cognitive computing models* generally describe the processes involved in executing cognitive computing tasks. The cognitive computing model, in this chapter, is defined to mimic the human brain processes mainly focusing on big data classification. The standard components of a cognitive model are the senses, thoughts, experiences, and memories, and they are defined below to describe cognitive computing model.

The component *senses* can be explained as computing ideas. They describe the awareness of certain events in the environment which are the stimuli of the environment that help the models detect, recognize, and classify objects (Weseley and McEntarffer, 2010) and their data domains. The knowledge of an object and its domain will lead to seen data settings.

The component *thoughts* can be defined as the connected computing ideas and they can be described as the concepts and mental pictures between the senses (Weseley and McEntarffer, 2010). In other words the thoughts may be considered as the connected graphs between the senses (or stimuli), where the senses are the nodes of the connected graph. The connected knowledge of the environment will lead to hypothetical (unseen) data settings.

The component *experiences* can be defined as the scored and connected computing ideas. The experiences are the knowledge about the environment gained from the thoughts and are scored according to cognitive rules used to create concepts and mental pictures in the thought processes. The scored knowledge and the connected graphs will lead to learned (seen) data or scored hypothetical data as shown in Fig. 2.

The component *memory* defines stored computing ideas that are scored and connected. In general, memorization is the stored experiences together with the thoughts and senses in memory. It can also be described as stored data—a set of connected senses with the hypothesis and their scores in memory. It will lead to informed data.

Anomalous events define the events that contribute highly to the proposed cognitive computing model. In other words, the anomalous events are considered as the unique events that are significantly different from the regular events. They can easily modify the senses, thoughts, experiences, and memories components of a cognitive computing model more than regular event. Therefore, these events will be detected and processed uniquely in the proposed cognitive computing approach.

Significant events define the events that are repeated or duplicated during the observation (or monitoring) of the environment; it is also called *cloning* of events in the proposed model. It also means that some of the events are considered as important events by the model; hence, it duplicates them by sampling with replacement. The statistical technique that may be adopted to simulate this concept is *bootstrapping* (Breiman, 2001; Hastie et al., 2009). The bootstrapping may be used to explain the importance of certain contextual information of the data while observing the environment.

2.2 Cognitive Computing Architectures

The *cognitive computing architecture* defines a framework that integrates multiple cognitive computing models to work together efficiently and produce effective approaches. In the proposed approach, the cognitive computing architecture is defined as the framework that integrates the proposed STE-M model, intra- and intercognitive computing models (denoted by *i*C*i*C), and multiple random forest classifiers (mixture of classifiers is also possible). Fig. 3 illustrates the proposed architecture. It reflects the model presented in Fig. 2, but it has added the STE-M model, *i*C*i*C model, and the random forest model as illustrated. We have selected the standard random forest classifier as the machine learning technique along with the standard machine learning processes known as the training and testing to support continuous learning.

An *intracognitive computing model* is defined as the computer model and its associated learning algorithms that describe the cognitive computing processes between similar data domains and feature spaces. In the intracognitive computing model, anomalous events of a class are identified and used to fuse the same class-type data to create the thoughts concept defined earlier in this section.

An *intercognitive computing model* is defined as the computer model and its associated learning algorithms that describe the cognitive computing processes between dissimilar data domains and feature spaces. In the intercognitive computing model, the anomalous events of the other class are used in the event fusion of a class.

2.3 Cognitive Computing System

The *cognitive computing system* defines an environment (or computing infrastructure) that can help implement, understand, and validate the proposed

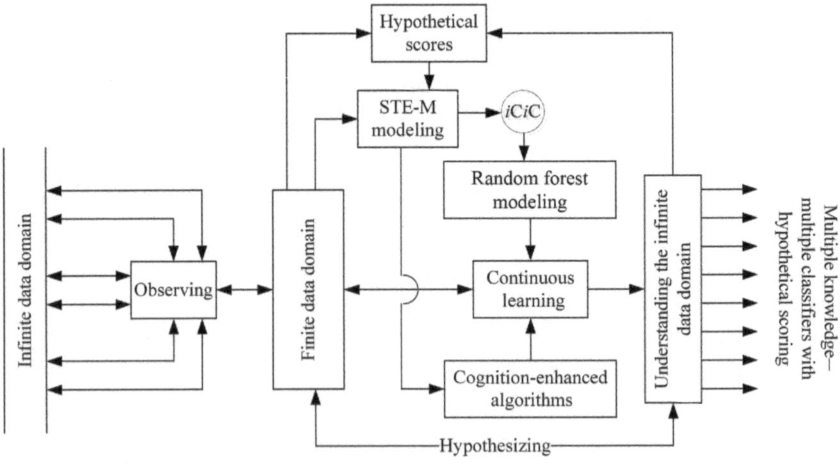

FIG. 3 The proposed cognitive computing architecture for big data classification.

cognitive computing architecture. In the proposed cognitive computing approaches, a robotic navigation scenario and associated computing environment are defined as the cognitive computing system. It involves big data classification under cune conditions and provides training and testing algorithms with an adaptive and continuous learning modules.

Big data environment is a computing environment which generates data that are uncertain, unpredictable, and unstructured in nature. It contains computing techniques and technologies that are capable of generating, processing, storing, and analyzing data that satisfy big data characteristics commonly defined as volume, velocity, and variety.

Cune condition is a condition in which the collected data is perturbed with multiple degradation sources, which include environmental noise (e.g., Gaussian noise or mixture of Gaussian), label noise (e.g., switching of labels), incomplete data, inaccurate data, and imbalanced data which are the general data characteristics of a big data environment resulted from uncertain, unpredictable, and unstructured data types.

Machine learning is a technique that provides parameterized learning models and associated optimization algorithms to tune the parameters through training and testing processes using labeled data sets. The general purpose of this technique is to divide data domain such that the classification of different varieties of objects through domain divisions. As stated in Suthaharan (2015), it may be divided into mathematical, hierarchical, and layered machine learning techniques. The technique selected in this paper is hierarchical, especially the random forest classifiers.

3 RANDOM FOREST CLASSIFIERS

The core unit of random forest classifiers is the decision tree. The decision tree is a hierarchical structure that is built using the features (or the independent variables) of a data set. Each node of the decision tree is split according to a measure associated with a subset of the features. The random forest is a collection of decision trees that are associated with a set of bootstrap samples that are generated from the original data set. The nodes are split based on the entropy (or Gini index) of a selected subset of the features. The subsets that are created from the original data set, using bootstrapping, are of the same size as the original data set. A detailed information on random forest classifiers can be found in the papers by Breiman (Breiman, 1996, 2001). In the standard random forest approach, the bootstrapping technique helps the development of random forest with a set of required number of decision trees in order to improve classification accuracy through the concept of overlap thinning as mentioned in Suthaharan (2015). Then an approach called bagging (bootstrap aggregate) technique is used to select the best trees with a voting scheme. This standard random forest approach is the one adopted in the proposed cognitive computing architecture.

4 THE STE-M MODEL

This section explains the proposed STE-M model. The three steps involved in this model are the detection of anomalous events, the extraction of significant events, and the cloning of events. These processes transform the original data domain to a cognitive computing-enabled data domain. The steps are illustrated in Fig. 4, which shows n observations that are created by observing an object (e.g., a human face) and extracting p features from it. Hence the first step creates a data domain $D=\{o_1,o_2,\ldots,o_n\}$ and a feature space $F=\{f_1,f_2,\ldots,f_p\}$. In the second step an anomalous event a_k is detected, and a set of significant events, $B=\{b_1,b_2,\ldots,b_n\}$, are extracted. These two categories of events are then fused to form a cloned data domain $C=\{c_1,c_2,\ldots,c_n\}$ in the final step. Although the detection of a single anomalous event is shown, it is possible to detect multiple anomalous events which will lead to multiple cloned data domain. It will support the integration of multiple classifiers goal of cognitive computing approaches.

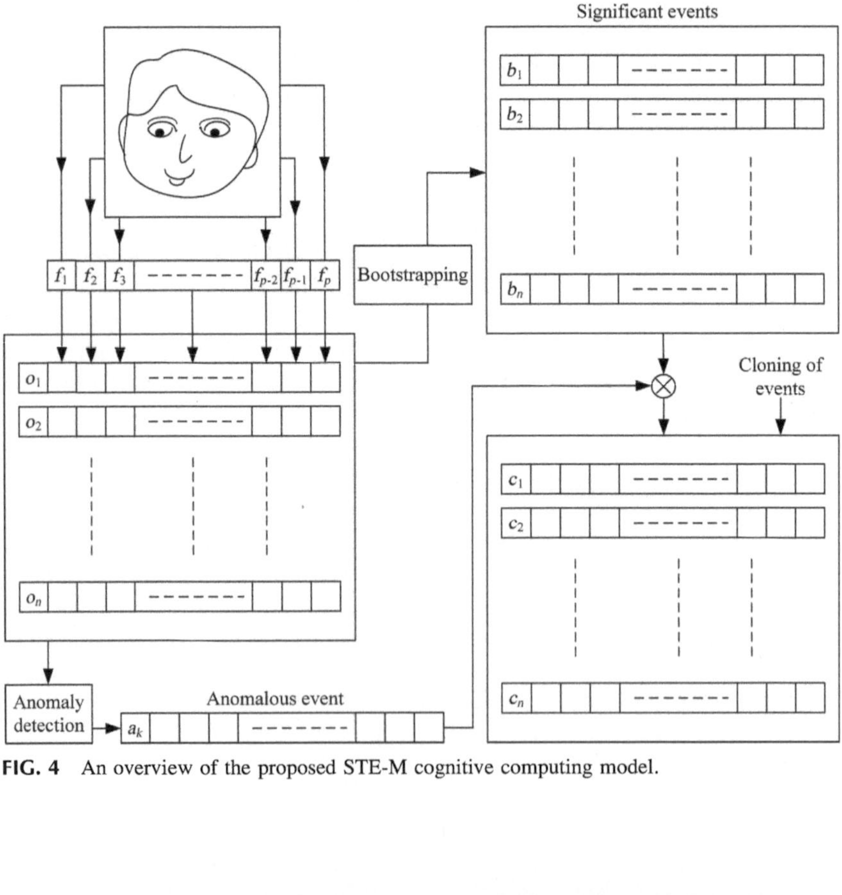

FIG. 4 An overview of the proposed STE-M cognitive computing model.

4.1 Modeling of Computing as Senses

The oscillation properties of a *sine* function are used to characterize and detect anomalous events. In general, if an event is observed from a big data environment, then we can have two possible cognitive computing characteristics, and they are "within features" characteristic and "between features" characteristic. Note that the term "features" represents the independent variables. The first characteristic enables the enhancement of *between-class* classification. That is, it is possible to recognize two observed values between classes that may give the same meaning or an opposite meaning. Similarly, the second characteristic enables the enhancement of *single-class* recognition or identification. That is, it is possible to recognize two distinct observations within a class that may give the same meaning or an opposite meaning.

These two characteristics are important for cognitive computing and its application to big data classification. These characteristics are considered as the main cause for the anomalous events in the proposed concept. They may be considered as *senses* and should not be ignored as outliers; instead they must be used for altering other events. Hence, anomalies are detected using the following *sine* function:

$$d_1 = sin(ae_1) \tag{1}$$

$$d_2 = sin(ae_2) \tag{2}$$

where ae_1 and ae_2 are the anomalous events detected from object 1 (i.e., class 1) and object 2 (i.e., class 2), respectively. The events that display high oscillation in the entire feature space are considered as the anomalous events. In other words the more alternating +1 and -1 sequence as illustrated in Fig. 5 for an event means that event are considered as the anomalous event. Similarly, the concept of significant event is defined as follows: if an event is significant, then it is repeatedly observed than an insignificant event. For example, this concept can be easily described by the statistical process, bootstrapping. The significant event can be extracted using bootstrapping that performs sampling with replacement. The bootstrapping technique plays an important role in the standard random forest technique. Hence, the component *senses* is associated with anomalous events and significant events that are represented by signal (or sine waves), and it can act as stimuli from the observed scene or environment.

4.2 Modeling of Computing as Thoughts

The component *thoughts* is to connect the sensed events to provide concepts and mental picture of the entire environment. In the STE-M model, thoughts are considered as a fusion of the significant events with the anomalous events. It forms several sets of connected graphs where the nodes are the related senses (signal representations) and edges are the relationships between these signals (or stimuli).

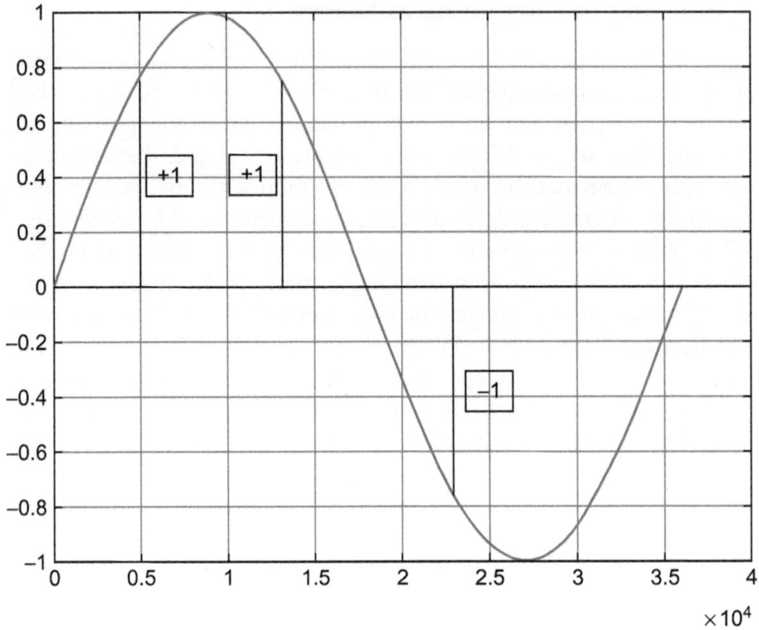

FIG. 5 A sine function that illustrates the sensing component of the proposed STE-M model.

4.3 Modeling of Computing as Experiences

The component *experiences* is defined as the scoring of thoughts (or connected senses) using comparative analysis between two or more objects as illustrated in Fig. 6. This is where the machine learning models can play a major role in the proposed cognitive computing approach. However, it requires a bridging technique to integrate the STE-M model with a machine learning technique. We have proposed intra- and interdomain technique to address this requirement and selected random forest as the machine learning technique. This becomes the cognitive computing architecture and it is explained in the next section.

4.4 Modeling of Computing as Memory

Memory is another important component in the proposed cognitive computing technique. Hence, it is added to the proposed STE-M model. The current big data techniques and technologies, such as Hadoop and MapReduce (Suthaharan, 2015), can help store the information that should be memorized and processed using a distributed computing environment for big data classification. The scored thoughts and associated information (e.g., anomalous events) are stored (or "remembered") in the big data system for future retrieval for testing phase of the learning algorithm.

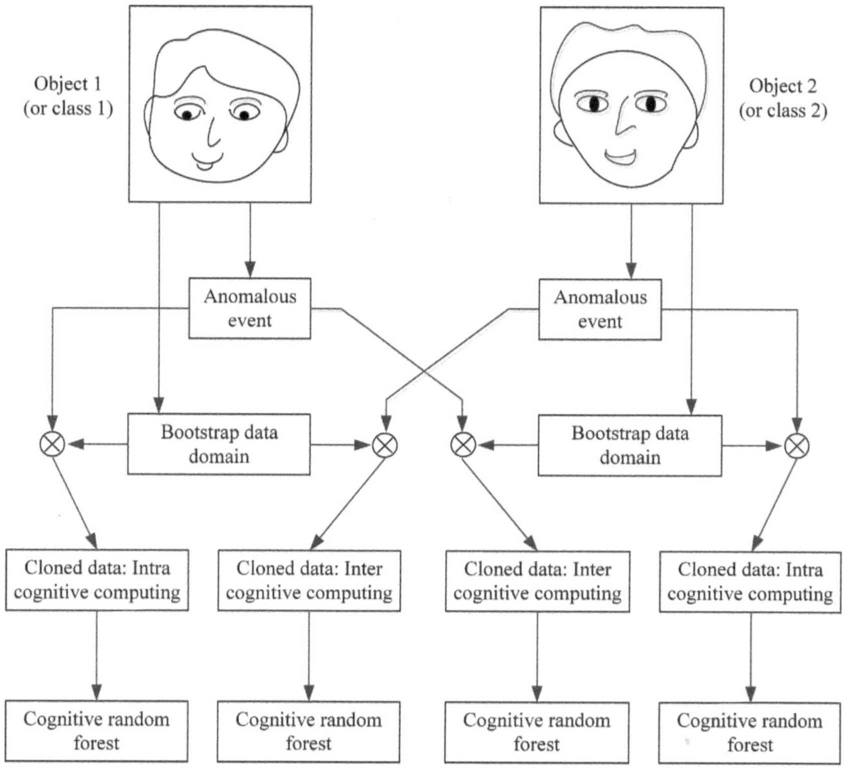

FIG. 6 An overview of the proposed cognitive computing architecture–cognitive random forest.

5 COGNITIVE RANDOM FOREST

The STE-M model is integrated with the standard random forest technique to make the random forest cognitive computing capable. According to Weseley and McEntarffer (2010), the thoughts may be categorized into concepts and mental pictures. We have adopted this idea and developed the STE-M model to represent the concepts, and the cognitive computing architecture to represent mental pictures. The proposed architecture is named cognitive random forest and illustrated in Fig. 6. It clearly shows the integration of the STE-M model, the intra- and intercognitive computing models (*iCiC*) for the domain division, and multiple random forest classifiers for improving classification accuracy with continuous learning.

5.1 Integration of Domain Divisions

It requires spatial and temporal reasoning (or cognitive rules), and we have selected fusion of events to reflect intracognitive domain division and intercognitive domain division. The models involved are the intracognitive computing and intercognitive computing, respectively. In intracognitive computing, the anomalous events of a class are used to fuse the same class data to create the concept of cognitive computing thoughts. In intercognitive computing, the anomalous events of the other class are used in the event fusion. These fused events will be transported to multiple random forest as illustrated in Fig. 6. These event fusion processes will transform the cloned data domain to a domain where the classes are highly separable (e.g., Fig. 12). We called this domain a *iCiC* data domain and divided it using classifiers for classification.

5.2 Random Forest for Continuous Learning

The random forests that are created for mental pictures are named as intracognitive and intercognitive random forests; hence, the proposed architecture is called the cognitive random forest. These random forests help the model create cognitive computing thoughts in testing phase later to establish classification cognitive rules.

6 COGNITIVE COMPUTING SYSTEM

The cognitive computing system that consists of the techniques and technologies is presented in Fig. 7. It shows two inner-boxes and one out-box to describe the proposed model, architecture, the system. The first inner-box describes the cognitive computing mode and it contains the proposed STE-M model. The second inner-box describes the cognitive computing architecture and it contains the STE-M model, the *iCiC* model, and the random forest models. The out-box describes the cognitive computing system and it contains the cognitive computing architecture and the machine learning techniques and the big data technologies labeled as natural, mobile, and computing environments.

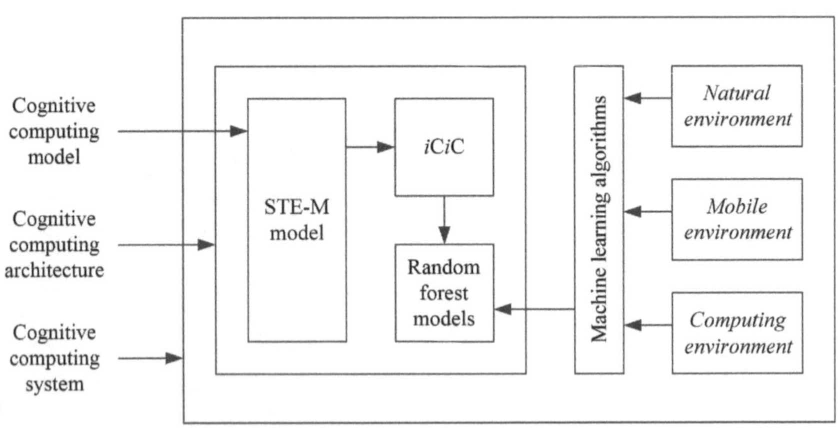

FIG. 7 The proposed cognitive computing system.

The same as a machine learning system, the proposed cognitive computing system also has a machine learning technique along with the cognitive computing architecture. Hence, it follows the training and testing phases to develop classifiers. The proposed architecture uses the random forest as the machine learning technique; hence, it follows the standard training and testing processes adopted in the random forest technique. The main difference is that the standard random forest divides the original data domain, but the cognitive random forest divides the *iCiC* data domain. The *iCiC* domain enhances the separability of classes with four different domains; therefore, it provides multiple random forest classifiers. The next section presents the results of a simulation conducted with cune conditions. The cune conditions are more complex and can lead to several scenarios to study. In this chapter, a simple case of cune conditions is studied. The analysis with more complex cune conditions will be our future research.

7 EXPERIMENTAL VALIDATION

One of the applications that can benefit from cognitive computing is mobile robot navigation. Suppose a robot is navigated in an environment which has grasslands, concrete slabs, and many other obstacles, then the settings can be described as a big data environment with cune condition. The robot is first navigated on the grassland with a certain speed, and then it suddenly encounters a concrete slab, which requires a rapid change in the rpm (revolution per minute) of the wheels. This situation then imposes the robot to classify the ground conditions momentarily and adjust the speed accordingly to avoid collisions with other objects. In addition, the cune condition brings complexity to the big data classification problem domain. In this simulation, the cune conditions are created using two Gaussian noises with $\mu=0$ and $\sigma=80$ for the training set and $\mu=0$ and $\sigma=90$ for the test set—note that the magnitude of the data ranges from 0 to 255. The big data environment with cune conditions considered in this section is to evaluate the standard random forest and the proposed cognitive random forest.

Two image data sets called "grassland" and "concrete slab" are used for the purpose of evaluating these approaches. The original color versions of these images are presented in Fig. 8 (left) and (right), respectively; however, their gray scale versions with the intensity values 0 (black) to 255 (white) are used in the analysis to reduce computational cost. The size of these images is of 256 $\times$ 256 pixels, and they are divided into 1024 blocks of 8 $\times$ 8 pixels. This process provides us with 1024 observations where each observation has 64 features, representing the pixel locations and their intensity values. Therefore, we have 2048 observations altogether with 64 features and two class labels—the labels 0 and 1 are used to represent the grassland and concrete slab classes, respectively.

For evaluation, the machine learning techniques require at least a training set and a test set; therefore, the data sets are divided into two sets with the

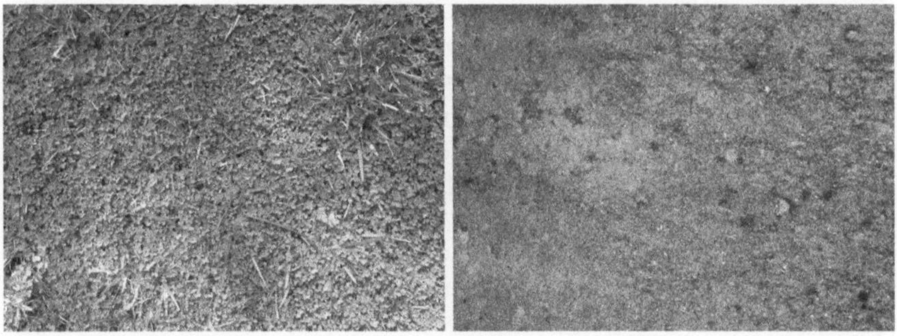

FIG. 8 Grassland image is on *left* and concrete slab image is on *right*.

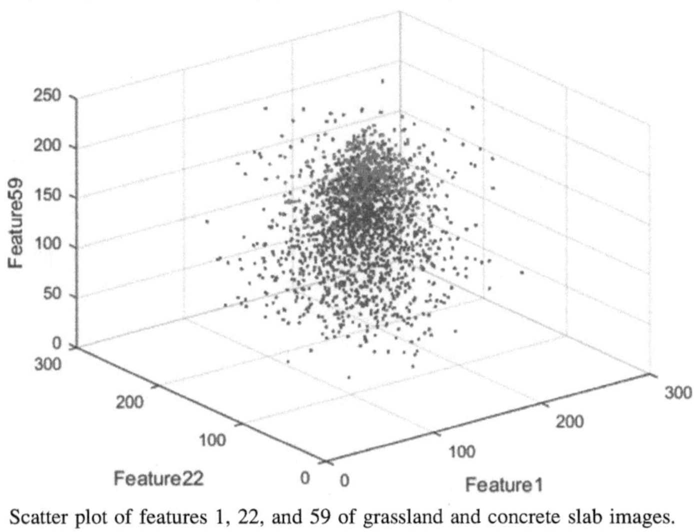

FIG. 9 Scatter plot of features 1, 22, and 59 of grassland and concrete slab images.

allocation of 1365 observations to the training set and 683 observations to the test set. It means that two-third of the entire data is used for training and one-third is used for testing. First, to illustrate classification characteristics of the images on a data domain, three features 1, 22, and 59 are selected and presented as a scatter plot in Fig. 9. We can clearly see the complexity in separating the two classes (grassland and concrete slab) with traditional classification techniques. The complexity can easily be increased by cune conditions.

7.1 Selected Cune Conditions

This simulation requires cune conditions. The development of cune conditions is a separate research problem in its own, especially when big data environment gets more complex and compound. In this experiment, several cune conditions are created by scaling the Gaussian noise only with $\mu=0$ and $\sigma=80$ for the training set and $\mu=0$ and $\sigma=90$ for the test set, and the techniques are evaluated. However, the results of three cune conditions are presented in this section, which are sufficient to compare the classification performances of the proposed cognitive random forest and the standard random forest using the grassland and concrete slab data sets.

7.1.1 Level-1 Cune Condition

This is a mild cune condition. It is called a level-1 cune condition because it provides a mild reshaping of both training and test data domains with the same strength. The strengths of the Gaussian noises are obtained using the scaler 0.001.

7.1.2 Level-2 Cune Condition

The level-2 cune condition reshapes the training and test data domains with a little stronger effect than that of level-1 cune condition. In this case, the strengths of the Gaussian noise are scaled by 0.5 in both training and testing phases.

7.1.3 Level-3 Cune Condition

It provides very strong cune condition to the big data environment. It is incorporated to reshape the data domains severely in both training and testing objectives for machine learning. In this case, the strengths of the Gaussian noise are scaled by 1 in both training and testing phases, that is, the same as the original Gaussian noise sets.

7.2 Standard Random Forest

The classification performance of the standard random forest is evaluated using grassland and concrete slab data sets with three levels of cune conditions in this section. The results and findings are also presented. Under the level-1 cune condition and with the use of all 1365 observations for the training, the standard random forest was able to grow 60 trees and achieve about 94% training accuracy. The out-of-bag error and the misclassification errors of labels 0 and 1 are calculated, and they are 0.10769, 0.10089, and 0.11433, respectively. The corresponding confusion matrix is:

$$M = \begin{bmatrix} 606 & 68 \\ 79 & 612 \end{bmatrix} \tag{3}$$

Where the out-of-bag error is calculated as (68/1365)+(79/1365), misclassification of 0 is calculated as 68/(68+606) and misclassification of 1 is calculated as 79/(79+612). One of the important things to note is that if the number of trees is not grown to above 60, then not all the observations are

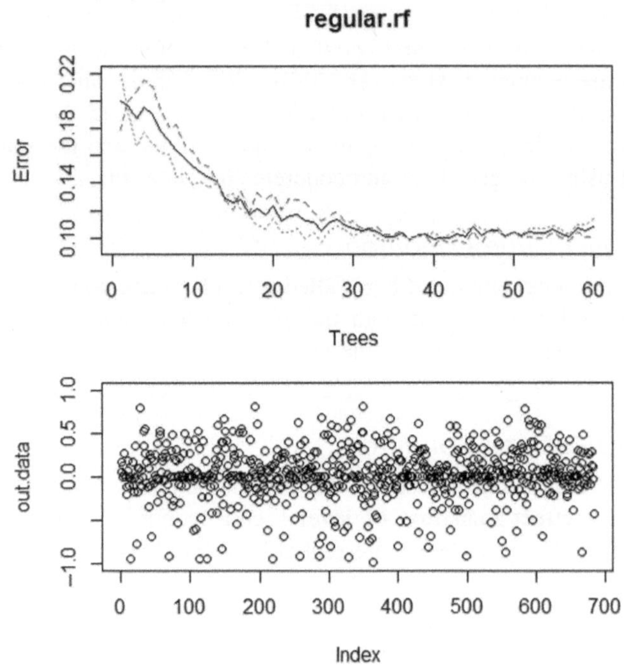

FIG. 10 Out-of-bag error (*black*) and misclassification of 0—*red* (*gray* in the print version) and 1—*green* (*light gray* in the print version) of the random forest are shown in the *top figure. Bottom figure* shows the prediction errors of random forest—level 1 cune condition.

used. The experiment also showed that it has used 52 features out of 64 to reach the best accuracy with 60 trees. The top graph in Fig. 10 shows the changes in three errors with respect to the number of trees grown in the random forest, and it illustrates that it is not feasible to reach 0 error.

In testing under the level-1 cune condition, the test data set of 683 observations was used to evaluate the prediction accuracy of the standard random forest classifier developed in the training phase. The predicted values for the class labels are shown in the bottom scatter plot in Fig. 10. In this plot the deviation from 0 indicates the error in the prediction. For the purpose of classification, the errors are associated with the points that fall out side the range (−0.5, 0.5), which gives about 88% testing accuracy.

Fig. 11 shows the training error relationships and the prediction errors of the level-2 cune condition. As we can see in the top graph, it requires more trees (i.e., 160 trees) to be grown in order to achieve about 88% training accuracy with minimum errors: 0.16263, 0.19436, and 0.13169, respectively. The bottom figure shows higher errors (deviation from 0) than in the previous case. The classification accuracy that is determined by the predictions based on the range (−0.5, 0.5) is 80%. Similarly, the level-3 cune condition is tested

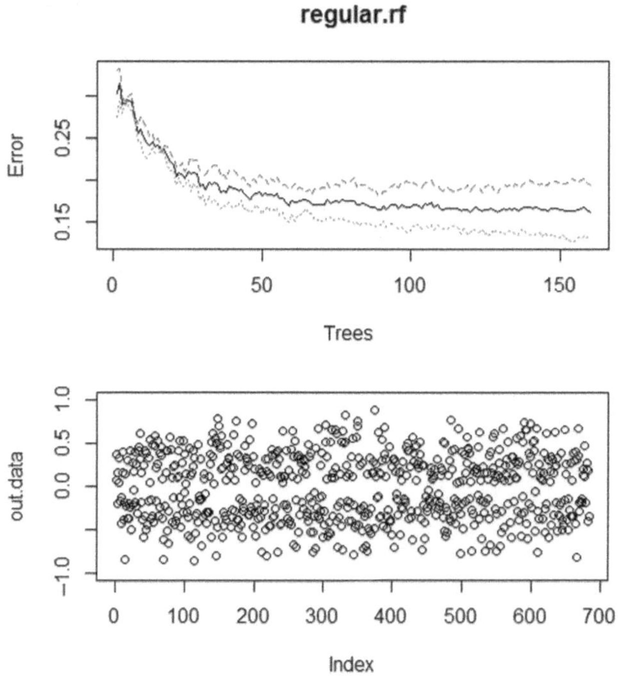

FIG. 11 Out-of-bag error (*black*) and misclassification of *0—red* (*gray* in the print version) and *1—green* (*light gray* in the print version) of the random forest are shown in the *top figure*. *Bottom figure* shows the prediction errors of random forest—level 2 cune condition.

and it gives about 83% training accuracy and 75% testing accuracy. The standard random forest cannot classify the two classes considered under the level-3 cune condition efficiently; therefore, the cognitive random forest is evaluated under this condition and the results are presented in the next section.

7.3 Cognitive Random Forest

In this simulation the proposed cognitive random forest is evaluated using the same training and test data sets with three levels of cune conditions. The cognitive random forest performed superior in both level-1 and level-2 cune conditions; therefore, the denouements of level-3 cune condition are sufficient to explain its efficiency. In cognitive random forest, the intra- and intercognition data domains were first created by selecting an anomalous event from each class and applying bootstrap sampling as described in Sections 4 and 5. Examples of these domains, using the same features 1, 22, and 59, are presented in the top two scatter plots of Fig. 12. They show significant separation between classes and the separability helps the random forest to be trained by growing a

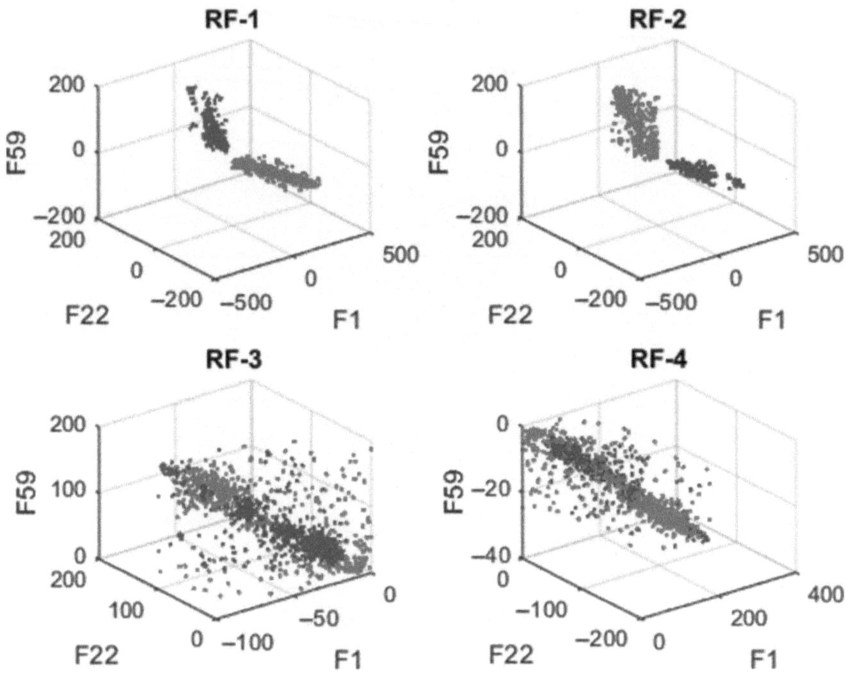

FIG. 12 Simulation results that show the effect of anomalous and significant events.

smaller number of trees and achieving high classification accuracy. In both domains, 100% training accuracy can be achieved, but with intracognition data domain, this training accuracy was achieved with 16 grown trees and with intercognition data domain, the accuracy was achieved with 20 trees.

The top plot in Fig. 13 shows the three-error relationships in intracognition data domains, but the use of intercognition data domains also gives a similar result. One of the grown trees in the cognitive random forest is extracted and presented in Fig. 14. It is a much smaller tree than the one grown in the standard random forest. It indicates the proposed approach is highly efficient than the standard approach. The tree also shows the features that are used at the nodes of the tree.

In testing, the same test data set (the one used for the validation of the standard random forest) is used and fused with the anomalous event stored in the memory before applying cognitive random forest classifiers. There are 683 observations in the test set and all of them are fused with the anomalous event stored in the memory first. The intra- and interdomain analysis (similar to the one adopted in training) is performed and prediction errors are calculated. The results are presented in the bottom plot of Fig. 13. A comparison between this figure and the corresponding results of the standard random forest in Fig. 10 (bottom plot) suggests the efforts put forward by the cognitive random forest have led to an excellent denouements.

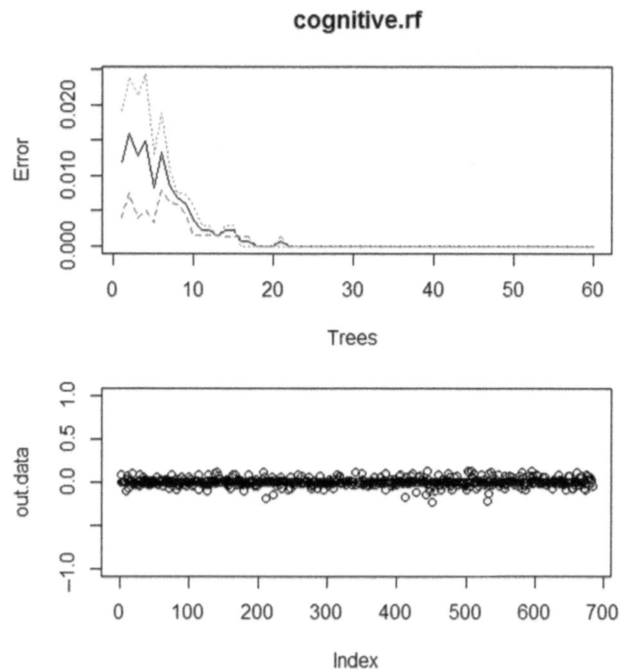

FIG. 13 Out-of-bag error (*black*) and misclassification of 0—*red* (*gray* in the print version) and *1*—*green* (*light gray* in the print version) of the cognitive random forest are shown in the *top figure*. *Bottom figure* shows the prediction errors of cognitive random forest. These results are from the application of cognitive random forest on the intracognitive data domain generated.

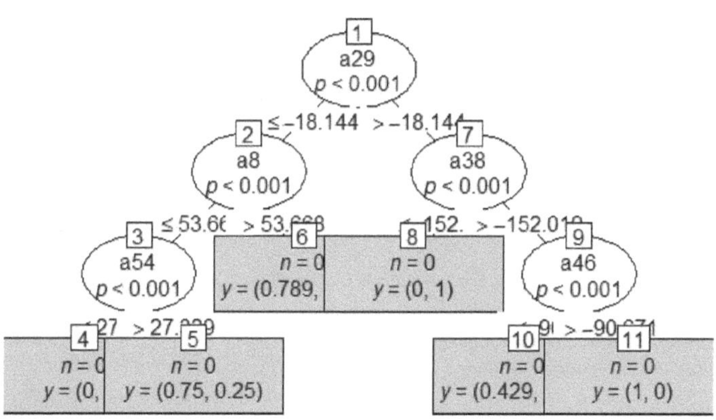

FIG. 14 This is the 18th decision tree of the cognitive random forest that was generated using *cforest* function of the *party* R-package—the corresponding tree of the standard random forest is large and hence not provided.

8 CONCLUSIONS

One of the conclusions is that the inclusion of cognitive computing in random forest machine learning techniques can significantly enhance its classification capabilities. Another conclusion is that the use of cognitive computing makes the random forest techniques produce a collection of scored classifiers than the original random forest; thus, it can help big data classification. We can also conclude that the cognitive random forest does not require pruning based on the data analyzed. Finally this research provides a novel approach and it not only helps enhance the random forest technique but also can be adopted with other machine learning techniques, making them suitable for big data classification. Getting the 100% training accuracy is not surprising and this is because the cognitive computing has already separated the classes for random forest.

More data sets with more cune conditions will be studied in the future research. The development of cune conditions for big data environment is not straightforward; therefore, this will also be the focus of immediate research. The data sets considered with the cune condition are complex enough to compare the performances of the standard random forest and cognitive random forest; however, the research results presented have demonstrated that the cognitive random forest is capable of handling more complex big data problems. Therefore, in the near future, streaming big data environment will be adopted and the techniques will be evaluated.

REFERENCES

Akbani, R., Kwek, S., Japkowicz, N., 2004. Applying support vector machines to imbalanced datasets. Machine Learning: ECML 2004, Springer, Berlin, Heidelberg, pp. 39–50.

Baranyi, P., Csapo, A., 2012. Definition and synergies of cognitive infocommunications. Acta Polytech. Hung. 9 (1), 67–83.

Biggio, B., Nelson, B., Laskov, P., 2011. Support vector machines under adversarial label noise. In: Asian Conference on Machine Learning, JMLR: Workshop and Conference Proceedings. 20, pp. 97–112.

Bishop, C.M., 2006. Pattern Recognition and Machine Learning. Springer Science+Business Media, LLC, New York.

Breiman, L., 1996. Bagging predictors. Mach. Learn. 45, 123–140.

Breiman, L., 2001. Random forests. Mach. Learn. 45, 5–32.

Breiman, L., Friedman, J., Stone, C.J., Olshe, R.A., 1984. Classification and Regression Trees. CRC Press, Boca Raton, FL.

Frenay, B., Verleysen, M., 2014. Classification in the presence of label noise: a survey. IEEE Trans. Neural Netw. Learn. Syst. 25 (5), 845–869.

Gustavo, E.A., Batista, P.A., Monard, M.C., 2003. Analysis of four missing data treatment methods for supervised learning. Appl. Artif. Intell. 17, 519–533.

Hastie, T., Tibshirani, R., Friedman, J., 2009. The Elements of Statistical Learning. Springer, New York.

Hearst, M.A., Dumais, S.T., Osman, E., Platt, J., Scholkopf, B., 1998. Support vector machines. IEEE Intell. Syst. Appl. 13 (4), 18–28.

Hinton, G.E., Srivastava, N., Krizhevsky, A., Sutskever, I., Salakhutdinov, R.R., 2012. Improving neural networks by preventing co-adaptation of feature detectors. Technical Report, arXiv:1207.0580, pp. 1–18.

Hurwitz, J., Kaufman, M., Bowles, A., 2015. Cognitive Computing and Big Data Analytics. Wiley Publishing, Indianapolis, IN.

Jeyakumar, V., Li, G., Suthaharan, S., 2012. Support vector machine classifiers with uncertain knowledge sets via robust optimization. Optimization 63, 1–18.

Kotipalli, K., Suthaharan, S., 2014. Modeling of class imbalance using an empirical approach with spambase dataset and random forest classification. In: Proceedings of the 3rd Annual Conference on Research in Information Technology, ACM, pp. 75–80.

Murthy, S.K., 1998. Automatic construction of decision trees from data: a multi-disciplinary survey. Data Min. Knowl. Disc. 2 (4), 345–389.

Ramoni, M., Sebastiani, P., 2001. Robust learning with missing data. Mach. Learn. 45, 147–170.

Suthaharan, S., 2000. Image and edge detail detection algorithm for object-based coding. Pattern Recogn. Lett. 21 (6), 549–557.

Suthaharan, S., 2014. Big data classification: problems and challenges in network intrusion prediction with machine learning. ACM SIGMETRICS Perform. Eval. Rev. 41 (4), 70–73.

Suthaharan, S., 2015. Machine Learning Models and Algorithms for Big Data Classification: Thinking with Examples for Effective Learning. Integrated Series in Information Systems, vol. 36. Springer US, New York, pp. 1–356.

Wan, L., Zeiler, M., Zhang, S., Cun, Y.L., Fergus, R., 2013. Regularization of neural networks using dropconnect. In: Proceedings of the International Conference on Machine Learning, pp. 1058–1066.

Weseley, A.J., McEntarffer, R., 2010. Barron's AP Psychology. Barron's Educational Series, Hauppauge, NY.

Chapter 7

Bayesian Additive Regression Tree for Seemingly Unrelated Regression with Automatic Tree Selection

S. Chakraborty[1]
University of Missouri, Columbia, MO, United States
[1]*Corresponding author: e-mail: chakrabortys@missouri.edu*

ABSTRACT

In this chapter, we propose a flexible Bayesian regression tree model when the response variable is a vector and the components of the vector are highly correlated. Our proposed Bayesian additive tree model can jointly model the correlation structure among the related response variables and provide a highly flexible and nonlinear regression structure for each of the individual regression functions. The number of trees in our multivariate Bayesian additive regression tree (seemingly unrelated regression) model (BART-SUR) is selected adaptively by treating it as a model parameter and assigning a prior distribution on it. We have designed an efficient Bayesian backfitting algorithm with reversible jump MCMC for our BART-SUR model. Our BART-SUR can jointly model the correlated response vector and at the same time it can adaptively select the number of trees required to fit the data set. The adaptive tree selection makes our model extremely fast and efficient. We demonstrate the superiority of our BART-SUR model over several out of the shelve popular methods like random forest, neural network, wavelet regression, and support vector machine through two simulation studies and three real data applications.

Keywords: Bayesian model, Correlated response, Ensemble model, Reversible jump MCMC, Regression tree, Seemingly unrelated regression

1 INTRODUCTION

In modern scientific experiments, data generated on an object tend to be in the form of a vector of responses rather a scalar quantity. Furthermore, several response variables on the same object are often very highly correlated. For example in chemometric studies (Osborne et al., 1984; Park et al., 2002) with

Handbook of Statistics, Vol. 35. http://dx.doi.org/10.1016/bs.host.2016.07.007

near-infrared (NIR) spectroscopy, the predictor variables arise from an essentially continuous curve at equally spaced points of reflectance. On the other hand there are multiple response variables measuring the composition of certain basic materials or tissue samples. Therefore in such cases we have one set of covariates but the response is a vector of correlated responses (Geladi and Dabakk, 2009). This is very common when people analyze composition of food, drink, pharmaceutical products, and petrochemicals (Ingle and Crouch, 1988) . The seemingly unrelated regressions (SUR) model, proposed by Zellner (1962), can be viewed as a more general case of this problem. The SUR model has been extensively applied in agricultural economics (O'Donnell et al., 1999; Wilde et al., 1999) and aquaculture economics (Samonte-Tan and Davis, 1998). In forestry it is used for modeling tree growth (Hasenauer et al., 1998). Its application and use in the natural sciences is rapidly increasing with the influx of mountain of data gathered through satellite imaginary and on field sensor networks.

In an SUR model (Denison et al., 2002; Zellner, 1962), we can have several individual regression relationships between the set of predictor variables and a set of response variables. The predictor variables corresponding to each components of the response vector can be completely different. Although apparently they look like independent set of relationship but they are linked by the fact that their error distributions are correlated. Let us consider we have a data set $D=\{(\boldsymbol{y}_1, \boldsymbol{X}_1), (\boldsymbol{y}_2, \boldsymbol{X}_2), \ldots, (\boldsymbol{y}_n, \boldsymbol{X}_n)\}$, where $\boldsymbol{y}_i=(y_{i1}, \ldots, y_{id})^T$ is the vector of multiple responses corresponding to the i-th observation. The $\boldsymbol{X}_i = \left(\boldsymbol{x}_{i1}^T; \ldots; \boldsymbol{x}_{id}^T\right)$, where $\boldsymbol{x}_{ij}$ is the set $(p_j \times 1)$ of the covariates corresponding to the j-th component in the response vector $\boldsymbol{y}_i$. Then following Zellner (1962) we can write the SUR model as follow:

$$\boldsymbol{y}_i = \boldsymbol{X}_i\boldsymbol{\beta} + \boldsymbol{e}_i, \tag{1}$$

where $\boldsymbol{X}_i = \left[\boldsymbol{x}_{i1}^T, \boldsymbol{0}, \ldots, \boldsymbol{0}; \boldsymbol{0}, \boldsymbol{x}_{i2}^T, \ldots, \boldsymbol{0}; \ldots; \boldsymbol{0}, \ldots, \boldsymbol{x}_{id}^T\right]$ (is a $d \times p$ matrix; $p=\sum_{j=1}^{d} p_j$) and $\boldsymbol{e}_i=(e_{i1}, \ldots, e_{id}) \sim N_d(\boldsymbol{0}, \Sigma)$. Therefore, solving the system of equations (1) for the correlated errors solves the problem of fitting an SUR model.

We notice here that the original SUR model (Zellner, 1962) is linear where as in many modern practical examples the linear model assumption is not good enough to explain the complex relationship between the set of the predictors and the corresponding response variables. To handle the problem of nonlinearity, richer class of models are proposed that uses basis expansion technique (Holmes et al., 2002). For example Denison et al. (2002) used nonlinear transformation of the predictor variables based on a set of basis function and formulated a Bayesian SUR model. More recently Ando and Zellenr (2010) developed a hierarchical Bayesian analysis of the SUR model based on importance sampling.

Other common approach is to apply the univariate response regression technique separately on each component of the response vector using separate nonlinear regression functions and ignore any existing relationship among the components of the response vector. Following this approach several popular nonlinear univariate regression methods like the support vector machine (SVM; Schölkopf and Smola, 2002; Wahba, 1999), random forest (RF; Breiman, 2001), neural networks (Ripley, 1996), and Gaussian process regression (Rasmussen and Williams, 2006) are widely used. However, all these mentioned approaches always ignore the underlying relationship between the responses and thus produces suboptimal predictive outcome.

In the Bayesian paradigm in recent years there is an explosion of several nonlinear flexible models which can be used to model the complex functional relationship between the predictor and the response. Neal (1996) developed Bayesian neural network model as a probabilistic model for the regular neural network. Tipping (2000) developed relevance vector machine (RVM) as a Bayesian extension of the celebrated SVM. Although all the mentioned Bayesian models can be well utilized for modeling nonlinear regression functions but none of them can model jointly all the components of the response vector. Brown et al. (2001) developed a Bayesian wavelet regression (BWR) on curves that can model jointly all the components of the response vector using wavelet coefficient and matrix variate normal distribution. A competing and more accurate prediction model is proposed by Chakraborty et al. (2012) that is based on the reproducing kernel Hilbert space theory.

Bayesian additive regression trees (BART) is proposed by Chipman et al. (2010) as a nonparametric Bayesian model to estimate underlying complex regression functions. The BART approach produces highly flexible nonlinear regressions, and it has been shown to be highly competitive with the existing modern predictive methods like bagging and boosting (Breiman, 1996). The original BART model is extended and improved upon by several other independent researchers. For example, Zhang et al. (2007) extended the BART to SBART for spatial data analysis. Bonato et al. (2011) developed a BART-based model for analyzing right censored survival data with accelerated failure time model. A faster and improved version of BART that can handle "big data" problems is developed by Kapelner and Bleich (2014).

In this chapter we extended the BART for multiple response regression model, otherwise known as SUR model. The regression function corresponding to each component of the response vector is modeled using an ensemble of simple trees. The error vector is assumed to follow a multivariate normal distribution with a variance covariance matrix. We put an inverse Wishart prior distribution on the covariance matrix of the error distribution. The number of trees needed for each component is kept random and chosen adaptively by putting a prior distribution on it. This greatly enhances the speed and efficiency of our BART-SUR model since most of the times only a hand full of trees are good enough to build the best predictive model. Our Bayesian formulation allows us to fully characterize the uncertainty in predictions in a coherent manner without requiring any cross-validation. In our BART-SUR model as the number of trees is kept random the dimension of the parameter space is changing within each MCMC iteration. To handle that problem in this chapter we propose a fast reversible jump MCMC (RJMCMC) algorithm (Green, 1995) to fit our model. For easy implementation of our BART-SUR model we have created a stand alone R package called *BartSur()* with detail documentation which can be downloaded from our website.

The remainder of this chapter is organized as follows. In Section 2, we introduce our BART-SUR model. In this section we discuss in detail the prior structures on the parameter space and the number of tree selection through a specially designed prior MCMC. In Section 3, we propose a Bayesian backfitting RJMCMC algorithm. In Section 4, we demonstrate the effectiveness of our BART-SUR model based on two simulation studies. In Section 5, we to demonstrate the potential of out BART-SUR based on three real data sets. In Section 6, we conclude with a discussion and some future possible extensions.

2 BART FOR SUR WITH AUTOMATIC TREE SELECTION

In this section, we will discuss in detail our proposed BART model for the seemingly unrelated regression (BART-SUR), choice of our prior distributions, and some justification behind our prior choices.

2.1 BART-SUR Model

Let us consider we have a data set $D=\{(\boldsymbol{y}_1, \boldsymbol{X}_1), (\boldsymbol{y}_2, \boldsymbol{X}_2), \ldots, (\boldsymbol{y}_n, \boldsymbol{X}_n)\}$, where $\boldsymbol{y}_i=(y_{i1}, \ldots, y_{id})^T$ is the vector of multiple responses corresponding to the i-th observation ($i=1,\ldots,n$). The $\boldsymbol{X}_i = \left(\boldsymbol{x}_{i1}^T; \ldots; \boldsymbol{x}_{id}^T\right)$, where $\boldsymbol{x}_{ij}$ is a $p_j \times 1$ set of covariates corresponding to the j-th response in the response vector $\boldsymbol{y}_i$. It is possible that all $\boldsymbol{x}_{ij}$ for $j=1, \ldots, d$ are same. For example in NIR spectroscopy data we get a single set of light reflectance intensity and from there we want to predict the composition of more than one materials (Geladi and Dabakk, 2009). On the other hand more generally $\boldsymbol{x}_{ij}$ for $j=1, \ldots, d$ can be different, then we have a generalization of Zellner's SUR model (Zellner, 1962). Therefore our nonlinear SUR model can be written as

$$\begin{pmatrix} y_{i1} \\ y_{i2} \\ \vdots \\ y_{id} \end{pmatrix} = \begin{pmatrix} f_1(\boldsymbol{x}_{i1}) \\ f_2(\boldsymbol{x}_{i2}) \\ \vdots \\ f_d(\boldsymbol{x}_{id}) \end{pmatrix} + \begin{pmatrix} e_{i1} \\ e_{i2} \\ \vdots \\ e_{id} \end{pmatrix}, \tag{2}$$

where $i=1, \ldots, n$, the function $f_j()$ is the unknown regression function to predict the j-th component of the response vector, and $\boldsymbol{e}_i=(e_{i1}, \ldots, e_{id}) \sim N_d(\boldsymbol{0}, \Sigma)$. The d components of our response vector $\boldsymbol{y}_i$ are modeled jointly with the help of the correlated error term. The basic assumption of the correlated error arises from the fact that since the d response observations are collected on the same individual or object they may be related through an underlying process. However, the error vectors are assumed to be independent between two different samples, or $\boldsymbol{e}_i \sim^{iid} N_d(\mathbf{0}, \Sigma)$ for $i = 1, \ldots, n$. The regression function $f_j()$ is unknown to us and we would like to avoid any strong parametric linear or nonlinear model assumption on it. To mitigate this problem we decide to assign a nonparametric representation of the regression function $f_j()$. In this chapter, we approximate each regression function $f_j()$ by a sum of m_j regression trees or an ensemble of trees (Chipman et al., 2010) as follow

$$f_j(\boldsymbol{x}_{ij}) \approx h_j(\boldsymbol{x}_{ij}) = \sum_{k=1}^{m_j} g_k^j(\boldsymbol{x}_{ij}), \quad j = 1, \ldots, d. \tag{3}$$

Therefore each regression function (corresponding to each component of the response vector) is modeled using separate set of sum of regression trees. However, the dependence among the components of the response vector is established through the correlated error structure $\boldsymbol{e}_i = (e_{i1}, \ldots, e_{id}) \sim N_d(\mathbf{0}, \Sigma)$. The model presented in (3) represent a "sum-of-trees model." It is basically an additive model with m_j number of small trees. Each tree for the j-th component of the response is denoted by $g_k^j(\boldsymbol{x}_j)$ and for the j-th component we have in total m_j number of trees. In later part of this section, we will introduce a better and easy to understand notation for each of the individual tree model through which can explicitly show the underlying parameters in a tree model. Combining (2) and (3) we get the following formulation of our BART-SUR model

$$\begin{pmatrix} y_{i1} \\ y_{i2} \\ \vdots \\ y_{id} \end{pmatrix} = \begin{pmatrix} \sum_{k=1}^{m_1} g_k^1(\boldsymbol{x}_1) \\ \sum_{k=1}^{m_2} g_k^2(\boldsymbol{x}_2) \\ \vdots \\ \sum_{k=1}^{m_d} g_k^d(\boldsymbol{x}_d) \end{pmatrix} + \begin{pmatrix} e_{i1} \\ e_{i2} \\ \vdots \\ e_{id} \end{pmatrix}, \tag{4}$$

$$\boldsymbol{e}_i = (e_{i1}, \ldots, e_{id}) \sim N_d(\mathbf{0}, \Sigma).$$

Before we start describing the concept of the individual trees and the associated parameters we would like to give a proper justification of using trees in our additive model. Predictors like linear or polynomial regression are global models, where a single predictive formula is supposed to hold over the entire data space. However, when the data have lots of features which interact in nonlinear and complex ways, a single global model hardly succeeds in accurately predicting the response outcome. An alternative approach to nonlinear regression is to subdivide the feature space into small subregions, where the interactions are easy to manage and model. We then partition the subpartitions again and keep doing that until we get to a section of space which can be modeled using a very simple model. This process is also known as recursive partitioning. Therefore the full global model has two distinct components: one is the recursive partition and other is the simple model in each of the final partitions. The global model we get by this process is also known as "tree model" (Breiman et al., 1984). In our BART-SUR model we have not one single tree but a collection or ensemble of many small trees. A nice analogy of that in the frequentist literature is the RF (Breiman, 2001).

In a single tree model, we have a set of binary rules with which the recursive partitioning is done on the predictor space known as the internal nodes. The final partitions are known as the terminal nodes, where simple models are assigned. In this chapter, T will represent the binary tree with all splitting rules and $M=\{\mu_1, \ldots, \mu_b\}$ represents the set of parameter values corresponding to each of the b terminal nodes. As the main idea in any tree model is to use a very simple estimator in the final partition often we use only the mean based on the data points in the particular subspace or the terminal node as the "simple model."

For the sake of simplicity and easy parameter interpretation from now onward we are going to denote the k-th single tree for the j-th component by $g_k^j\left(\boldsymbol{x}_j; T_k^j, M_k^j\right)$. Rewriting (4) with this new notation we get

$$\begin{pmatrix} y_{i1} \\ y_{i2} \\ \vdots \\ y_{id} \end{pmatrix} = \begin{pmatrix} \sum_{k=1}^{m_1} g_k^1\left(\boldsymbol{x}_1; T_k^1, M_k^1\right) \\ \sum_{k=1}^{m_2} g_k^2\left(\boldsymbol{x}_2; T_k^2, M_k^2\right) \\ \vdots \\ \sum_{k=1}^{m_d} g_k^d\left(\boldsymbol{x}_d; T_k^d, M_k^d\right) \end{pmatrix} + \begin{pmatrix} e_{i1} \\ e_{i2} \\ \vdots \\ e_{id} \end{pmatrix}, \tag{5}$$

$$\boldsymbol{e}_i = (e_{i1}, \ldots, e_{id}) \sim N_d(\boldsymbol{0}, \Sigma); \quad i = 1, \ldots, n.$$

Our BART-SUR model as presented in Eq. (5) is a direct extension of the original BART (Chipman et al., 2010). With a large number of trees (m_j), BART-SUR model can encompass a large class of linear and nonlinear functions with excellent out of sample predictive performance. It is important to notice that for each component we have m_j number of small trees. The value of m_j can be same or different for all the components. We believe that to have better flexibility m_j should be different for each of the component of our BART-SUR model. Different set of predictor value may require different degree of model complexity to predict the corresponding response component. For example, some component may need fewer trees (small value of m_j) whereas for others we may need to resort to large number of trees. In the original BART model the number of trees needed is fixed and is supplied by the user. Chipman et al. (2010) recommended using large number of trees. That means for an SUR model we need to fix the number of trees for each of the components. However, in this chapter we treat m_j (the number of trees for the j-th component) as random parameter of the model and put a prior distribution on it. This helps us adaptively select the number of trees required for each components.

In our BART-SUR model the model parameters are denoted by $\Theta=\{\Theta_1, \ldots, \Theta_d, \Sigma\}$. The $\Theta_j=\left\{\left(T_1^j, M_1^j\right), \ldots, \left(T_{m_j}^j, M_{m_j}^j\right), m_j\right\}$ is the set of tree parameters for the j-th component ($j=1, \ldots, d$). To be noted here that since the

TABLE 1 Model Parameters of BART-SUR

m_j:	The number of trees for modeling the *j*-th response component.
T_k^j:	Parameters for the binary regression trees for the *k*-th tree of the *j*-th component.
M_k^j:	Parameter values for the terminal nodes for the *k*-th tree of the *j*-th component.
Σ:	Covariance matrix for the error vector **e**.

number of trees m_j is kept random the dimension of the parameter space Θ_j is changing conditional to the variable m_j. Since our BART-SUR model contains a large number of parameters to avoid any confusion and increase clarity and interpretation, we provided in Table 1.

2.2 Prior Distribution

In this section we specify priors distribution on the model parameters as described in Table 1.

Prior distributions on $\left(T_k^j, M_k^j\right)$: We will follow the prior distribution for $\left(T_k^j, M_k^j\right)$ as suggested by Chipman et al. (2010). The prior distribution of the tree parameters for the *j*-th component conditional on the number of trees m_j are assigned as

$$\pi\left(\left(T_1^j, M_1^j\right), \ldots, \left(T_{m_j}^j, M_{m_j}^j\right) | m_j\right) = \prod_{k=1}^{m_j} \pi\left(M_k^j | T_k^j\right) \pi\left(T_k^j\right)$$
$$\text{and } \pi\left(M_k^j | T_k^j\right) = \prod_l \pi\left(\mu_{kl}^j | T_k^j\right). \tag{6}$$

This prior structure implies independent prior on each individual trees. The priors for $\pi\left(T_k^j\right)$ in (6) are assigned following the suggestion in Chipman et al. (2010) such that (i) the distribution on the assignment of the splitting variable at each internal node is a uniform distribution over all available variables corresponding to the particular response component. That means we pick one variable out of p_j variables corresponding to the *j*-th component for splitting with equal probability; (ii) the splitting rule in each interior node (conditional on the splitting variable) follows a discrete uniform distribution over the set of available splitting values corresponding to the particular response component; and (iii) On all the nodes across all *j*-components we assign iid probability that a node at depth h is nonterminal is given by $a(1+h)^{-c}$, where $0 < a < 1$ and $0 \leq c < \infty$ are the parameters that controls the size of the tree. Conditional on T_k^j, on the μ_{kl}^j parameter we assign conjugate normal distribution $N\left(0, \frac{0.25}{k^2 m_j}\right)$. This particular prior induces the fact that the leaf nodes will be shrunk to zero and as we increase the number of trees m_j the contribution from one individual tree decreases. This mimics the concept of weak leathering strategy (Schapire, 1990). In our chapter we fixed $k=2$, $a=0.5$, $c=2$. This ensures each individual tree in our BART-SUR model is small and have moderate shrinkage.

Prior distributions on Σ: On the error variance–covariance matrix Σ we put an inverse Wishart distribution as $\Sigma \sim IW(r, R)$. Where $r > d$ and R is a $d \times d$ positive definite matrix. A good default choice for the hyperparameters is to set $r=d+1$ and $R=r\boldsymbol{I}$, where $\boldsymbol{I}$ is the identity matrix. This choice of the prior parameter produces the least informative prior structure on Σ.

Prior distribution on m_j: In the original BART model (Chipman et al., 2010) the authors kept the number of trees fixed and argued that the BART model fits well as long as m is large enough. Increasing m further caused only a gradual increase in the out-of-sample mean squared error (MSE). Since the amount of shrinkage on the leaf-node values depends on m_j, increasing the number of trees m_j lessens the effect of over-fitting. While it may be relatively safe to set m_j to a very large value as far as model fitting and prediction is concerned, unfortunately, a value too large will unnecessarily cause the algorithm to run much slower. To overcome that problem we put prior distributions on the number of trees m_j for each of the j-components of our BART-SUR model as follow

$$P\left[m_j = k\right] = p_{jk}, \tag{7}$$

for $k = m_{j,\min}, \ldots, m_{j,\max}$. Sufficiently large support for m_j ensures that the model can grow freely as required from the data. In this chapter, we recommend to use the default prior for the number of trees as $m_{j,\min} = \max\left\{2, \left\lceil \frac{10}{d-1} \right\rceil\right\}$ and $m_{j,\max} = \max\left\{2, \left\lceil \frac{300}{d-1} \right\rceil\right\}$ with the prior probabilities p_{jk} being proportional to 2^{-m_j} to aggressively shrink the number of trees down. This encourages our BART-SUR model to select small ensemble of trees and thus produces much faster result than the original BART model. Nevertheless, if the data have a strong need of large number of trees, the prior shrinkage on the number of trees will be overwhelmed and the model will end up selecting sufficiently large number of trees.

3 FITTING BART-SUR MODEL THROUGH MCMC

In this section we develop an efficient RJMCMC (Green, 1995) algorithm to generate samples from the posterior distribution of our BART-SUR model. Here, we require the RJMCMC because the dimension of our parameter space is changing with different values of the number of trees m_j. The original

BART model is fitted using a novel Gibbs sampling scheme called Bayesian backfitting. In this chapter, we used the Bayesian backfitting method (Chipman et al., 2010) as a template to develop our own algorithm. The step by step algorithm to fit our BAT-SUR model is as follow:

Step 1. Start with an initial value of our model parameters Θ^0.

Step 2. Select the j-th component of the response vector.

Step 3. Since the dimension of our parameter space is changing the Gibbs sampler is replaced with a reversible jump step that proposes to (1) add a tree, (2) delete a tree, or (3) update the current set of trees for the j-th component. To increase m_j, a new tree, u, is drawn from the prior distribution and is inserted into the end of current list of trees for the j-th component, θ_j. Decreasing m_j removes the current last tree in the list. The deterministic transform functions for these two steps are then:

$$g_{m_j \to m_j+1}\left(\theta_j^{(t)}, u^{(t)}\right) = \left(\left(\theta_j^{(t)}, u^{(t)}\right), \varnothing\right)$$
$$g_{m_j \to m_j-1}\left(\theta_j^{(t)}, \varnothing\right) = \left(\theta_{-m_j}^{(t)}, \theta_{m_j}^{(t)}\right).$$

Both of the above transform function have an absolute value of the determinate of the Jacobin equal to $\left|J_{m_j}\right| = 1$. By drawing the u from the prior, the tree priors cancel out in the acceptance probability, and we are left with a likelihood ratio of the fits under the two set of trees times the ratio of the priors on the number of trees. The remaining step of updating the current set of trees proceeds exactly as in the case with m_j fixed, its transform function is:

$$g_{m_j \to m_j}\left(\theta_j^{(t)}, \theta^\star\right) = \left(\theta^\star, \theta_j^{(t)}\right)$$

Here, the Jacobin is a permutation matrix, so the absolute value of its determinate is 1 as well. The $\theta^\star$ is a proposal from the same full conditional as in the original BART; everything cancels in the acceptance probability as usual with full conditionals, so $\theta^\star$ is always accepted if we perform this step.

Note, the add and delete steps should have the same dimensions as the update step; u could be viewed as the very first tree in a new set of trees, but since those additional trees (1) do not appear in the likelihood, nor (2) change the absolute value of the determinate of the Jacobin, and (3) the tree priors cancels out, they disappear from the acceptance probability entirely, so it is unnecessary to actually draw them.

If the current number of trees reaches the maximum possible, then the add step occurs with probability 0, with the extra probability going to the update step. Similarly if the number of tree reaches the minimum, then the delete step never occurs. At the beginning of the

MCMC chain for a user-specified number of iterations, the number of trees is held fixed at the initial value so that the fit settles down a little before the add, delete, and swap move starts.

Step 4. Conditional on m_j and Σ the parameters related to the j-th component of our BART-SUR model $\Theta_j = \left\{ \left(T_1^j, M_1^j\right), \ldots, \left(T_{m_j}^j, M_{m_j}^j\right) \right\}$ is updated using the Bayesian backfitting MCMC algorithm described in Chipman et al. (2010). For detail description of our modification of Bayesian backfitting algorithm for our BART-SUR model please see Appendix A.

Step 5. Go back to Step 2 and cycle through all the d components of our response vector.

Step 6. Conditional on the rest of the parameters and the data the posterior distribution of Σ is an Inverse Wishart distribution. We sample directly from

$$\Sigma|\ldots \sim \mathrm{IW}_d(r+n, R^*), \tag{8}$$

where $R^* = R + \sum_{i=1}^{n} (\boldsymbol{y}_i - \boldsymbol{h}_i)(\boldsymbol{y}_i - \boldsymbol{h}_i)^T$, $\boldsymbol{h}_i^T = \left(\sum_{k=1}^{m_1} g_k^1\left(\boldsymbol{x}_{i1}; T_k^1, M_k^1\right), \ldots, \sum_{k=1}^{m_d} g_k^d\left(\boldsymbol{x}_{id}; T_k^d, M_k^d\right)\right)$.

Step 7. We keep on cycling through Step 2–6 until we attain convergence.

4 SIMULATION STUDIES

In this section we propose two simulation scenarios to simulate data sets and apply our BART-SUR model. We compare the performance of our BART-SUR model against several state of the art machine learning models to be discussed in later part of this section.

Simulation Study 1: Here, we simulate data sets from $d=3$ dimensional nonlinear SUR model. We assume that the each covariate set has $p_j = 10$ number of different predictor variables in it, where $j=1, 2, 3$. Therefore we have the following three equation SUR model.

$$y_j = \boldsymbol{x}_j \boldsymbol{\beta}_j + e_j, \quad j = 1, \ldots, 3,$$

where the covariate vector $\boldsymbol{x}_j = (x_{j1}, \ldots, x_{j10})^T$ is generated from a Gaussian process. The $x_{jk} \in (-2, 2)$ and the covariance function is given by $k\left(\boldsymbol{x}_{jl}, \boldsymbol{x}_{jr}\right) = \exp\left(\frac{-|\boldsymbol{x}_{jl} - \boldsymbol{x}_{jr}|^2}{c}\right)$ and $c = 2 * j$. Therefore we notice for each of the three response component the underlying Gaussian process is different. We generate the regression coefficients $\boldsymbol{\beta}_j = (\beta_{j1}, \ldots, \beta_{j10})^T$ from an *Uniform*$[0, 1]^{10}$, $j=1, 2, 3$. The error vector $\boldsymbol{e} = (e_1, e_2, e_3)$ is generated from multivariate normal distribution $N_3(\boldsymbol{0}, \boldsymbol{V})$, where the variance covariance matrix $\boldsymbol{V} = ((v_{rs}))$ has variance $v_{rr}=1$ and the covariances $v_{rs}=0.8$, r, $s=1$, 2, 3. This simulation setting is mimicking a typical SUR scenario where on a single individual we have a three variate response and for each component of the response we have separate set of the covariates or predictors. All three response components are connected through an underlying correlated error structure.

Simulation Study 2: Here, we simulate data sets from a $d=4$ dimensional nonlinear model. The nonlinear functions connecting the response variables are four different variations of the Friedman's 5 dimensional test function ($p_j=5$).

$$y_1 = 10\sin(\pi x_1 x_2) + 20(x_3 - 0.5)^2 + 10x_4 + 5x_5 + e_1$$

$$y_2 = 5\sin(\pi x_1 x_2) + 10(x_3 - 0.5)^4 + 8x_4 + 3x_5 + e_2$$

$$y_3 = 20\sin(\pi x_1 x_2) + 15(x_3 - 0.5)^2 + 30x_4^2 + 15x_5 + e_3$$

$$y_4 = 15\sin\left(\pi x_1^2 x_2\right) + 10(x_3 + 0.5)^2 + 5x_4 + 15\sqrt{x_5} + e_4,$$

where x_1, x_2, x_3, x_4, x_5 iid $U(0, 1)$ and $\boldsymbol{e}=(e_1, e_2, e_3, e_4)$ are generated from multivariate normal distribution $N_4(\mathbf{0}, \boldsymbol{V})$, where the variance covariance matrix $\boldsymbol{V}$ is chosen as $v_{rs}=0.8^{|r-s|}$, r, $s=1$, …, 4. To be noted here that under this simulation scenario the five covariates are same for all components of the response; however, the underlying functions that connects the predictor variables with the response are different for each component. Also, all components of the response vector are connected between each other via correlated normal error.

In both the simulation settings we scale the response variable and the covariates to have zero mean and unit variance. Under each simulation setting we have generated $n=150$ data points. We keep 100 data points for training and 50 data points in the validation or test set. We repeat each simulation scenario 100 times. In Tables 2 and 3, we report the average mean squared errors of prediction (MSEP) corresponding to each component of the response vector for Simulation Study 1 and Simulation Study 2, respectively. We compare the performance of our BART-SUR model against several modern nonlinear models like (i) RF, (ii) BART, (iii) SVM, (iv) Bayesian linear SUR model (BL-SUR), (v) BWR, and (vi) Gaussian process regression (GPR). Out of the mentioned six competing methods (iv), (v), and (vi) can jointly estimate all the components of the response. Whereas, for methods (i)–(iii) each component of our response vector is modeled separately. Therefore methods (i)–(iii) do not takes into account the underlying correlation structure among the responses. The competing methods are all very well documented procedures. To keep the chapter at a reasonable length we decided not to include any more explanation about the individual methods. However, interested readers can refer to Hastie et al. (2001) for detail description. The SVM is fitted using the *svm*() function in R with polynomial kernel, and RF is fitted using the *randomForest*() function in R. Each of the mentioned models requires some tuning parameters to be adjusted to get the best prediction result. For

TABLE 2 Simulation Study 1: Average MSEP (and SD) in the Test Set

	y_1	y_2	y_3
RF	0.23	0.18	0.16
	(0.03)	(0.04)	(0.03)
BART	0.14	0.15	0.11
	(0.04)	(0.05)	(0.03)
SVM	0.14	0.21	**0.10**
	(0.06)	(0.04)	(0.02)
BL-SUR	0.41	0.32	0.35
	(0.07)	(0.07)	(0.05)
BWR	0.15	0.11	0.18
	(0.03)	(0.05)	(0.04)
GPR	0.18	1.32	0.14
	(0.04)	(0.05)	(0.04)
BART-SUR	**0.10**	**0.09**	**0.10**
[1.02, 1.16]	(0.03)	(0.02)	(0.03)

The range of calculated Gelman–Rubin diagnostic scores is reported in [ ] for our BART-SUR. The number reported in ( ) is the standard deviations. The three columns correspond to three components of the response vector. The bold numbers are the lowest MSEP.

RF it is the number of generated trees, for SVM with polynomial kernel it is the degree of the polynomial. We determine these tuning parameters by doing a fivefold cross-validation on the training data. We have used the *mlegp*() package in R to fit GPR model for multidimensional outputs. For Bayesian models like BL-SUR, BART, and BWR we choose the priors as suggested by the authors in their respective original manuscripts. The BL-SUR model is fitted using the *rsurGibbs*() function in the *Bayesm* R library. The BART model is fitted with 300 trees using the *bart*() function in the *BayesTree* R library. The BWR model is fitted using the matlab code as provided by the authors Brown et al. (2001) from http://www.stat.rice.edu/~marina/publications.html. To maintain the continuity of this chapter and ease to follow the draft more detail, description of all chosen model parameters are relegated to Appendix B.

To fit our BART-SUR model we run 20,000 MCMC iterations and use four independent chains with different starting points to check for any possible problem with multimodality. We use the first half of our MCMC chain as the burn in and use the postburn in samples to calculate the predicted values of the response in the test set and other model parameters. The overall

TABLE 3 Simulation Study 2: Average MSEP (and SD) in the Test Set

	y_1	y_2	y_3	y_4
RF	0.14	0.13	0.18	0.14
	(0.05)	(0.04)	(0.05)	(0.04)
BART	0.09	0.08	0.09	0.08
	(0.03)	(0.02)	(0.02)	(0.03)
SVM	0.14	0.10	0.09	0.10
	(0.06)	(0.04)	(0.04)	(0.04)
BL-SUR	0.21	0.26	0.18	0.22
	(0.04)	(0.06)	(0.05)	(0.04)
BWR	0.13	0.10	0.11	0.09
	(0.03)	(0.04)	(0.03)	(0.03)
GPR	0.33	0.22	0.15	0.19
	(0.08)	(0.06)	(0.06)	(0.05)
BART-SUR	**0.07**	**0.06**	**0.07**	**0.06**
[1.04, 1.20]	(0.02)	(0.03)	(0.02)	(0.02)

The range of calculated Gelman–Rubin diagnostic scores is reported in [ ] for our BART-SUR. The number reported in ( ) is the standard deviations. The four columns correspond to four components of the response vector. The bold numbers are the lowest MSEP.

convergence is checked using Gelman–Rubin diagnostic in the *coda* package in R. It is evident that different data set requires different number of MCMC iterations to fit a model. Through the Gelman–Rubin diagnostic we found out that on average 20,000 MCMC iterations yields very good convergence. So in all our simulation studies and also real data analysis (in next section) we have ran 20,000 MCMC iterations. All prior parameters are chosen following the guideline mentioned in Section 2.2.

The results in Tables 2 and 3 indicate that our BART-SUR model has significantly lower average MSEP than all other standard methods. In Simulation Study 1 our BART-SUR model produced on average 28% and 18%, lower MSEP for y_1 and y_2 components, respectively, than the best competing method. For the component y_3 it has matched the best performing SVM model. In Simulation Study 2 the covariates are connected to the response vector through the Friedman's nonlinear functions. We can see in Simulation Study 2 also our BART-SUR resulted in lowest average MSEP (at least 22%, 25%, 22%, and 25% lower average MSEP) than the best standard method which is the BART. This strongly indicates the fact that when there are nonlinearity of the regression function BART and BART-SUR works much better than the existing other methods. This also testifies to the fact that our additive tree model can capture the complex nonlinear functional relationship between the covariates and the response vector. Comparing BART and our BART-SUR we see that in the presence of correlated multidimensional responses we gain significant prediction accuracy through our BART-SUR model. On the other hand BART which fits each of the response components independently cannot explore the underlying correlation structure and hence has poorer prediction performance than BART-SUR. Another major advantage of our BART-SUR is the automatic tree selection. Through our assigned

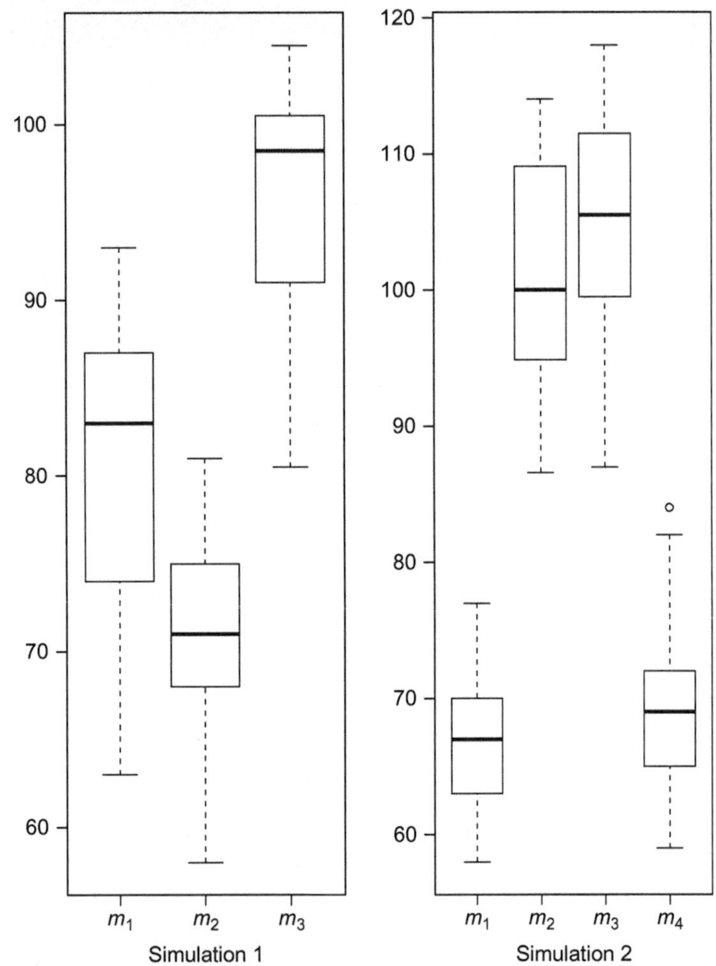

FIG. 1 Boxplot of the posterior means of number of trees m_j across 100 replications for Simulation Study.

prior on the number of trees (mj) we can adaptively select the required number of trees. In Fig. 1, we provide a boxplot of the posterior mean number of trees selected for each components of the response vector for Simulation Study 1 and Simulation Study 2 across all 100 replications. From Fig. 1, it is clear that different components require different number of trees in our BART-SUR model. In Simulation Study 2, the second and the third component require at least 100 trees, whereas in the first and the fourth component only 70 trees gave us the best prediction. This adaptive tree selection through prior adds to our model flexibility. Depending on the complexities of the underlying regression function our model can self-determine how many trees it really need to attain the best prediction accuracy.

5 DATA ANALYSIS

In this section, we implement our BART-SUR model on three real life data sets that have multiple correlated response variables. The first data set arise from a chemometric study (Osborne et al., 1984), the second data set is originated from a study in educational psychology (Timm, 1975), and the third is an econometric data set (Baltagi, 2005).

5.1 Biscuit Dough Data Set

This data example is obtained from a chemometric experiment done by Osborne et al. (1984). The objective of the study was to find out the composition of biscuit dough. The four compounds studied in this analysis are fat, sugar, flour, and water. Therefore here our response is a vector of length four or in our earlier model notation $d=4$. We want to accurately predict the percentage of these four compounds based on their near infra-red reflectance signature. On each sample $p=256$ spectral points are recorded as predictor variables over wavelengths ranging from 1380 nm to 2400 nm with a gap of 4 nm. Due to several factors like autofluorescence, scattering of light, and nonmonochromatic radiation the linearity assumption is violated and its is very difficult to predict the compositions based on the NIR spectrum. Moreover since all four components are based on the same material there is a strong correlation between the four compounds. In this data set, we have 78 samples in total. Of that we randomly chose two-third as the training set and one-third as the test set. We repeat this 100 times. To be noted here that for this chemometric data we have one set of NIR measurement as the covariate whereas the response is a vector of four components. So this is a special case of an SUR model. In Table 4, we report the average MSEP and the standard errors of the MSEP for the biscuit dough data across all 100 splits in the test set. Our proposed BART-SUR model outperformed all the other methods by a significant margin. Moreover from Fig. 2 it is clear that the four components fat, sugar, flour, and water have different degree of nonlinear

TABLE 4 Biscuit Dough Data: Average MSEP (and SD) in the Test Set

	Fat	Sugar	Flour	Water
RF	0.26	0.19	0.27	0.26
	(0.15)	(0.15)	(0.12)	(0.07)
BART	0.13	0.23	0.28	0.23
	(0.05)	(0.07)	(0.04)	(0.04)
SVM	0.36	0.34	0.23	0.12
	(0.12)	(0.11)	(0.06)	(0.03)
BL-SUR	0.42	0.39	0.60	0.33
	(0.14)	(0.16)	(0.12)	(0.12)
BWR	0.10	0.20	0.26	0.09
	(0.03)	(0.06)	(0.08)	(0.02)
GPR	0.27	0.38	0.32	0.26
	(0.04)	(0.06)	(0.08)	(0.06)
BART-SUR	**0.08**	**0.15**	**0.18**	**0.08**
[1.03, 1.28]	(0.01)	(0.03)	(0.03)	(0.03)

The range of calculated Gelman–Rubin diagnostic scores is reported in [ ] for our BART-SUR. The number reported in ( ) is the standard deviations. The four columns correspond to four components (fat, sugar, flour, water) of the response vector. The bold numbers are the lowest MSEP.

relationship with the NIR spectra and because of that our BART-SUR model has adaptively selected on average different number of trees for each of the components. For fat, sugar, and flower on average our BART-SUR model picked up over 100 trees, whereas for water it needed less than 80 trees. Adaptive selection of trees eliminate the problem of choosing large number of trees as done in the original BART model.

5.2 Rohwer Data Set

This data set is obtained from an experiment by William D. Rohwer (Timm, 1975) on 69 kindergarten children. We would like to investigate how well some performance measures of aptitude and achievement can be predicted with the help of the performance on a set of paired-associate (PA) tasks. In this data, the response variables are Student Achievement Test (SAT), Peabody Picture Vocabulary Test (PPVT), and Raven Progressive Matrices Test (RPMT). The predictor variables are socioeconomic status (a factor with levels high and low), performance on a "named" PA task, performance on a

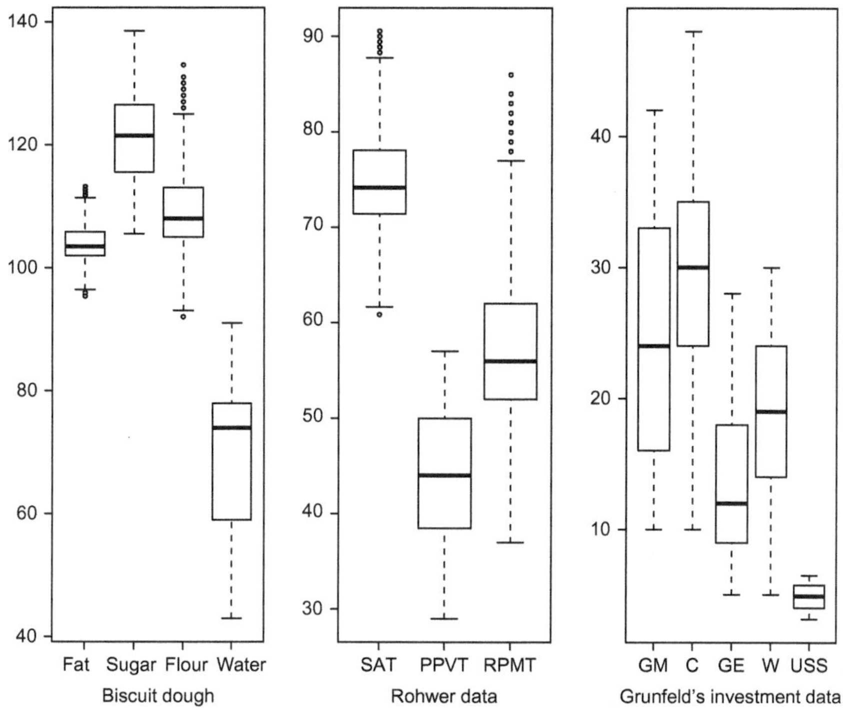

FIG. 2 Boxplot of the posterior means of number of trees m_j across 100 replications.

"still" PA task, performance on a "named still" PA task, performance on a "named action" PA task, and performance on a "sentence still" PA task. Therefore according our notation we have $d=3$ components for our response vector and we have collected same $p=6$ covariates on every individual. We have the same six covariate measurements corresponding to every component of the response vector. So this data set can also be considered as a special case of the SUR model. Before we fit the models we scale the predictor variables and the response variables. We split the data randomly using 2/3 of the samples as training set and the remaining 1/3 as the test set. We repeat the random splitting 100 times and the prediction accuracy of our BART-SUR model and other competing models are reported in Table 5. Based on the reported average MSEP in Table 5 we notice that our BART-SUR model has made significant improvement in predicting the SAT and RPMT scores over all competing models. For the component PPVT score the accuracy of BART-SUR is matched by the RF. From Fig. 2, we notice our model on average chose more than double the number of trees required for SAT as compared to the PPVT component of the response. This implicitly indicates the fact that the underlying relationship between SAT and the predictor variables is much more complex than that between PPVT and the same set of covariates.

TABLE 5 Rohwer Data: Average MSEP (and SD) in the Test Set

	SAT	PPVT	RPMT
RF	0.21	**0.18**	0.20
	(0.05)	(0.05)	(0.04)
BART	0.23	0.19	0.20
	(0.04)	(0.05)	(0.03)
SVM	0.29	0.31	0.22
	(0.03)	(0.06)	(0.03)
BL-SUR	0.43	0.28	0.25
	(0.05)	(0.04)	(0.06)
BWR	0.30	0.21	0.29
	(0.06)	(0.05)	(0.04)
GPR	0.22	0.25	0.20
	(0.03)	(0.02)	(0.04)
BART-SUR	**0.14**	**0.18**	**0.17**
[1.04, 1.12]	(0.03)	(0.04)	(0.04)

The range of calculated Gelman–Rubin diagnostic scores is reported in [ ] for our BART-SUR. The number reported in ( ) is the standard deviations. The three columns correspond to three components (SAT, PPVT, and RPMT) of the response vector. The bold numbers are the lowest MSEP.

5.3 Grunfeld's Investment Data

Here, we obtain a data set on five US manufacturing farms General Motors (GM), Chrysler (C), General Electric (GE), Westinghouse (W), and US Steel (USS) collected over a period of 20 years (1935–1954) (Baltagi, 2005). The response variables are the gross investment including the repairs and maintenance of each of these five companies. Therefore, according to our model notation here $d=5$ and $n=20$. The predictor variables are the market value of the firm and the firm's capital stock. Notice that for each of the component corresponding to a particular firm the set of predictor variable will be different as the market value and capital stock value will different for each of the individual company. Since all these five companies operate in United States and international market with similar goods and services there exists an underlying association between their gross investment values. So it is appropriate that the gross investment values of these five companies should be modeled jointly. Since in this data set we have only 20 observations instead of splitting it into test and training set we do leave one out cross-validation (LOOCV) and calculate the MSEP. In Table 6, we report the LOOCV error

TABLE 6 Grunfeld's Investment Data: LOOCV Error (and SD)

	GM	C	GE	W	USS
RF	0.10	0.08	0.05	0.07	**0.03**
	(0.009)	(0.02)	(0.01)	(0.01)	(0.007)
BART	0.09	0.08	0.10	**0.05**	**0.03**
	(0.02)	(0.02)	(0.03)	(0.008)	(0.005)
SVM	0.13	0.12	0.08	0.08	0.06
	(0.05)	(0.04)	(0.02)	(0.03)	(0.01)
BL-SUR	0.08	0.10	0.07	0.09	0.06
	(0.006)	(0.03)	(0.01)	(0.02)	(0.008)
BWR	0.12	0.09	0.10	0.13	0.05
	(0.04)	(0.03)	(0.03)	(0.02)	(0.01)
GPR	0.09	0.12	0.08	0.08	0.04
	(0.02)	(0.03)	(0.03)	(0.02)	(0.006)
BART-SUR	**0.05**	**0.07**	**0.05**	0.06	**0.03**
[1.01, 1.10]	(0.01)	(0.008)	(0.01)	(0.009)	(0.004)

The range of calculated Gelman–Rubin diagnostic scores is reported in [ ] for our BART-SUR. The number reported in ( ) is the standard deviations. The five columns correspond to the five components (GM, C, GE, W, and USS) of the response vector. The bold numbers are the lowest MSEP.

of our BART-SUR model and other competing models. From Table 6, it is clear that for the GM and C our BART-SUR gives significantly lower cross-validation error compared to all other methods. Whereas for GE our BART-SUR matches the same level of LOOCV error as the best performing standard model which is the RF. Same phenomenon is also observed for USS, our BART-SUR prediction is as good as the best performing BART and RF. Only in W component our BART-SUR MSEP is marginally above the BART model. However, it is to be noted here that BART-SUR through the adaptive tree selection scheme needs on average only 20 trees for the W component (Fig. 2), whereas in the original BART the number of recommended trees is set at 300.

6 CONCLUSION

In this chapter we have extended the BART for SUR model. In many modern analysis whenever the response is a vector quantity with the components (of the vector) somehow correlated among themselves it is appropriate to model and predict them jointly. In our proposed BART-SUR model we can jointly model for a multiple response regression problem. The regression function corresponding to each component of the response is modeled using ensemble of small trees. The dependency between each component of the multivariate response vector is established using a correlated random error structure and then putting an inverse Wishart prior on it. The prior structure on the tree parameters is chosen following Chipman et al. (2010). Our BART-SUR model can automatically select the number of required trees. The adaptive tree selection is done treating the number of trees as a model parameter and then putting a discrete prior distribution on it. In our model each component of the response variable can have separate number of trees. This adds to the model flexibility because depending on the complexity of the underlying true regression function our model can automatically decide the number of trees needed to fit each component. If for a specific component the underlying regression function does not show much nonlinear or local behavior BART-SUR will use less number of trees. On the other hand if for another component the underlying regression function shows many local and nonlinear behavior our BART-SUR will make sure that a large number of trees are added in the model to increase the prediction accuracy and overall model fit.

Our BART-SUR model is fitted using a Bayesian backfitting algorithm. Since the number of trees in our model is kept random the dimension of our parameter space is changing across different iterations of MCMC. In this chapter we proposed a novel RJMCMC scheme to navigate through different dimension of our parameter space. In the two presented simulation studies and three real data analysis our BART-SUR resulted in significantly lower prediction error compared to the available state of art linear and nonlinear models. Since we do not need to start with a very large number of trees our BART-SUR can run much faster than the standard BART where we have to assign a large number for the tree size. We have created a R package called *BartSur* based on our model which is easy to use. The *BartSur* package is currently works on LINUX platform and can be obtained from the authors upon request.

Due to vast potential of predictive modeling BART has recently received lots of attention. The automatic tree selection developed in our BART-SUR model can also be adopted for binary and multicategory data to solve classification problem. These are few open areas for future research.

APPENDICES

Appendix A Bayesian Backfitting Algorithm for BART-SUR Model

Conditional on Σ and m_j we can sample the $\Theta_j = \left\{\left(T_1^j, M_1^j\right), \ldots, \left(T_{m_j}^j, M_{m_j}^j\right)\right\} (j = 1, \ldots, d)$ parameters through the following steps.

Step (i) Select the BART regression function for the j-th component $h_j(\boldsymbol{x}_j)=\sum_{k=1}^{m_j} g_k^j\left(\boldsymbol{x}_j;T_k^j,M_k^j\right)$. Conditional on Σ, m_j, and $h_{l\neq j}$ we only need to update the $\Theta_j=\left\{\left(T_1^j,M_1^j\right),\ldots,\left(T_{m_j}^j,M_{m_j}^j\right)\right\}$ parameters.

Step (ii) Each individual tree parameters $\left(T_k^j,M_k^j\right)k=1,\ldots,m_j$ in the regression function for the j-th component $h_j(\boldsymbol{x}_j)$ is updated identically as mentioned in Chipman et al. (2010) through Bayesian backfitting. However, the only difference will be the posterior distribution of M_k^j conditional on all other parameters. Conditional on Σ, m_j, and $h_{l\neq j}$ the posterior distribution of M_k^j or $\mu_{kl}^j(l=1,\ldots)\overset{iid}{\sim}N\left(a_{kl}^j,b_{kl}^j\right)$. Here, the posterior mean a_{kl}^j and variance b_{kl}^j are calculated based on the collection of data points in the l-th node of the k-th tree for the j-th component using the normal-inverse gamma conjugate formula with priors $\mu_{kl}^j\sim N\left(0,\frac{0.25}{k^2m_j}\right)$. The conditional error variance is obtained from the conditional distribution of the j-th diagonal element of Σ given the all other remaining elements of the matrix Σ. This is effectively replacing the σ^2 component in BART algorithm by $\sigma_{j|l\neq j}^2$.

Appendix B Choice of the Tuning Parameters for the Competing Models

In this section, we explain in details what are the tuning parameters and how did we select those tuning parameters for the competing models like RF, SVM, GPR, BWR, BART, and Bayesian linear regression for seemingly unrelated regression (BL-SUR).

Random forest: For RF we fix the number of trees equal to 500. The *mtry* parameter which represents the number of variables randomly sampled as candidates at each split is selected using the *tuneRF*() function in R. Other parameters like *nodesize* and *maxnodes* are set to the default recommended values (Breiman, 2001).

Support vector machine: For SVM we use the *eps-regression* option for the *svm*() function in the R package *e*1071. We chose polynomial kernel since it gives the lowest prediction error. The polynomial kernel parameter like the degree of the polynomial is chosen based on fivefold cross-validation.

Gaussian process regression: We fit the GPR using the *mlegp*() function in R. We assume that the Gaussian process has a constant mean function. For our multiple response output we set the option in *mlegp*() function such that it fits independent Gaussian processes to principle component weights obtained through singular value decomposition of the output. The *nugget* parameter is set to 1 for stability of our computation.

Bayesian additive regression tree: We fit the BART using the *bart*() function in the *BayesTree* package in R. We set *ntree* = 300 to make sure sufficiently large number of trees are used in the BART model. The degrees of freedom for the error variance is set to 3 to have near diffuse proper prior. All remaining parameters are chosen as recommended in Chipman et al. (2010).

Bayesian wavelet regression: We fit the BWR using the matlab code as provided from http://www.stat.rice.edu/~marina/publications.html. For BWR we run four independent chains of 50,000 iterations and discarded the first 25,000 as burn in. All prior parameters are chosen such that we have near diffuse proper prior as recommended in Brown et al. (2001).

Bayesian linear regression for SUR: We fit the Bayesian Linear Regression For SUR model using the *rsurGibbs*() function in the *bayesm* package in R. In the *rsurGibbs*(), function the prior parameter for the regression coefficients are chosen such that it has a large variance. The inverse Wishart prior for the variance covariance matrix Σ is set such that the degrees of freedom is $d+3$ and the scale matrix is $(d+3)$ $*\boldsymbol{I}$. The *rsurGibbs*() is fitted with four independent chains of 50,000 iterations with the first half as burn in.

REFERENCES

Ando, T., Zellenr, A., 2010. Hierarchical Bayesian analysis of the seemingly unrelated regression and simultaneous equations models using a combination of direct Monte Carlo and importance sampling techniques. Bayesian Anal. 5, 65–96.

Baltagi, B.H., 2005. Econometric Analysis of Panel Data, third ed. John Wiley, Chichester.

Bonato, V., Baladandayuthapani, V., Broom, B.M., Sulman, E.P., Aldape, K.D., Do, K.A., 2011. Bayesian ensemble methods for survival prediction in gene expression data. Bioinformatics 27, 359–367.

Breiman, L., 1996. Bagging predictors. Mach. Learn. 24, 123–140.

Breiman, L., 2001. Random forests. Mach. Learn. 45, 5–32.

Breiman, L., Friedman, J., Olshen, R., Stone, C., 1984. Classification and Regression Trees. Wadsworth, Belmont, CA.

Brown, P.J., Fearn, T., Vannucci, M., 2001. Bayesian wavelet regression on curves with application to a spectroscopic calibration problem. J. Am. Stat. Assoc. 96, 398–408.

Chakraborty, S., Ghosh, M., Mallick, B., 2012. Bayesian non-linear regression for large p small n problems. J. Multivar. Anal. 108, 28–40.

Chipman, H., George, E., McCulloch, R., 2010. BART: Bayesian additive regression trees. Ann. Appl. Stat. 4 (1), 266–298.

Denison, D., Holmes, C., Mallick, B., Smith, A.F.M., 2002. Bayesian Methods for Nonlinear Classification and Regression. Wiley, London.

Geladi, P., Dabakk, E., 2009. Computational Methods and Chemometrics in Near Infrared Spectroscopy, Encyclopedia of Spectroscopy and Spectrometry. Academic Press, Oxford.

Green, P.J., 1995. Reversible jump Markov chain Monte Carlo computation and Bayesian model determination. Biometrika 82, 711–732.

Hasenauer, H., Monserud, R.A., Gregoire, T.G., 1998. Using simultaneous regression techniques with individual-tree growth models. For. Sci. 44 (1), 87–95.

Hastie, T., Tibshirani, R., Freidman, J., 2001. The Elements of Statistical Learning. Springer, New York, NY (Chapter 3).

Holmes, C.C., Denison, D.G.T., Mallick, B.K., 2002. Accounting for model uncertainty in seemingly unrelated regressions. J. Comput. Graph. Stat. 11, 33–51.

Ingle, J.D.J., Crouch, S.R., 1988. Spectrochemical Analysis. Prentice Hall, New York, NJ.

Kapelner, A., Bleich, J., 2014. Prediction with missing data via Bayesian additive regression trees. Can. J. Stat. 43, 224–239. ArXiv e-prints.

Neal, R.M., 1996. Bayesian Learning for Neural Networks. Springer-Verlag, New York, NY.

O'Donnell, C.J., Shumway, C.R., Ball, V.E., 1999. Input demands and inefficiency in US agriculture. Am. J. Agric. Econ. 81, 865–880.

Osborne, B.G., Fearn, T., Miller, A.R., Douglas, S., 1984. Application of near-infrared reflectance spectroscopy to compositional analysis of biscuits and biscuit doughs. J. Sci. Food Agric. 35, 99–105.

Park, E.S., Oh, M.S., Guttorp, P., 2002. Multivariate receptor models and model uncertainty. Chemom. Intell. Lab. Syst. 60, 49–67.

Rasmussen, C.E., Williams, C., 2006. Gaussian Processes for Machine Learning. MIT Press, Cambridge, MA.

Ripley, B.D., 1996. Pattern Recognition and Neural Networks. Cambridge University Press, Cambridge, UK.

Samonte-Tan, G.P.B., Davis, G.C., 1998. Economic analysis of stake and rackhanging methods of farming oysters (Crassostrea iredalei) in the Philippines. Aquaculture 160, 239–249.

Schapire, R.E., 1990. The strength of weak learnability. Mach. Learn. 5, 197–227.

Schölkopf, B., Smola, A., 2002. Learning with Kernels. MIT Press, Cambridge, MA.

Timm, N.H., 1975. Multivariate Analysis with Applications in Education and Psychology. Wadsworth (Brooks/Cole), Monterey, CA.

Tipping, M., 2000. The relevance vector machine. In: Solla, S., Leen, T., Muller, K. (Eds.), Neural Information Processing Systems—NIPS. 12. MIT Press, Cambridge, MA, pp. 652–658.

Wahba, G., 1999. Support vector machines, reproducing Kernel Hilbert spaces and the randomized GACV. In: Schölkopf, B., Burges, C., Smola, A. (Eds.), Advances in Kernel Methods. MIT Press, Cambridge, MA, pp. 69–88.

Wilde, P.E., McNamara, P.E., Ranney, C.K., 1999. The effect of income and food programs on dietary quality: a seemingly unrelated regression analysis with error components. Am. J. Agric. Econ. 81 (4), 959–971.

Zellner, A., 1962. An efficient method of estimating seemingly unrelated regressions and testing for aggregation bias. J. Am. Stat. Assoc. 57, 348–368.

Zhang, S., Shih, Y.C.T., Muller, P., 2007. A spatially-adjusted Bayesian additive regression tree model to merge two datasets. Bayesian Anal. 2, 611–634.

Section C

Applications

Chapter 8

Cognitive Systems for the Food–Water–Energy Nexus

V.P.A. Lonij[1] and J.-B. Fiot
IBM Research, Dublin, Ireland
[1]Corresponding author: e-mail: vincentl@ie.ibm.com

ABSTRACT

Providing for the food, water, and energy needs of a growing world population is a grand challenge. The way we choose to address this challenge as a society will have far-reaching impacts, for instance, on public health, national security, and the global climate. The interactions between the food, water, and energy systems offer us an opportunity to improve efficiency, but can also make the system more complex. Cognitive systems can help mitigate this complexity and thus improve efficiency.

What food, water, and energy have in common is that they are often not produced where they are consumed, they are costly to transport, and they are hard to store efficiently in large quantities. And this is where cognitive computing comes in: if you cannot store a resource you must have good forecasts of supply and demand. This requires handling large scale datasets from multiple sources, using machine learning methods to build forecasting models, and leveraging optimization techniques to help incorporate forecasting results into a decision-making process.

We will use energy as an example throughout the bulk of this chapter with the understanding that the same methods, challenges, and solutions can be applied more broadly to food, water, and other constrained resources.

Sense: We will first discuss methods to make the most of sensor data. For example, it is expensive to deploy large networks of ground sensors (e.g., weather stations), and it is therefore beneficial to make use of sensors that can cover large areas, like radar or satellite images. However, it is challenging to estimate ground conditions based on satellite measurements alone. We will discuss machine learning algorithms that can learn the mapping from wide area sensor data to local conditions based on only a few ground-truth measurements. Applications include, for example, to estimate rainfall based on radar measurements, or to estimate solar power generation based on satellite images of clouds.

Predict: Next, we will discuss forecasting methods ranging from a few minutes ahead to days or even years ahead. We will discuss forecasting methods for energy demand, solar energy generation, and wind generation. Different methods are effective for these different technologies as well as different forecasting horizons. Very short-range forecasts might use autoregressive models, while mid- to long-range forecasts

Handbook of Statistics, Vol. 35. http://dx.doi.org/10.1016/bs.host.2016.07.003

may require physical models. We also discuss hybrid methods that combine expert knowledge with machine learning.

React: Finally, we will discuss how to use the outputs of these analytics tools to serve decision-making processes. How optimization can help improve infrastructure planning and economic dispatch of power generation. And how cognitive systems can help system operators to make sense of messy data from multiple sources, provide recommended actions, and enable better decision making.

We will highlight several different mathematical methods, including autoregressive models, generalized additive models, fully connected neural networks, deep learning, convolutional neural networks, and nonlinear optimization in the context of energy. We will conclude the chapter with an outlook on how current trends in the cognitive computing will impact the broader challenge of managing constrained resources.

Keywords: Cognitive computing, Machine learning, Resource management, Forecasting, Neural networks, Energy, Water, Food

1 INTRODUCTION

Food, water, clean air, and energy: these are all resources that we cannot live without. On a foundational level they are needed for human survival, but their relative abundance in developed countries also forms the basis of our modern lifestyles. Many of us can hardly imagine not being able to turn on a light by simply flipping a switch or not having clean running water available in every household. This is, however, not the case in many parts of the world. And even in developed nations, managing a reliable supply of these constrained resources is becoming more costly and more complex.

In this chapter we will first discuss what makes a resource constrained and how the constrained nature of food, water, and energy leads to global grand challenges. We will then discuss why cognitive systems have the potential to help solve these challenges. We will use energy as an example throughout the bulk of this chapter with the understanding that the same methods, challenges, and solutions can be applied more broadly to other constrained resources.

This chapter is intended to be understandable both to those with a background in resource management looking for an inventory of cognitive tools, as well as those with a background in cognitive computing looking for a new fruitful field for new applications. We will, therefore, discuss methods and applications as well as the key challenges and what makes these challenges important to solve. We will review the state of the art in cognitive computing for energy applications and provide an outlook on ongoing and future work.

1.1 Constrained Resources

Demand for food, water, and energy will grow by approximately 35%, 40%, and 50%, respectively, owing to an increase in the global population and the consumption patterns of an expanding middle class (National Intelligence Council, 2012). What these constrained resources have in common is that they are often not produced where they are consumed, they are costly to transport, and they are hard to store efficiently in large quantities. The difficulty of storing a resource is what economists call an inter-temporal constraint. And this is where cognitive computing comes in: if you cannot store a resource you must have a good forecasts of supply and demand. With growing population not in the same location as growing resource availability, this challenge is likely to increase in the coming decades.

1.1.1 Food

The world population is expected to grow by more than 30% over the next 30 years (Alexandratos et al., 2012). The vast majority of this growth will be in developing nations and primarily in urban areas. By 2050, urban areas will account for 70% of the world population. This growing population, combined with growing per capita consumption, will lead to global food consumption of 27 trillion kcal/day in 2050, up from 19 trillion kcal/day in 2010 (Alexandratos et al., 2012; Godfray et al., 2010). This aggregate food consumption equates to roughly 2700 kcal/person/day, which means that in principle there is enough global food production to meet the nutritional needs of the global population. However, this is not the case in practice with about 795 million people that are undernourished globally according to the Food and Agriculture Organization of the United Nations (Food and Agriculture Organization (FAO), 2015). In other words, there is sufficient food, it is just not in the right place at the right time. Even in developed nations, the food system exhibits major inefficiencies. The US Department of Agriculture estimates 31% or 133 billion pounds of the 430 billion pounds of food produced in the United States was lost before it was consumed (Buzby et al., 2014).

1.1.2 Water

Access to clean, fresh water is essential for survival and to prevent disease. About 10% of the globally available surface and groundwater supply and about 30% of rainwater is used by humans (Oki and Kanae, 2006). Despite this, about 748 million people lack access to improved drinking water (World Health Organization, 2014). The reason for this is the high variability of water resource availability both in time and in space. Of global water resources, a large fraction is available where human demands are small, such as in the Amazon basin, Canada, and Alaska (Cosgrove and Rijsberman, 2014), this water is therefore unavailable for consumption without transporting it first. In addition, flow during floods and wet seasons cannot be used during the low-flow seasons unless storage systems are in place (Oki and Kanae, 2006). Again, developed nations are not immune to the challenges of supplying their populations with clean water. For example, in California, groundwater levels continue to drop due to an extended period of drought and high consumption (Howitt et al., 2014).

1.1.3 Energy

The International Energy Agency estimates that, in 2013, total world energy consumption was 13,541 Mtoe (Megaton oil equivalents), equal to an average power consumption of 18.0 terawatts (International Energy Agency, 2015a). This demand for energy is expected to rise dramatically, by about 70% by 2040, largely in response to rapid economic growth in the developing world (International Energy Agency, 2015c). Solar capacity was over 177GW in 2014 with an increase in 2014 of 40GW (International Energy Agency, 2015b). Worldwide, this accounts for a small fraction at present, though the rapid growth of renewable capacity may lead to renewables becoming the largest source of electricity by the early 2030s (National Intelligence Council, 2012). The spatially localized and intermittent nature of renewable resources such as solar and wind will have to be addressed through a combination of transmission, energy storage, and demand response. This drive toward renewables combined with rapid growth in distributed generation will require a change in business models of utility companies. This change will be further driven by an increase in the availability of data, including smart meters (Institute for Electric Innovation, 2014).

1.1.4 Food–Water–Energy Nexus

One of the major challenges in managing food, water, and energy resources is that they are strongly connected and subject to competing interests (Bazilian et al., 2011; National Intelligence Council, 2012). Fig. 1 shows a schematic

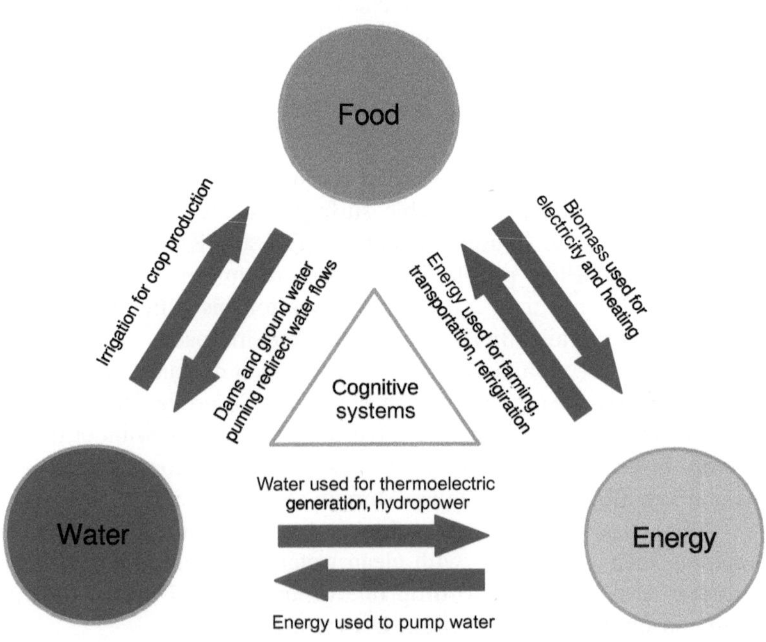

FIG. 1 The food–water–energy nexus.

summary. Energy is used to pump water, and water is used to generate electricity. Each kilowatt-hour (kWh) of a thermoelectric power generator uses approximately 100 L of water, primarily for cooling purposes. Aggregate fresh water use for thermoelectric generation accounts for 41% of all freshwater withdrawals (Kenny et al., 2009). In the meantime, collection, treatment, transportation, and end-use of water accounts for up to 20% of electricity consumption (California Department of Water Resources, 2008).

The agricultural sector is a large user of water resources, accounting for 37% of all freshwater withdrawals in the United States (Kenny et al., 2009) and roughly 70% globally (United Nations World Water Assessment Programme, 2015).

The end-use energy demand of the global food sector is around 32% of current global energy demand (Food and Agriculture Organization (FAO), 2011). Though biofuels are responsible for only a small fraction of global energy consumption, they make up the largest share of renewable energy generation (mostly used for heating) (International Panel on Climate Change, 2011).

These three resources also share infrastructure: river dams sometimes serve the dual purpose of generating hydroelectric power and storing water for irrigation or human consumption (United Nations World Water Development Programme, 2014).

There is an opportunity here for improved efficiency. For example, the power use for water transportation could be used to mitigate the intermittent nature of renewable generation. In this example, a water system operator could choose to pump water at times of high solar power or wind power generation. Of course, coupling these three components comes at the cost of greater complexity. We will discuss how cognitive computing can be used to alleviate some of this complexity by helping decision makers focus on the most important choices.

1.2 The Value of Cognitive Systems in Managing Constrained Resources

Cognition, from a human perspective, is the ability to sense the world around us, reason based on what we observe, and make decisions as a result of this reasoning. A key skill is the ability to make sense of large amounts of knowledge acquired from an environment that is often too complex to exactly understand or predict.

Cognitive computing has a dual interpretation. First, it can refer to cognitive algorithms which simulate the human brain; artificial neural networks (ANNs) are an example of this. These algorithms can provide high accuracy predictions for complex data-intensive problems, but can also shed light on the way the human brain itself works (Valiant, 1995). Second, it can refer to cognitive systems, which exhibit cognitive abilities, including the ability to process unstructured data and interact naturally with humans, irrespective of the underlying algorithmic implementation. In this chapter we will primarily take the second perspective, but we will also review how cognitive algorithms have made cognitive systems perform better.

It noteworthy that these three processes map almost directly onto some of the most popular areas in computer science at present: big data, machine learning, and decision support. Recent advances in computing power, machine learning methods (including advances in ANNs and deep learning), and an explosion in the quantity of available data have made it feasible to actually build real cognitive systems (Kelly III and Hamm, 2013).

Take, for example, the simple question of whether or not to take an umbrella when leaving home. To make this decision one needs to take into account an expectation about the weather which might be based on a weather report combined with looking out the window combined with previous experience regarding the weather in the region. A probability of rain should then be weighed against the inconvenience of having to carry an umbrella that may not be needed.

As we will see in this chapter, cognitive systems can not only help improve weather forecasts themselves, they can also be used to combine data from multiple sources and help make decisions. The decision to take an umbrella might not sound important enough to warrant the use of such complex tools; however, the same challenges are faced by food producers deciding when to sow and when to reap, or by power plant operators deciding when and how much energy to produce.

Cognitive systems have been deployed in diverse areas like health care, online advertising, and children's toys. In this chapter we will focus on how cognitive systems have been and could be used to manage constrained resources such as food, water, and energy.

Besides the critical importance of these resources to the foundation of our modern lifestyle, the systems that manage them have properties that make this a particularly promising field for the application of cognitive methods.

First, there is a large and increasing volume of data available about the operation of these systems and the factors that drive them. For example, in the energy field data are available from weather stations, satellites, numerical weather predictions, power flow telemetry, and smart meters. These data are related through geographical location, network connectivity, and ownership. Cognitive methods are most effective in cases where large amounts of data are available from multiple sources, and when solving problems requires decisions to be made under uncertainty. Indeed, the management of food, water, and energy supplies fit that bill; input data consist of a mix of time-series, images, network graphs, and public policy constraints.

Second, constrained resources are either hard to store efficiently, like electricity, or difficult to transport, like water, or both, as is the case for food. These difficulties can be alleviated by good forecasts which can in turn drive better decisions of when to produce, when to consume and what infrastructure to put in place. Good forecasts of solar energy generation, for example, can reduce the cost of reserve generation (Gowrisankaran et al., 2011). Cognitive algorithms excel at finding the nonlinear relationships between driving factors and variables of interest. This applies to time-series data, as we will see in Section 3, as well as image and video data, as we will describe in Section 4.

Finally, managing any one resource individually is already a complex task. Taking into account the interactions between them offers an opportunity to gain efficiency, but also makes the task of optimally managing them even more complex. We will discuss in Section 5 how cognitive systems can help to mitigate the complexity. For example, to make the output of any analysis consumable to decision makers it needs to be presented in an easily consumable, human readable way. As resource management systems become more interconnected, cognitive systems can help hide some of the complexity though decision support; this includes generating recommended actions, such as power plant scheduling or market bidding strategies.

The confluence of these three factors has led to the recent emergence of the first operational cognitive systems to manage energy (Lonij et al., 2016; Treinish et al., 2016).

2 INVARIANCE, CORRELATION, AND DATA

Cognitive systems learn from data. The larger the quantity of available data to train an algorithm, the better the accuracy of the predictions and decisions made by the system. Two properties of the data that are essential to make this work are invariance and correlation. We will explain why they are important to enable cognitive computing and show how they hold at least approximately, for many problems in resource management. And in those instances where invariance or correlation is violated we will show how problems can be reduced to an invariant one. This explains recent successes in applying cognitive algorithms to constrained resource management problems, and it is the reason why we expect these successes to increase in the future.

2.1 Invariance

One major challenge in applying cognitive algorithms to physical systems is that "more data" often means "older data." So the more data one includes in the training dataset the larger the chance that some of that data are no longer representative of the current dynamics of the system. For cognitive algorithms, and more broadly, machine learning algorithms, to be successful, the signal to be modeled must exhibit some degree of invariance. The example of time-invariance boils down to the requirement that "what is true today is till true tomorrow." Time-invariance is also referred to as stationarity.

Invariance is a property of a system: invariance to a parameter means that the system's dynamics do not depend explicitly on that parameter.

For example, in the case of time-invariance, we can write a signal $s(t)$ as some forecast operator (F) applied to a set of variables ($\{p(t)\}$).

$$s(t) = F(\{p(t)\}_p) \tag{1}$$

Note that the $\{p(t)\}$ may include lagged variables (e.g., $p_i(t) = p_j(t - \delta)$). Invariance is then given by the condition that at a future or past time ($t' = t + \Delta$) the same expression still holds.

$$s(t+\Delta) = F(\{p(t+\Delta)\}_p) \tag{2}$$

We can generalize the concept of time-invariance to other dimensions as well. For example, a signal is spatially invariant if it does not explicitly depend on location. The signal may vary as a function of location but this variation is really a function of an underlying property that depends on location. Take, for example, the height of snow on the ground: such a quantity may depend on the water vapor density in the atmosphere as well as the altitude of the earth surface. Snowfall may more likely at high altitude and would therefore appear to depend on location. However, the true relationship is between snowfall and altitude and does not explicitly dependent on location.

Explicit dependence on location is undesirable because it leads to poor generalizations, i.e., it implies that the extension of the model to new locations that were not part of the training set will be either impossible or result in low accuracy. Conversely, a model that depends explicitly on location may need data at each location of interest. The assumption of spatial invariance is analogous to the concept of shift-invariance in image processing which forms the foundation of the convolution layer in convolutional neural networks.

We do not argue that invariance is a requirement for machine learning to work, instead, we argue only that those problems which can be phrased in an invariant way will yield greater success.

2.2 Correlation

The second property of importance is correlation. We use the word correlation here in the general sense of there being a relationship between two quantities, without specifying any particular way of determining the existence of correlation (like Pearson's or Spearman's correlation).

This argument also applies to traditional statistical forecasting methods. Compare, for example, temporal autocorrelation of the return over time of a financial asset, like a stock, and the temporal autocorrelation of energy demand, shown in Fig. 2. While stock market returns appear random, energy demand shows a rich structure in the autocorrelation data.

Autocorrelation, as used in this example, is a linear measure and does not prove the absence of a nonlinear relationships. Therefore, the low autocorrelation as a function of lag of the financial asset does not imply that it is

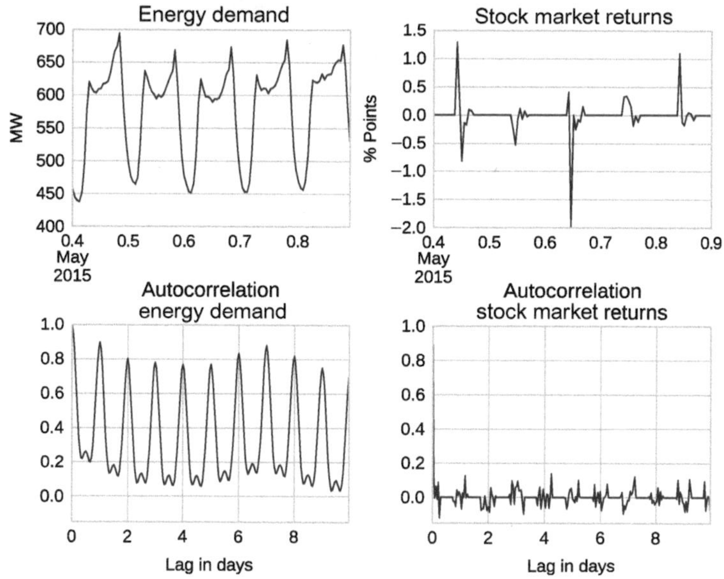

FIG. 2 Comparison of autocorrelation of stock market returns and energy demand. While stock market returns appear random, energy demand shows a rich structure in the autocorrelation data. *Top left*: hourly energy demand in the state of Vermont as a function of time (data published by ISO New England). *Top right*, hourly returns on a financial asset (IBM). *Bottom left and right*: autocorrelation of energy demand and stock market returns, respectively. Note, in particular, the peaks in autocorrelation of energy demand at lags of 1 day and 7 days, indicating daily and weekly seasonality. Autocorrelation of financial asset returns is consistently and significantly smaller.

impossible to forecast future returns based on past returns. It does, however, suggest that it may be much more difficult. Of course, this has not stopped many people from attempting to forecast stock prices anyway.

Energy consumption signals on the other hand (as well as weather data and renewable generation signals) exhibit much stronger autocorrelation and cross-correlation. There is clear evidence that there are relationships between the predictor variables and the variables we want to predict (target variables). Examples of such relationships include the relationship between irradiance and photo-voltaic (PV) generation, and the relationship between time of day and power demand. This is reflected in the state of the art accuracy forecasts of power demand, with mean absolute percentage errors of less than 2% (Lonij et al., 2016).

What remains is to find the specific relationships between the many available predictor variables and the target variable of interest. Such relationships can be found based on prior knowledge of the system, and we will see several examples of such physics-based models. But the quest to model ever more complex systems with higher accuracy, necessitate that these predictor functions be approximated. ANNs have been shown to be effective at being universal approximators (Sutskever and Hinton, 2008). Because of this, ANNs have been applied to a wide range of problems, and we will explore how they have been used in the energy field.

2.3 Sparse Data, Big Data, and Data Types

The remainder of this chapter is organized following the different data types that are available in resource management. The sources and types of data also go hand in hand with the methods that are used to analyze them. The type of data matters, e.g., time-series vs images vs video, but so does the amount of available data; when the dataset is sparse, e.g., power consumption during heat waves, it may be necessary to use physical models or hybrid models that incorporate strong priors. When a dataset is large and complex, such as a series of satellite images, it becomes necessary to use neural networks and deep learning.

The primary data in resource management are time-series data representing the production, transportation, and consumption of resources. In particular the subsequent sections will focus on energy, and we will be dealing with power generation, transmission, and consumption data. In addition, a major driver of the power system is the environment. Temperature, wind, irradiance, snow, and humidity all affect both consumption and generation from renewable sources such as solar and wind power. This environmental data have a spatial component which, as we will see, can be exploited to get more accurate and complete information about the world around us. Combining image data from satellites with ground sensor data for irradiance can extend the range of ground sensors. And when sequences of images are interpreted as video, they can be used to extrapolate and make some of the most accurate short-term forecasts of renewable generation.

Data-driven approaches to model time-series involve the estimation of the parameters of an assumed underlying model. Examples include hidden Markov models (Rabiner and Juang, 1986), autoregressive models (Valipour et al., 2013), and linear dynamical systems (Luenberger, 1979). Methods based on neural networks (NN) include traditional multilayer perceptrons, restricted Boltzmann machines (Hinton and Salakhutdinov, 2006), and deep learning (Längkvist et al., 2014).

Data from numerical weather models, radar stations, and satellites can be interpreted as images and video. We will see that techniques from object tracking, such as Kalman filtering (Welch and Bishop, 2006), optical flow (Corpetti et al., 2002), and video compression (Furht et al., 2012) help enhance the forecasts obtained from time-series models.

3 TIME-SERIES DATA

In the energy domain the most ubiquitous data are power production and consumption data. Measured in the form of currents, voltages, active power, and reactive power, these data are acquired as series of time-value pairs (time-series). These time-series represent household energy consumption or aggregates at different points on the power grid. With the recent adoption of advanced metering infrastructure, i.e., "smart meters," the volume of this data has sky-rocketed (Institute for Electric Innovation, 2014). A second important source of time-series data comes from ground-based weather sensors, including temperature, wind, irradiance, precipitation, etc. These data can be used to understand the drivers of power consumption and generation.

Broadly speaking there have been two branches of study in handling time-series data: data-driven models and physics-based models. These two branches span a spectrum of problems. On one end of the spectrum lie purely physical models, suitable when only sparse data is available and interpretability is key. On the other end lie problems where datasets are huge and the patters too complex to capture with models built from first-principles. In addition, some interesting recent developments have shown the success of combining these two approaches.

The strong temporal autocorrelation of power consumption data also makes it a good candidate for autoregressive forecasting models. In addition, there are correlations between power consumption and environmental factors such as temperature. The challenge is to map out these relationships in a way that generalizes to data-points that have not yet been observed.

For example, the power consumption of a household has strong autocorrelation: power consumption at a given level tends to persist for some time. However, the power consumption of two neighboring households does not necessarily have a correlation, i.e., there should not be any a priori assumption of spatial continuity. Furthermore, because these time-series are driven by human behavior as well as local environmental factors there tend to be strong multifrequency seasonalities: patterns that repeat daily, weekly, and yearly. If data covering a sufficiently long period of time is available, these seasonalities can be learned by cognitive algorithms or accounted for by seasonal autoregressive models. If insufficient data are available, prior knowledge can be used in the form of physics-based models or in the form of constraints or regularizations of machine learning algorithms.

3.1 Physics-Based Methods

Physics-based models lie on one end of the spectrum, they have relatively few free parameters and are usually highly interpretable. In fact, one of the principal reasons to build physical models to gain understanding about the workings of a system. For this reason they are popular when modeling systems that are driven by physical processes that we understand (as opposed to, say, human behavior).

Physics-based models focus on the underlying process that generated the data. An example of this is the output of a solar power generator. The physical process that drives the conversion of solar radiation to electrical current is well understood. Once the irradiance incident on a solar module is known, its power output can be determined with accuracy better than a few percent.

One of the challenges, however, is the difficulty of determining the incident irradiance. Even if a ground measurement of irradiance near a PV power plant is available, converting from the time-series of measured irradiance (usually on a horizontal plane) to a time-series of produced power is challenging due to several detailed effects that influence irradiance absorbed by the PV modules. For example, shade, soiling of the modules, orientation of the modules, and incident angle of light (Marion et al., 2005). This is where data-driven models come in.

3.2 Data-Driven Methods

The impact of weather conditions on PV power production has been studied using support vector machines (Shi et al., 2012), wavelet transforms (Lave et al., 2012b), and ANNs (Giraud and Salameh, 1999). Similarly, in wind forecasting the mapping between wind speed and wind generator output has been studies using a number of physical models (Ackermann et al., 2005; Costa et al., 2008) as well as an array of neural network-based models (Li et al., 2001; Marvuglia and Messineo, 2012).

These models provide a mapping from environmental and calendar variables onto power generation. These models are used to evaluate the potential of new locations for power generators (Palma et al., 2008) and to monitor the operational performance of power plants (Marvuglia and Messineo, 2012). Of course, once the mapping from environmental factors to power consumption and production is known, the same models can be used for forecasting if an exogenous forecast of environmental variables is available, e.g., from numerical weather forecasting models (Mathiesen and Kleissl, 2011; Perez et al., 2010).

Modeling power load is more challenging because it is, at least in part, driven by human behavior. While the impact of environmental factors, such as temperature, on power consumption of, for example, air-conditioning could be modeled using fundamental physical principles, the choices of people regarding when to use their air-conditioning is more challenging to model. This is why load modeling and forecasting is more amenable to a data-driven approach. Given the nonlinear nature of the relationship between load and environment, there has been longstanding and significant interest in applying ANN approaches to demand forecasting (Jebaraj and Iniyan, 2006; Metaxiotis et al., 2003; Park et al., 1991). Variations of these methods focus on identifying similar days (Mandal et al., 2006). A parallel but equally fruitful direction focuses on interpretability of models. Generalized additive models in particular have demonstrated high accuracy while maintaining the interpretability that is valuable for decision making (Fan and Hyndman, 2012; Wood, 2006).

Since the relationship between environment and power load is less well understood, and may be nonlocal, there has been work to develop power forecasting models that have no exogenous variables, i.e., purely autoregressive models (Suganthi and Samuel, 2012). Autoregressive ARIMA models have been combined with ANN to account for the diurnal cycle of solar power generation (Ji and Chee, 2011; Voyant et al., 2013).

However, the greatest success has been achieved by combining concurrent and autoregressive features (Alfares and Nazeeruddin, 2002; Lonij et al., 2016).

3.3 Hybrid Methods

What about the case where little data are available but the physical drivers of the system's dynamics are not well known? Take, for example, the modeling of power usage during extreme weather events like heat waves, or modeling the output of a newly installed PV system for which physical parameters are not known. Strong seasonality and nonstationarity, i.e., a lack of invariance can limit the effectiveness of autoregressive models. Longer-term seasonality and trends with timescales of a year or greater can limit the usefulness of data-driven method if only a small amount of historical training data is available.

Fortunately, hybrid models, that can incorporate prior knowledge of the system into data-driven methods, can remedy this. Take again as an example, a PV generator. The output of a PV generator consisting of stationary flat plate PV modules depends on time of day and time of year. Learning this behavior entirely from data would require many years of data. Fortunately, the primary cause of this behavior is well understood: it is due to the day–night cycle and the changing incident angle of sunlight. This effect can be modeled with physical models that have only a few free parameters. The residuals of this physical model can then be input to data-driven methods, even if only a few months of training data are available.

The simplest way to incorporate this information is to model clear sky index instead of power output (Perez et al., 2010). More recent efforts have focused on simultaneously optimizing parameters of a set of physical models and the parameters of a data-driven model to jointly predict the same system (Lonij et al., 2016). More general methods can impose shape constraints, e.g., monotonicity, convexity, concavity, to the relationships between predictor variables and target variables (Chen and Samworth, 2015).

In power load forecasts, prior knowledge can be used in multiple ways. For example, knowledge of holidays or special events can be incorporated into machine learning algorithms as categorical variables (Fan and Hyndman, 2012). Furthermore, if certain exogenous variables, such as temperature, are known to be significant, different models can be chosen depending on the range of these exogenous variables (Song et al., 2006). The PV contribution to net power flow on a power-line can also be modeled using a physical model. The parameters of this physical model can then help understand the drivers of this power flow, such as installed PV capacity connected to the power-line. Operator expertise can be incorporated through expert systems (Hsu and Ho, 1992), and these expert systems can in turn be combined with novel machine learning methods, including ANN (Kim et al., 1995). Numerical weather predictions can be combined with ANN to produce better irradiance forecasts (Voyant et al., 2012).

3.4 Hierarchical Time-Series and Missing Data

Power networks are unique in the sense that telemetry can be both over-complete and under-complete depending on how many sensors are deployed. In the over-complete case, multiple sensors measure the same or closely related quantities, like the power flow into a transformer and the power flow out. The power flow through one node on a network can be calculated as a weighted sum of power flows through connected nodes. Whether or not there are sufficient data to calculate the flow through a power-line can be determined through observability analysis (Monticelli and Wu, 1985). If there is an over-complete set of measurements, then the measurements can be used for fault detection and anomaly detection. This type of situational awareness is essential in an operational cognitive energy forecasting system.

The set of available measurements is under-complete, for example, if the sum of several power flows into a transformer is measured, but the flow on the individual lines is not. In this case it is still possible to detect anomalous data-points, by comparing data to a model prediction for the same data point. The process can be iterative: train a model on a dataset, identify anomalous data-points as those points where the model produces large errors, and remove those points from the dataset and repeat.

4 IMAGES, VIDEO, AND SPATIO-TEMPORAL DATA

While the number and the quality of images have exploded in the recent years, understanding and making sense of image data still raise multiple challenges. From a computer's point of view, an image is nothing more than a multidimensional array of numbers. Efficiently capturing, transmitting, storing, and analyzing large quantities of image data unlock new applications in multiple industries. With a decreased cost of computing power, several image processing and computer vision techniques also regain strong interests in the scientific and industrial communities. In this section, we analyze a few recent cognitive computing success stories that rely on image data.

4.1 Making the Most of Sensor Data

One of the challenges in quantifying our environment is that the environment itself is spatially and temporarily continuous, while weather sensors on weather stations typically provide discrete point measurements. For example, a rain gauge measures rainfall only in one location, an irradiance sensor measures irradiance at a single point. However, the quantity of interest is total rainfall, say, over the area of a crop field, or the total irradiance over the area of a solar farm.

Traditional sampling theory, pioneered by Kotelnikov, Nyquist, Shannon, and Whittaker, states that a signal can be exactly recovered from a set of uniformly spaced samples taken at the so-called *Nyquist rate* of twice the highest frequency present in the signal of interest. Recently, compressed sensing (CS) has attracted a lot of attention, by suggesting one might be able to reconstruct a high-dimensional signal with a sparse set of measurements which could be more compact than the limit from the traditional sampling theory (Donoho, 2006).

In the field of energy, CS has been applied to weather radar data, for example, to estimate remote precipitation using radar measurements. In this example, where precipitation samples are highly correlated in space and time, matrix completion techniques can be applied (Mishra et al., 2014).

Imaging sensors can fill in the gaps between point sensors. Images can be obtained from satellites, radar stations, or ground-based sky imagers. In the case of satellites, a number of different quantities can be obtained, including radiation in the visible spectrum, infrared radiation, and radar soundings. In the case of ground-based radar stations the principal quantity obtained is reflectivity and polarization of reflected radar waves.

The challenge in making optimal use of these imaging sensors is that they do not make a direct measurement of the quantity of interest. For example, an image of the visible spectrum obtained from a satellite would show the location of clouds, not the ground-level irradiance. To calculate ground-level irradiance one would have to infer cloud altitude, depth, and optical density from the image.

In the case of estimating precipitation from radar reflectivity measurements we see a similar challenge. Radar reflectivity measurements give us a measure of the moisture content of the atmosphere. This is not the same as the amount of liquid that makes it to ground level, which is the quantity we are ultimately interested in.

The structure of this problem, i.e., having a large dataset with a limited set of labels alligns very well with the methods known from supervised and semi-supervised machine learning. Which is why there has been a significant amount of recent work invested in making the most of radar data, satellite data, and data from ground-based sky imagers.

Radar-based estimates of precipitation can be used for agriculture, hydrology, waste water management, and wireless communications (Govindaraju and Rao, 2013; Mounce et al., 2014; Tokar and Johnson, 1999). Finding a correspondence between radar measurements and ground-based rain gauges can be hindered by a range of data quality challenges, and the first step in addressing this problem is to apply quality control techniques (Harrison et al., 2000).

To address the remaining gap between real world data and physics-based models, there has been a long history of applying data-driven methods and ANN to convert radar measurements to ground precipitation (Govindaraju and Rao, 2013; Huang et al., 2015; Nayak et al., 2013). Recent preliminary results suggest that recurrent neural networks can be used to integrate an irregular sequence of instantaneous measurements in order to estimate total rainfall over a longer period of time (Kaggle, 2016). Seasonality can be addressed by using adaptive neural networks (Chiang et al., 2007).

Similarly, data from satellite radiometers can be used to estimate rainfall (Chandrasekar et al., 2014; Tapiador et al., 2004). The advantage of satellite imagery is that it covers a much larger area. More recently, due to the rapid adoption of solar power as a renewable energy source, there has been a growing interest in using satellite data to estimate and forecast ground-level irradiance (Marquez et al., 2013; Perez et al., 2004). The primary interest for remote measurements of ground-level irradiance is resource assessment for the planning of new generator locations (Hammer et al., 2003; Mellit and Pavan, 2010). In planning the location of new generators, it is important for the plant owner to maximize the output of a plant over its lifetime; however, it is important for grid operators to consider the effect of geographic dispersion of system locations on the variability of aggregate generation. As a result, there has been significant study of spatio-temporal correlations of irradiance, and the implication of geographic smoothing (Hoff and Perez, 2012; Lave et al., 2012a). A drawback of using satellite data is the relatively limited spatial and temporal resolution.

Higher spatial and temporal resolution forecasts are especially important for predictions of PV power generation, which is impacted by local irradiance which can fluctuate by more than 80% in several seconds due to changing cloud cover. An example of this is shown in Fig. 3. In the search

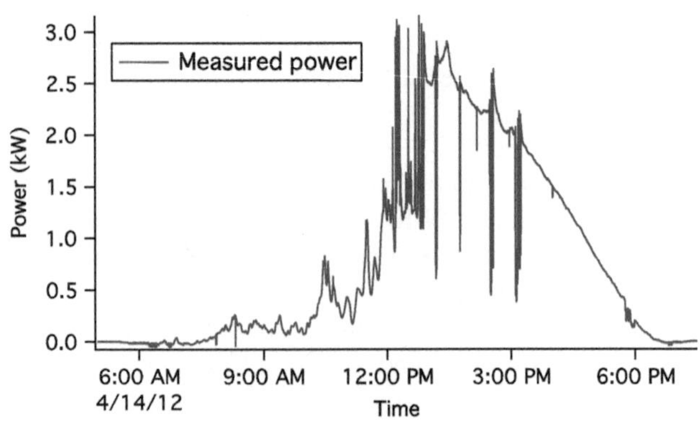

FIG. 3 Generation of photo-voltaic (PV) generators can fluctuate rapidly due to intermittent cloud cover.

for higher temporal and spatial resolution, researchers turned to ground-based sky imagers. These sky images are particularly useful for irradiance forecasting because they are well suited to detect clouds and because even commodity cameras can record multiple high-resolution frames per second. There are two challenges in interpreting images from sky imagers: identifying the difference between cloud pixels and clear sky pixels under varying conditions and the estimation of cloud height. Identifying cloud height can be achieved using a set of cameras for stereoscopy (Nguyen and Kleissl, 2014).

Though the majority of the literature to date has used simple manually designed features and thresholds for cloud segmentation. Recent advancements have used unsupervised clustering (Heinle et al., 2010) and adaptive thresholding (Jayadevan et al., 2015) to improve accuracy. Motion of clouds, detected from a sequence of images, can also be used to improve cloud segmentation (Jayadevan et al., 2012; Papin et al., 2002). However, this motion can be put to use in and even more valuable context: forecasting.

4.2 Forecasting and Deep Learning

Atmospheric data have a spatio-temporal structure. In other words, there is a relationship between forecasts at time t and location $\vec{x}$ and measurements at time $t' = t - \delta$ and a different (usually "upwind") set of locations $\{\vec{x}'\}$.

This implies that a sequence of images, or more generally, a set of spatially dispersed measurements taken over an extended period of time can be used to make short-term forecast of the same measure quantity (Lonij et al., 2012, 2013).

The simplest example of a spatio-temporal relationship assumes a linear translation with fixed velocity:

$$p(\vec{x}, t) = p(\vec{x} - \delta \vec{v}, t - \delta) \tag{3}$$

where $\vec{v}$ is the drift velocity. The velocity in this equation can be obtained from an independent source (Bosch et al., 2013) or learned from the data itself. The latter case is often used for image data because high spatial and temporal resolution are needed to match features in one or more consecutive frames.

The operating principle of these methods is to analyze two consecutive frames in a sequence in order to find matching configurations of pixels in both images (Leese et al., 1971). The assumption is that the object corresponding to those pixels moved from one location to another. In the context of clouds, an additional difficulty is the fact that clouds can change shape in addition to their translation. To address this challenge, cloud dynamics can be modeled as an optical flow problem (Corpetti et al., 2002). That is, motion vectors can be assigned to each pixel in the image, usually with some constraint on the

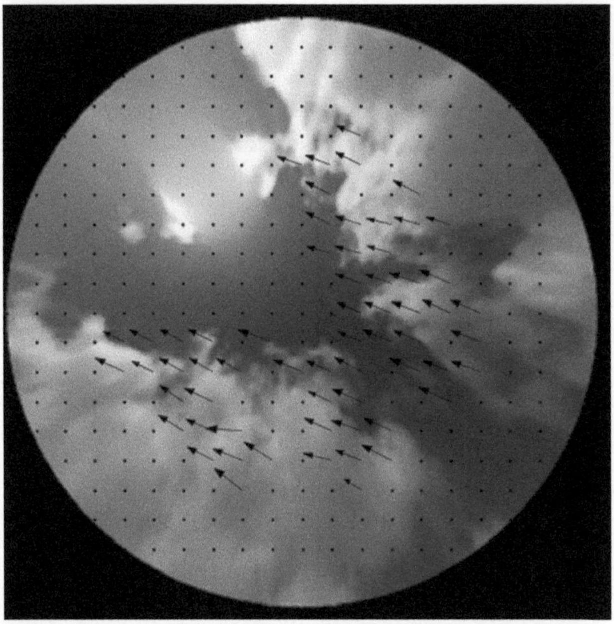

FIG. 4 Cloud motion vectors obtained through block-motion estimation after quality control filtering. *Figure reproduced, with permission, from Chow, C.W., Urquhart, B., Lave, M., Dominguez, A., Kleissl, J., Shields, J., Washom, B., 2011. Intra-hour forecasting with a total sky imager at the UC San Diego solar energy testbed. Sol. Energy 85 (11), 2881–2893. doi:10.1016/j.solener.2011.08.025.*

variation of motion vectors assigned to neighboring pixels (Chow et al., 2015). Algorithms from video compression, such as block-motion estimation, have been shown to be an effective compromise maintaining some of the granularity of optical flow algorithms, while requiring fewer computational resources (Chow et al., 2011; Furht et al., 2012). Fig. 4 shows the result of a block-motion estimation algorithm.

Subsequent images in the sequence can then be predicted by translating each pixel, or set of pixels, in the direction of the estimated motion vector. When combined with the segmentation algorithms, these predictive image sequences can be used to make forecasts of irradiance, or solar power generation. This methodology has been applied successfully to both images from ground-based cameras (Chow et al., 2015; Jayadevan et al., 2012) as well as images from satellites (Bedka and Mecikalski, 2005; Perez et al., 2010).

This same principle has been applied to extend the reach of point sensors. The spatial correlation of, say, irradiance can be estimated from a point measurement. Past upwind conditions and future downwind conditions can then be estimated from the measurements at a single point, effectively creating an array of virtual irradiance sensors (Perez et al., 2012).

With the recent success of deep learning methods in analyzing and classifying images (Bengio, 2009), it is encouraging to see that these methods are also being applied to atmospheric forecasting (Klein et al., 2015; Xingjian et al., 2015). What is particularly interesting is that these problems have motivated the development of new deep neural network to policies as well as new types of layers, including letters specifically designed to capture spatial translations in images (Klein et al., 2015).

To evaluate the benefit of not whole algorithms to forecast cloud motion, we must first discuss the metrics we need to compare forecasts.

4.3 Forecasting Horizon vs Accuracy

Despite efforts to incorporate ground-truth data into physics-based models, like numerical weather models, purely data-driven models outperform physics-based models in short-term forecasts. For forecasting horizons greater than several hours, physics-based predictions perform better. This may be due to the ability of physics-based models to accommodate strong nonlinearity, provided that the form of the nonlinearity is known. It should also be noted that the relationship between forecast error and forecasting horizon depends on the detailed measurement process, the metric that is chosen and the averaging interval. For example, for irradiance forecasting, it matters how measurement points are aligned with wind direction, how measurement points are spatially distributed and how measurements are averaged over time. Under mild assumptions this relationship can be modeled (Lonij et al., 2013). Fig. 5 shows this theoretical relationship compared to measurements from multiple authors. The relationship holds over three orders of magnitude.

One of the key assumptions of this model is the linear relationship between forecasts at time t and location $\vec{x}$ and measurements at time $t' = t - \delta$ and a different (usually "upwind") set of locations $\{\vec{x}'\}$. It will be very interesting to see if future cognitive algorithms, for example, based on deep NN topologies can beat this benchmark.

5 AUTONOMOUS SYSTEMS TO MANAGE COMPLEXITY

In the previous sections we discussed how forecasts related to constrained resources can be produced. The challenges associated with the cost to store and transport constrained resources can be mitigated with these forecasts. Of course this is predicated on the assumption that actions be taken in response to these forecasts. In the case of electric utilities, short-term actions may consist of dispatching power plants, activating demand response, and bidding into markets. Longer-term actions can include planning of infrastructure like transmission lines, storage and renewable generation capacity as well as entering into power purchase agreements, or investing in power futures contracts that often cover several years into the future.

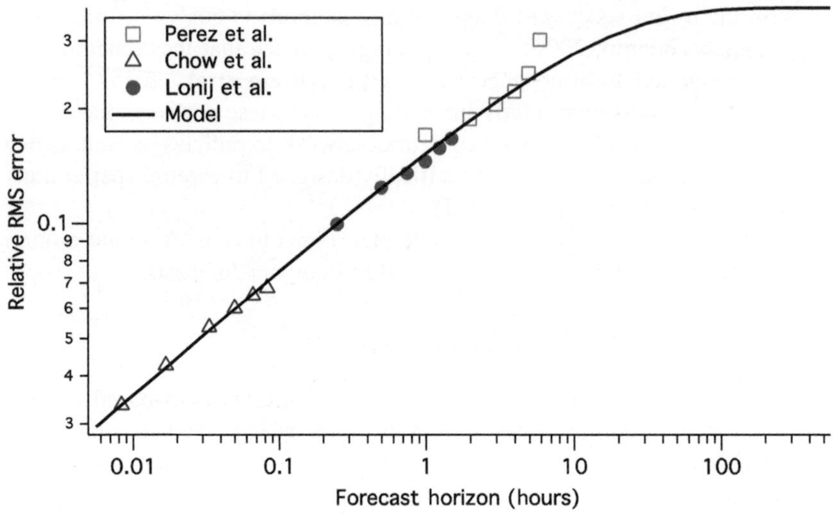

FIG. 5 Forecast accuracy vs forecasting horizon. Forecast accuracy for three different methods of short-term irradiance forecasts based on cloud motion detection. Results, using a ground-based sky imager (Chow et al., 2011), are shown as *blue* (*gray* in the print version) *triangles*. Results using a network of ground-based irradiance sensors are shown as *red* (*gray* in the print version) *circles* (Lonij et al., 2013). Results using satellite images are shown as *green* (*gray* in the print version) *squares* (Perez et al., 2010). *Figure reproduced, with permission, from Lonij, V.P., Brooks, A.E., Cronin, A.D., Leuthold, M., Koch, K., 2013. Intra-hour forecasts of solar power production using measurements from a network of irradiance sensors. Sol. Energy 97, 58–66. doi:10.1016/j.solener.2013.08.002.*

Managing a system with such complex behavior is a grand challenge for artificial intelligence. When we consider the interactions that the power system has with systems used to manage other constrained resources such as food and water, the problem becomes more complex, but there is also an opportunity to build greater efficiency into the system if the problem of complexity can be addressed. Complexity can be reduced by the use of intelligent systems and by building human–computer interfaces that can present a user with the most pertinent information that facilitates decision making.

Making use of forecasts and other analytics can be done at three different levels of sophistication. First, cognitive systems can summarize messy data from multiple sources and present conclusions in formats that are easy to ingest for human decision makers. Second, cognitive systems can provide recommendations for optimal actions. Finally, it is possible to build fully autonomous systems that achieve predefined goals without human intervention.

Natural human–computer interfaces are an essential part of any cognitive system. This starts with the interpretation of weather data. An interesting line of research has focused on generating human readable text describing the output of a numerical weather model. These methods use fuzzy quantifiers to aggregate trends and other features in time-series data (Kacprzyk et al., 2008), or by translating summary statistics into natural language (Sripada et al., 2003). Similarly, speech recognition can be used to input natural language queries about data or the state of the weather in a particular location. These human–computer interfaces can help deliver summarized versions of complex information to decision makers.

The next level of sophistication is to present recommended actions to decision makers. There is a long history of using linear and nonlinear optimization methods to help decide the most economic dispatch of power plants (Wood and Wollenberg, 2012). Power dispatch has to take into account the topology of the power network, including transmission constraints (Kumar et al., 2005). This is a very computationally intensive problem and methods using neural networks as approximators have been used (Siano et al., 2012). The reduced computational requirements of the trained neural network allows optimization algorithms to run on embedded devices.

With the growth of renewable capacity, optimization methods have been applied to mitigate the intermittent nature of wind and solar power through the optimal use of battery storage (Teleke et al., 2009; Wang et al., 2008). The extension of this concept is the collective optimization of multiple power generating an consuming entities which are not necessarily colocated. This concept is referred to as a virtual power plant (VPP). For example, the combination of a (nondispatchable) solar generator, a flexible load, such as a heating system, and battery storage could be used to generate or consume power on demand. Such a VPP can be used to make the power grid more reliable or to optimize revenue for the owner of the hardware subject to fluctuating power prices. Such systems could operate fully independently, without human intervention.

Similarly, a home owner who has installed PV modules on his roof and owns an electric vehicle could in principle be a power market participant (Farhangi, 2010). However, such an individual is unlikely to have sufficient expertise to find the optimal charging behavior of his vehicle subject to fluctuating external power price signals and his personal need for flexibility in his transportation. These problems could be solved with cognitive algorithms that learn both the behavior of the consumer as well as the impact of environmental factors (Farhangi, 2010). Such systems are already available in the form of smart thermostats that optimize power consumption based on learned behavior of the resident of a house and the thermal properties of the house. This is an autonomous cognitive system that is connected to the power grid.

Autonomous interaction between these systems requires machine to machine communication. This communication is in itself a challenge because of bandwidth limitation, and power consumption constraints. Again in this case the communications system could be optimized using cognitive methods that adjust which part of the spectrum is used and what the communication network topology is (Zhang et al., 2012).

What is especially interesting in this case is that the challenges of optimizing power generation and consumption play out in a miniature setting inside some of the devices connected to the power grid. For example an electric vehicle has to balance between energy efficiency and acceleration in the presence of multiple power sources (battery, capacitor, fossil fuel). Since vehicles have to respond quickly and do not have powerful computers, it is beneficial to train an ANN to make decisions rather than to embed a complex solver (Moreno et al., 2006).

6 CONCLUSION

In this chapter we discussed the emergence of cognitive systems for energy management. We showed how different methods, ranging from physics-based methods to hybrid methods to fully data-driven ANNs go hand in hand with the quantity and type of available data. Going forward, data will be a driving force of innovation. The energy field is currently ahead in terms of the amount and quality of telemetry that is collected about the state of the power system. The widespread availability of weather data from multiple sources already enables the use of cognitive systems in water management and agriculture. However, better data collection in these fields is likely needed to unlock the full potential of cognitive systems applied to the food–water–energy nexus.

The continuing trend toward a more distributed operation of the power grid, in the form of distributed generation, and distributed energy storage, combined with demand response and smart home appliances, will eventually create a need for cognitive solutions. These cognitive systems will empower end-users to efficiently manage their energy production and generation, and help grid operators make better, more informed decisions about an increasingly complex system.

The connections between the food, water, and energy systems already exist. However, these connections are not yet fully exploited in the management of these systems. As more data are collected about each of these systems it will become more feasible to find globally optimal ways of managing the food–water–energy nexus. Cognitive systems will be essential in building these solutions.

REFERENCES

Ackermann, T., 2005. Wind Power in Power Systems. vol. 140. John Wiley & Sons, Ltd, Chichester, UK.

Alexandratos, N., Bruinsma, J., et al., 2012. World agriculture towards 2030/2050: the 2012 revision. Tech. rep., ESA working paper.

Alfares, H.K., Nazeeruddin, M., 2002. Electric load forecasting: literature survey and classification of methods. Int. J. Syst. Sci. 33 (1), 23–34.

Bazilian, M., Rogner, H., Howells, M., Hermann, S., Arent, D., Gielen, D., Steduto, P., Mueller, A., Komor, P., Tol, R.S., et al., 2011. Considering the energy, water and food nexus: towards an integrated modelling approach. Energy Policy 39 (12), 7896–7906.

Bedka, K.M., Mecikalski, J.R., 2005. Application of satellite-derived atmospheric motion vectors for estimating mesoscale flows. J. Appl. Meteorol. 44 (11), 1761–1772.

Bengio, Y., 2009. Learning deep architectures for AI. Found. Trends Mach. Learn. 2 (1), 1–127.

Bosch, J.L., Zheng, Y., Kleissl, J., 2013. Deriving cloud velocity from an array of solar radiation measurements. Sol. Energy 87, 196–203.

Buzby, J.C., Farah-Wells, H., Hyman, J., 2014. The estimated amount, value, and calories of post-harvest food losses at the retail and consumer levels in the United States, 121. USDA-ERS Economic Information Bulletin.

California Department of Water Resources, 2008. Managing an uncertain future: climate change adaptation strategies for California's water. Tech. Rep., California Department of Water Resources.

Chandrasekar, V., Ramanujam, K.S., Chen, H., Le, M., Alqudah, A., 2014. Rainfall estimation from spaceborne and ground based radars using neural networks. In: Geoscience and Remote Sensing Symposium (IGARSS), 2014 IEEE International, pp. 4966–4969.

Chen, Y., Samworth, R.J., 2015. Generalized additive and index models with shape constraints. J. R. Stat. Soc. Ser. B (Stat Methodol.) 78 (4), 729–754.

Chiang, Y.M., Chang, F.J., Jou, B.J.D., Lin, P.F., 2007. Dynamic ANN for precipitation estimation and forecasting from radar observations. J. Hydrol. 334 (1), 250–261.

Chow, C.W., Urquhart, B., Lave, M., Dominguez, A., Kleissl, J., Shields, J., Washom, B., 2011. Intra-hour forecasting with a total sky imager at the UC San Diego solar energy testbed. Sol. Energy 85 (11), 2881–2893.

Chow, C.W., Belongie, S., Kleissl, J., 2015. Cloud motion and stability estimation for intra-hour solar forecasting. Sol. Energy 115, 645–655.

Corpetti, T., Memin, E., Perez, P., 2002. Dense estimation of fluid flows. IEEE Trans. Pattern Anal. Mach. Intell. 24 (3), 365–380. ISSN 0162-8828. http://dx.doi.org/10.1109/34.990137.

Cosgrove, W.J., Rijsberman, F.R., 2014. World Water Vision: Making Water Everybody's Business. Routledge, Abingdon-on-Thames.

Costa, A., Crespo, A., Navarro, J., Lizcano, G., Madsen, H., Feitosa, E., 2008. A review on the young history of the wind power short-term prediction. Renew. Sust. Energ. Rev. 12 (6), 1725–1744.

Donoho, D.L., 2006. Compressed sensing. IEEE Trans. Inf. Theory 52 (4), 1289–1306.

Fan, S., Hyndman, R.J., 2012. Short-term load forecasting based on a semi-parametric additive model. IEEE Trans. Power Syst. 27 (1), 134–141.

Farhangi, H., 2010. The path of the smart grid. IEEE Power Energ. Mag. 8 (1), 18–28.

Food and Agriculture Organization (FAO), 2011. 'Energy-Smart' food for people and climate. FAO, Rome, Italy.

Food and Agriculture Organization (FAO), 2015. The State of Food Insecurity in the World. FAO, Rome.

Furht, B., Greenberg, J., Westwater, R., 2012. Motion Estimation Algorithms for Video Compression. vol. 379. Springer Science & Business Media, New York.

Giraud, F., Salameh, Z.M., 1999. Analysis of the effects of a passing cloud on a grid-interactive photovoltaic system with battery storage using neural networks. IEEE Trans. Energy Convers. 14 (4), 1572–1577.

Godfray, H.C.J., Beddington, J.R., Crute, I.R., Haddad, L., Lawrence, D., Muir, J.F., Pretty, J., Robinson, S., Thomas, S.M., Toulmin, C., 2010. Food security: the challenge of feeding 9 billion people. Science 327 (5967), 812–818.

Govindaraju, R.S., Rao, A.R., 2013. Artificial Neural Networks in Hydrology, vol. 36. Springer Science & Business Media, Dordrecht.

Gowrisankaran, G., Reynolds, S.S., Samano, M., 2011. Intermittency and the value of renewable energy. NBER Working Paper No. 17086.

Hammer, A., Heinemann, D., Hoyer, C., Kuhlemann, R., Lorenz, E., Müller, R., Beyer, H.G., 2003. Solar energy assessment using remote sensing technologies. Remote Sens. Environ. 86 (3), 423–432.

Harrison, D., Driscoll, S., Kitchen, M., 2000. Improving precipitation estimates from weather radar using quality control and correction techniques. Meteorol. Appl. 7 (2), 135–144.

Heinle, A., Macke, A., Srivastav, A., 2010. Automatic cloud classification of whole sky images. Atmos. Meas. Tech. 3 (3), 557–567.

Hinton, G.E., Salakhutdinov, R.R., 2006. Reducing the dimensionality of data with neural networks. Science 313 (5786), 504–507.

Hoff, T.E., Perez, R., 2012. Modeling PV fleet output variability. Sol. Energy 86 (8), 2177–2189.

Howitt, R., Medellín-Azuara, J., MacEwan, D., Lund, J., Sumner, D., 2014. Economic analysis of the 2014 drought for California agriculture. Center for Watershed Sciences, University of California, Davis.

Hsu, Y.Y., Ho, K.L., 1992. Fuzzy expert systems: an application to short-term load forecasting. IEE Proc. C 139 (6), 471–477.

Huang, B.J., Tseng, T.H., Tsai, C.M., 2015. Rainfall estimation in weather radar using support vector machine. In: Intelligent Information and Database Systems. Springer, Switzerland, pp. 583–592.

Institute for Electric Innovation, 2014. Utility scale smart meter deployments. Tech. Rep., The Edison Foundation.

International Energy Agency, 2015a. Key world energy statistics.

International Energy Agency, 2015b. Photovoltaic power system programme, trends 2015.

International Energy Agency, 2015c. World energy outlook 2015 factsheet. Tech. Rep., International Energy Agency.

International Panel on Climate Change, 2011. IPCC special report on renewable energy sources and climate change mitigation. Technical Report.

Jayadevan, V.T., Lonij, V.P.A., Rodriguez, J.J., Cronin, A.D., 2012. Forecasting Solar Power Intermittency Using Ground-Based Cloud Imaging. World Renewable Energy Forum, Denver.

Jayadevan, V.T., Rodriguez, J.J., Cronin, A.D., 2015. A new contrast-enhancing feature for cloud detection in ground-based sky images. J. Atmos. Ocean. Technol. 32 (2), 209–219.

Jebaraj, S., Iniyan, S., 2006. A review of energy models. Renew. Sust. Energ. Rev. 10 (4), 281–311.

Ji, W., Chee, K.C., 2011. Prediction of hourly solar radiation using a novel hybrid model of (ARMA) and (TDNN). Sol. Energy 85 (5), 808–817.

Kacprzyk, J., Wilbik, A., Zadrożny, S., 2008. Linguistic summarization of time series using a fuzzy quantifier driven aggregation. Fuzzy Sets Syst. 159 (12), 1485–1499.

Kaggle, 2016. How much did it rain? II. https://www.kaggle.com/c/how-much-did-it-rain-ii.

Kelly III, J., Hamm, S., 2013. Smart Machines: IBM's Watson and the Era of Cognitive Computing. Columbia University Press, New York, NY.

Kenny, J.F., Barber, N.L., Hutson, S.S., Linsey, K.S., Lovelace, J.K., Maupin, M.A., 2009. Estimated Use of Water in the United States in 2005, U.S. Geological Survey Circular 1344. US Geological Survey, Virginia.

Kim, K.H., Park, J.K., Hwang, K.J., Kim, S.H., 1995. Implementation of hybrid short-term load forecasting system using artificial neural networks and fuzzy expert systems. IEEE Trans. Power Syst. 10 (3), 1534–1539.

Klein, B., Wolf, L., Afek, Y., 2015. A dynamic convolutional layer for short range weather prediction. In: Proceedings of the IEEE Conference on Computer Vision and Pattern Recognition, pp. 4840–4848.

Kumar, A., Srivastava, S., Singh, S., 2005. Congestion management in competitive power market: a bibliographical survey. Electr. Power Syst. Res. 76 (1), 153–164.

Längkvist, M., Karlsson, L., Loutfi, A., 2014. A review of unsupervised feature learning and deep learning for time-series modeling. Pattern Recogn. Lett. 42, 11–24. http://www.sciencedirect.com/science/article/pii/S0167865514000221.

Lave, M., Kleissl, J., Arias-Castro, E., 2012a. High-frequency irradiance fluctuations and geographic smoothing. Sol. Energy 86, 2190–2199. http://dx.doi.org/10.1016/j.solener.2011.06.031.

Lave, M., Kleissl, J., Stein, J., 2012b. A wavelet-based variability model (WVM) for solar PV power plants. IEEE Trans. Sustainable Energy 4 (2), 501–509.

Leese, J.A., Novak, C.S., Clark, B.B., 1971. An automated technique for obtaining cloud motion from geosynchronous satellite data using cross correlation. J. Appl. Meteorol. 10 (1), 118–132.

Li, S., Wunsch, D.C., O'Hair, E., Giesselmann, M.G., 2001. Comparative analysis of regression and artificial neural network models for wind turbine power curve estimation. J. Sol. Energy Eng. 123 (4), 327–332.

Lonij, V.P.A., Jayadevan, V.T., Brooks, A.E., Rodriguez, J.J., Koch, K., Leuthold, M., Cronin, A.D., 2012. Forecasts of PV power output using power measurements of 80 residential PV installs. In: 2012 38th IEEE Photovoltaic Specialists Conference (PVSC).

Lonij, V.P., Brooks, A.E., Cronin, A.D., Leuthold, M., Koch, K., 2013. Intra-hour forecasts of solar power production using measurements from a network of irradiance sensors. Sol. Energy 97, 58–66.

Lonij, V.P.A., Fiot, J.-B., Chen, B., Fusco, F., Pompey, P., Gkoufas, Y., Sinn, M., Tougas, D., Coombs, M., Stamp, A., 2016. A scalable demand and renewable energy forecasting system for distribution grids. In: Proceedings of the IEEE PES General meeting.

Luenberger, D., 1979. Introduction to Dynamic Systems: Theory, Models, and Applications. Wiley, New York, NY.

Mandal, P., Senjyu, T., Urasaki, N., Funabashi, T., 2006. A neural network based several-hour-ahead electric load forecasting using similar days approach. Int. J. Electr. Power Energy Syst. 28 (6), 367–373.

Marion, B., Adelstein, J., Boyle, K., Hayden, H., Hammond, B., Fletcher, T., Canada, B., Narang, D., Kimber, A., Mitchell, L., Rich, G., Townsend, T., 2005. Performance parameters for grid-connected PV systems. In: Photovoltaic Specialists Conference, 2005. Conference Record of the Thirty-first IEEE, pp. 1601–1606.

Marquez, R., Pedro, H.T., Coimbra, C.F., 2013. Hybrid solar forecasting method uses satellite imaging and ground telemetry as inputs to ANNs. Sol. Energy 92, 176–188.

Marvuglia, A., Messineo, A., 2012. Monitoring of wind farms' power curves using machine learning techniques. Appl. Energy 98, 574–583.

Mathiesen, P., Kleissl, J., 2011. Evaluation of numerical weather prediction for intra-day solar forecasting in the continental United States. Sol. Energy 85, 967–977.

Mellit, A., Pavan, A.M., 2010. A 24-h forecast of solar irradiance using artificial neural network: application for performance prediction of a grid-connected PV plant at Trieste, Italy. Sol. Energy 84 (5), 807–821.

Metaxiotis, K., Kagiannas, A., Askounis, D., Psarras, J., 2003. Artificial intelligence in short term electric load forecasting: a state-of-the-art survey for the researcher. Energy Convers. Manag. 44 (9), 1525–1534.

Mishra, K.V., Kruger, A., Krajewski, W.F., 2014. Compressed sensing applied to weather radar. In: Geoscience and Remote Sensing Symposium (IGARSS), 2014 IEEE International, 1832–1835.

Monticelli, A., Wu, F.F., 1985. Network observability: theory. IEEE Trans. Power App. Syst. PAS-104 (5), 1042–1048.

Moreno, J., Ortúzar, M.E., Dixon, J.W., 2006. Energy-management system for a hybrid electric vehicle, using ultracapacitors and neural networks. IEEE Trans. Ind. Electron. 53 (2), 614–623.

Mounce, S., Shepherd, W., Sailor, G., Shucksmith, J., Saul, A., 2014. Predicting combined sewer overflows chamber depth using artificial neural networks with rainfall radar data. Water Sci. Technol. 69 (6), 1326–1333.

National Intelligence Council, 2012. Global trends 2030: alternative worlds. National Intelligence Council, Washington, DC.

Nayak, D.R., Mahapatra, A., Mishra, P., 2013. A survey on rainfall prediction using artificial neural network. Int. J. Comput. Appl. 72 (16), 32–40.

Nguyen, D.A., Kleissl, J., 2014. Stereographic methods for cloud base height determination using two sky imagers. Sol. Energy 107, 495–509.

Oki, T., Kanae, S., 2006. Global hydrological cycles and world water resources. Science 313 (5790), 1068–1072.

Palma, J., Castro, F., Ribeiro, L., Rodrigues, A., Pinto, A., 2008. Linear and nonlinear models in wind resource assessment and wind turbine micro-siting in complex terrain. J. Wind Eng. Ind. Aerodyn. 96 (12), 2308–2326.

Papin, C., Bouthemy, P., Rochard, G., 2002. Unsupervised segmentation of low clouds from infrared METEOSAT images based on a contextual spatio-temporal labeling approach. IEEE Trans. Geosci. Remote Sens. 40 (1), 104–114.

Park, D.C., El-Sharkawi, M., Marks, R., Atlas, L., Damborg, M., et al., 1991. Electric load forecasting using an artificial neural network. IEEE Trans. Power Syst. 6 (2), 442–449.

Perez, R., Ineichen, P., Kmiecik, M., Moore, K., Renne, D., George, R., 2004. Producing satellite-derived irradiances in complex arid terrain. Sol. Energy 77 (4), 367–371. ISSN 0038-092X. http://dx.doi.org/10.1016/j.solener.2003.12.016. http://www.sciencedirect.com/science/article/pii/S0038092X03004687.

Perez, R., Kivalov, S., Schlemmer, J., Hemker Jr., K., Renne, D., Hoff, T.E., 2010. Validation of short and medium term operational solar radiation forecasts in the US. Sol. Energy 0038-092X84 (12), 2161–2172. http://dx.doi.org/10.1016/j.solener.2010.08.014.http://www.sciencedirect.com/science/article/pii/S0038092X10002823.

Perez, R., Kivalov, S., Schlemmer, J., Hemker Jr., K., Hoff, T.E., 2012. Short-term irradiance variability: preliminary estimation of station pair correlation as a function of distance. Sol. Energy 0038-092X86 (8), 2170–2176. http://dx.doi.org/10.1016/j.solener.2012.02.027.http://www.sciencedirect.com/science/article/pii/S0038092X12000928.

Rabiner, L.R., Juang, B.H., 1986. An introduction to hidden Markov models. IEEE ASSP Mag. 3 (1), 4–16.

Shi, J., Lee, W.J., Liu, Y., Yang, Y., Wang, P., 2012. Forecasting power output of photovoltaic systems based on weather classification and support vector machines. IEEE Trans. Ind. Appl. 48 (3), 1064–1069.

Siano, P., Cecati, C., Yu, H., Kolbusz, J., 2012. Real time operation of smart grids via FCN networks and optimal power flow. IEEE Trans. Ind. Inf. 8 (4), 944–952.

Song, K.B., Ha, S.K., Park, J.W., Kweon, D.J., Kim, K.H., 2006. Hybrid load forecasting method with analysis of temperature sensitivities. IEEE Trans. Power Syst. 21 (2), 869–876. ISSN 0885-8950. http://dx.doi.org/10.1109/TPWRS.2006.873099.

Sripada, S., Reiter, E., Davy, I., 2003. SumTime-Mousam: configurable marine weather forecast generator. Expert Update 6 (3), 4–10.

Suganthi, L., Samuel, A.A., 2012. Energy models for demand forecasting—a review. Renew. Sust. Energ. Rev. 16 (2), 1223–1240.

Sutskever, I., Hinton, G.E., 2008. Deep, narrow sigmoid belief networks are universal approximators. Neural Comput. 20 (11), 2629–2636.

Tapiador, F., Kidd, C., Hsu, K.L., Marzano, F., 2004. Neural networks in satellite rainfall estimation. Meteorol. Appl. 11 (1), 83–91.

Teleke, S., Baran, M.E., Huang, A.Q., Bhattacharya, S., Anderson, L., 2009. Control strategies for battery energy storage for wind farm dispatching. IEEE Trans. Energy Convers. 24 (3), 725–732.

Tokar, A.S., Johnson, P.A., 1999. Rainfall-runoff modeling using artificial neural networks. J. Hydrol. Eng. 4 (3), 232–239.

Treinish, L.A., Cipriani, J.P., Praino, A., Singhee, A., Wang, H., Sinn, M., Lonij, V., Fiot, J.-B., Chen, B., 2016. Enabling coupled models to predict the business impact of weather on electric utilities. IBM J. Res. Dev. 60 (1), 5:1–5:10. ISSN 0018-8646. http://dx.doi.org/10.1147/JRD.2015.2489478.

United Nations World Water Assessment Programme, 2015. The United Nations world water development report. Water for a sustainable world.

United Nations World Water Development Programme, 2014. The United Nations world water development report. Water and energy.

Valiant, L.G., 1995. Cognitive computation. In: FOCS. IEEE, Washington, DC, p. 2.

Valipour, M., Banihabib, M.E., Behbahani, S.M.R., 2013. Comparison of the ARMA, ARIMA, and the autoregressive artificial neural network models in forecasting the monthly inflow of Dez dam reservoir. J. Hydrol. 476, 433–441.

Voyant, C., Muselli, M., Paoli, C., Nivet, M.L., 2012. Numerical weather prediction (NWP) and hybrid ARMA/ANN model to predict global radiation. Energy 39, 341–355. http://dx.doi.org/10.1016/j.energy.2012.01.006.

Voyant, C., Muselli, M., Paoli, C., Nivet, M.L., 2013. Hybrid methodology for hourly global radiation forecasting in Mediterranean area. Renew. Energy 53, 1–11. http://dx.doi.org/10.1016/j.renene.2012.10.049.

Wang, J., Shahidehpour, M., Li, Z., 2008. Security-constrained unit commitment with volatile wind power generation. IEEE Trans. Power Syst. 23 (3), 1319–1327.

Welch, G., Bishop, G., 2006. An introduction to the Kalman filter. Technical report.

Wood, S., 2006. Generalized Additive Models: An Introduction with R. CRC Press, Boca Raton, FL.

Wood, A.J., Wollenberg, B.F., 2012. Power Generation, Operation, and Control. John Wiley & Sons, Hoboken, NJ.

World Health Organization, 2014. UN-water global analysis and assessment of sanitation and drinking-water (GLAAS).

Xingjian, S., Chen, Z., Wang, H., Yeung, D.Y., Wong, W.K., Woo, W.C., 2015. Convolutional LSTM network: a machine learning approach for precipitation nowcasting. In: Advances in Neural Information Processing Systems, 802–810.

Zhang, Y., Yu, R., Nekovee, M., Liu, Y., Xie, S., Gjessing, S., 2012. Cognitive machine-to-machine communications: visions and potentials for the smart grid. IEEE Netw. 26 (3), 6–13.

Chapter 9

Cognitive Computing Applications in Education and Learning

M.T. Irfan* and V.N. Gudivada[†,1]
*Bowdoin College, Brunswick, ME, United States
[†]East Carolina University, Greenville, NC, United States
[1]Corresponding author: e-mail: gudivadav15@ecu.edu

ABSTRACT

Education and learning stand out among many application areas of cognitive computing due to their practical appeal as well as their research challenge. We can gauge the level of research challenge once we recognize the broad spectrum of human learning, the complex and not fully understood human learning process, and various influencing factors of learning, such as pedagogy, technology, and social elements. We survey some of the key research problems and recent efforts to solve them. In particular, we examine the important roles played by the Educational Data Mining and Learning Analytics communities in these research efforts. We find evidence of significant progress over the last few years. We also find evidence of concerted efforts by academia and industry in bringing some of these research results to the public in the form of learning and educational software.

Keywords: Cognitive learning, Learning Analytics, Educational Data Mining, Cognitive systems, Cognitive computing, Personalized learning, Data science

1 INTRODUCTION

Much of the excitement about cognitive computing is spurred by its enormous potential in learning, only a small fraction of which has so far been realized. The overarching goal here is to devise computational frameworks to help us learn better by exploiting data about our learning processes and activities. There are two important aspects of it—the mechanisms or insights about *how* we actually learn and the external manifestations of our learning activities. The state of the art at this moment reveals that a lot of energy and focus have gone into the latter front, while a lot remains to be done on the former front. An example will help distinguish these two.

Handbook of Statistics, Vol. 35. http://dx.doi.org/10.1016/bs.host.2016.07.008

It is well known that humans learn from a very small number of examples. For instance, it takes us only one example of a stop sign and one example of a lobster to differentiate one from the other, even though both are red. It is an extremely complex question how our brains process these two different objects in terms of their representations as well as their generalizations. The representation of a stop sign may encode some visual clues about a stop sign, such as its polygonal shape and its color, whereas its generalization would enable us to recognize stop signs from various distances as well as angles. Devising a computational framework to learn from a very small number of examples is currently at a nascent stage and obviously has an enormous potential for future work (Lake et al., 2015).

In contrast, we can devise a computational framework for the same task *if* we are allowed to use a large number of examples. More specifically, we can "train" a machine learning algorithm with a large number of labeled examples of both stop signs and lobsters (Murphy, 2012). Typically, these examples are represented by various features. In some applications these features are handcrafted while in others they are automatically discovered by the algorithm itself. The important requirement here is a large amount of data. In education and learning, we often generate a huge amount of data through our learning activities, which presents us with both an enormous potential and a great challenge. The reader may readily relate to this as *Big Data*.

Big Data are typically characterized by five V's—volume (data are large-scale), variety (data come in many different forms), velocity (data are generated over time, often times in a "streaming" fashion), veracity (data have some elements of uncertainty), and value (extracting actionable intelligence). In the application settings of education and learning, the first four V's are prominently visible. For example, a Learning Management System (LMS) like Moodle (2016) or any Massive Open Online Course (MOOC) generates a huge volume of data (see Section 2). Moreover, the data come in a variety of forms, such as a student's answers to quantitative questions as well as essay-type questions, which are very different in many respects. It is also obvious that the data are generated continually with great velocity and that there will be uncertainty in the data if we collect data about students' level of engagement or their understanding of materials. The cognitive computing community has embraced the emergence of Big Data, thereby shaping up several subfields relevant to education and learning. The ongoing efforts have led to many success stories.

In the next section, we discuss Educational Data Mining (EDM) and Learning Analytics (LA)—two major areas of cognitive computing dealing with education and learning. Next, we present a survey of some recent results in EDM and LA. We conclude this chapter by touching upon some success stories and giving pointers to several challenging research directions. The following acronyms are used in this chapter.

ACT-R	Adaptive Control of Thought-Rational
AFM	Additive Factor Model
EDM	Educational Data Mining
ISPeL	Interactive System for Personalized eLearning
ITS	Intelligent Tutoring Systems
LA	Learning Analytics
LAK	Learning Analytics and Knowledge (Conference)
LMS	Learning Management System
MOOC	Massive Open Online Course
PFA	Performance Factors Analysis

2 EDM AND LA

EDM and LA are two emerging areas that embody the major applications of cognitive computing to education and learning. These two areas are not separated by any hard boundary and to the contrary, they often share common objectives and stakeholders. However, they fundamentally differ in several dimensions, including research methodology. These differences have been beautifully explained by Siemens and Baker (2012). In brief, humans play a central role in LA, whereas the bulk of the work in EDM is performed through automation. Humans do play a role in EDM, but that is peripheral in comparison to LA where they are the key components of the apparatus. Furthermore, LA models and analyzes the learning system as a whole, in contrast to EDM's divide-and-conquer approach. Next, we address these two areas separately. We also give an illustrative example for each.

2.1 Educational Data Mining

EDM has been inspired by advances in data mining and machine learning. The overarching goal is to help teachers, students, and other stakeholders achieve their respective objectives by utilizing Big Data in education. This critically hinges on understanding and modeling how students learn. As a result, The International Educational Data Mining Society (2016) has defined the field as "concerned with developing methods for exploring the unique and increasingly large-scale data that come from educational settings, and using those methods to better understand students, and the settings which they learn in." We should note here that this society organizes the annual International Conference on Educational Data Mining, which debuted in 2008.

The recurrent research themes in EDM are: mining educational data from various sources that include MOOCs and Intelligent Tutoring Systems (ITS); mining human–computer interaction (HCI) data in the educational setting; assessing students' affective state using multimodal sensory data; modeling students and predicting their performance; modeling, curating, and evaluating learning pathways; and connecting EDM with other related disciplines, including emerging areas like LA (see Section 2.2) as well as more mature areas like learning theory and learning sciences.

Although EDM started as a subfield of data mining, it now boasts an interesting and varied research landscape. As we will see in Section 3, it has welcomed tools and techniques from HCI and computer vision. In fact, many of the recent works in EDM can be thought of as data-centric applications of cognitive computing in education and learning.

For the most part, the methodologies in EDM research are rooted in several major areas of computer science, such as data mining and machine learning. Although EDM research typically keeps humans in the loop, the major focus is on automation through computational modeling, tools, and techniques. Next, we present a simplistic example to illustrate this point.

2.1.1 An EDM Example

For this fictitious example, imagine that a tutor has access to multidimensional data about her students' learning behavior on an online learning system. In particular, the system has recorded each student's number of sign-ins and the percentage of lesson completion as a measure of the student's progress. The input data points are shown as black dots in Fig. 1A. This figure is generated using an online visualization tool (Mohan, 2016). The tutor's goal is to *classify* her students according to their behavior to better assist them. However, the labeling for this classification is not clear a priori. This is a typical unsupervised learning problem in machine learning, also known as clustering (Murphy, 2012).

Among many different approaches to clustering, we pick the well-known k-means clustering algorithm for this example. The parameter k represents the number of clusters desired. An illustration is shown in Fig. 1 for $k = 4$. The first step is to choose k points for cluster centers (aka centroids), often times randomly. The initial centroids are shown as triangles in Fig. 1A.

Assuming that the points lie in a 2D Cartesian coordinate plane, the x coordinate of the centroid is the average of the x coordinates of all the points that belong to the cluster. The y coordinate of the cluster is calculated similarly. The same principle applies for points in an n-dimensional space. However, a suitable distance metric needs to be defined to measure distance between two points. The centroids need not be actual data points. Next, each data point is classified to its closest centroid based on a distance metric. This is shown in Fig. 1B, where different classes of points are shown in different colors. This classification is not an ideal grouping of points since cluster boundaries are not crisp. However, this completes one iteration of the algorithm.

In the next iteration, for each class, its new centroid is recalculated. The recalculated centroids are shown in Fig. 1C. Since the centroids have changed,

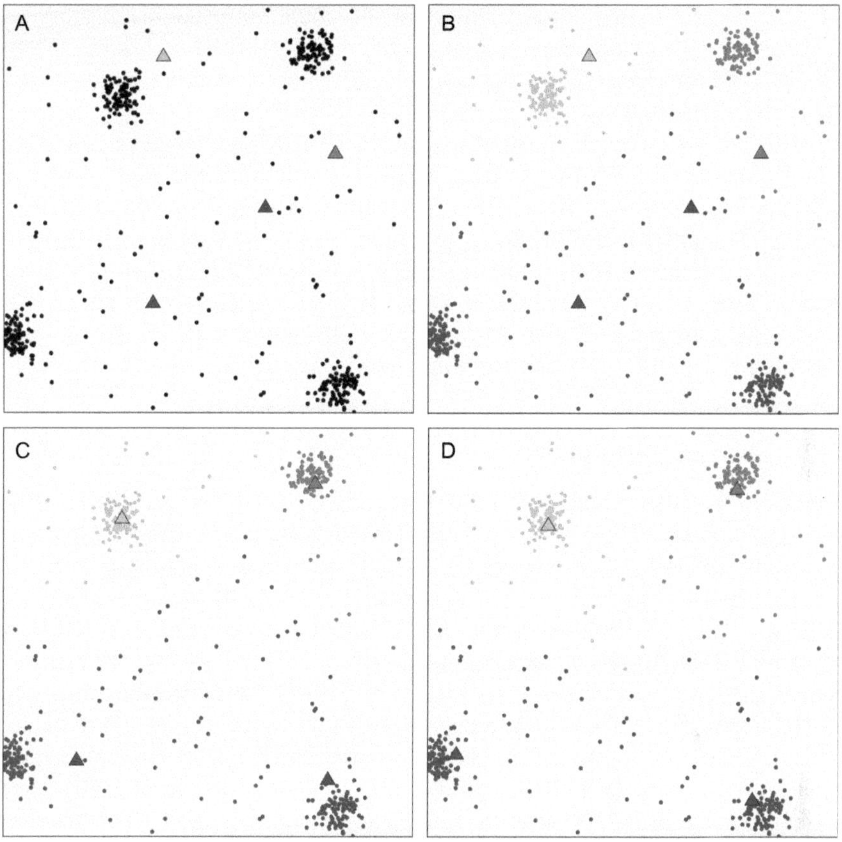

FIG. 1 (A) Input data points and four initial centroids (*triangles* filled with *different colors* (different *gray shades* in the print version)) are shown. The *x*-axis represents the number of times a student has signed in and the *y*-axis the percentage of lesson completion. Each data point corresponds to one student. (B) The data points are classified according to their proximity to the centroids. (C) The centroid of each of the four classes is recalculated by taking the average of the *x* and *y* coordinates of the data points in that class. (D) Successive iterations of centroid (re-)calculation and data point classification ultimately converge to this clustering. An interpretation of these clusters requires human experts with domain knowledge.

some of the points will now move into a different class. The recalculation of the centroids and the classification of the points are repeated, ultimately converging to the scenario shown in Fig. 1D.

The interpretation of clustering is most often left to human experts with domain knowledge—in this case, the tutor. The tutor may interpret the bottom left cluster shown in Fig. 1D as the group of students who made very little progress and also lacked in their level of efforts as evident from the small number of sign-ins. She may interpret the bottom right cluster as the group of students who signed in many times but still could not make much progress, perhaps giving an indication of short attention spans. Based on her interpretation of the four clusters, the tutor may design separate learning paths for different types of students.

This simple example illustrates many of Siemens and Baker's points on EDM (Siemens and Baker, 2012). First, it shows that the main focus of EDM is automated discovery. Human expertise is a means to that end. Second, EDM's approach of breaking up a problem into smaller parts is obvious here, since we are dealing with one small part of what we may call ITS. More interestingly, we can extend this example to model students with the goal of predicting learning outcomes. Lastly, Fig. 1 illustrates some of the common tools used in EDM, such as clustering and visualization.

2.2 Learning Analytics

LA is a relatively new field with the overarching goal of synergistically combining human judgment with technological advances for a wide range of learning activities, spanning from classroom learning to informal learning. The first International Conference on Learning Analytics and Knowledge (LAK) was held in 2011. At that time, the LA community identified LA field as: "Learning analytics is the measurement, collection, analysis and reporting of data about learners and their contexts, for purposes of understanding and optimising learning and the environments in which it occurs" (Long and Siemens, 2011). Since then the field has evolved significantly. For example, the sixth LAK conference held in 2016 Gašević and Lynch (2016) has identified five major branches of LA: (1) analytic approaches, methods, and tools; (2) theories and theoretical concepts; (3) measures of learning, change, and success; (4) learning activities, applications, and interventions; and (5) data sources.

Compared to EDM, LA embodies a broader umbrella of learning settings, tools, techniques, human and social factors, and applications areas. EDM is focused on automation and often times concerns with unearthing hidden patterns from data, whereas LA is more holistic and human-centric (Siemens and Baker, 2012). In a similar vein, as we will see in Section 3, predicting student performance is a central problem in EDM. In contrast, LA is more concerned with predicting the outcome of a learning approach or system. Also, LA is inherently an area of analytics, and therefore various analytical tools, such as discourse analysis and social network analysis, are more prominent in LA than in EDM. However, the border between these two fields is sometimes blurred due to their shared objectives and stakeholders. Furthermore, since both fields deal with predictive models (although applied to different domains), many of the computational techniques pertaining to machine learning and data mining are common between them. Next, we present an example of a cognitive model for LA.

2.2.1 An LAs Example

Given LA's holistic approach to an overall system, it is hard to come up with simple examples like the one we presented for EDM in Section 2.1.1. Here, we present a recent application architecture proposed by Gudivada (2016) for personalized learning. He notes that learning applications typically generate huge volumes of structured and unstructured data. Structured data can be easily obtained from a LMS's logs, whereas unstructured data come from a variety of sources such as blogs, emails and other forms of course messages, and discussions. Due to the ease of working with structured data, a vast majority of research on LA deals with only that type of data. However, if we consider building an effective system for personalized learning, dealing with unstructured data becomes essential. Gudivada proposes a comprehensive framework for personalized learning that can work with both structured and unstructured data. The framework is named Interactive System for Personalized eLearning (ISPeL).

Gudivada's architecture is hierarchical, roughly consisting of the following layers: network and hardware layer, databases and data analytics layer, cognitive analytics layer, and personalization layer. An illustration of the application level architecture is shown in Fig. 2. Some preliminaries are due before we describe this architecture.

First, Darwin Information Typing Architecture (DITA) is an XML data model for single source authoring and multichannel publishing (Bellamy et al., 2011; White, 2013). A DITA content (e.g., computer algorithms) is composed of DITA topics (e.g., time and space complexity, greedy algorithms, dynamic programming). It is desirable that each topic is minimal but self-contained. Second, a Resource Description Framework Schema is a general-purpose ontology language (Broekstra et al., 2001), typically used for storing domain information and topic hierarchies. Third, LMSs are software applications to facilitate education. Well-known examples of LMSs are Blackboard, Moodle, and D2L.

We are now ready to describe the ISPeL architecture in a bottom-up fashion. The bottom layer of ISPeL (Fig. 2) consists of nine databases. These databases dynamically keep track of information from various sources in order to provide the students a personalized learning experience. For example, the *DITA Topics* database keeps information about the topics being taught. The *Assessment Data* consists of each student's scores, time to complete tests, performance level on tests, and other related information. The *User Models* database keeps information about individual students, their preferences and progress, and learning styles.

The Question Generation Models, Scaffolding & Feedback Models, and Personalization Models are the main components for generating personalized questions, feedback, and lesson plans. The *Question Generation Models* module has capability to automatically generate questions for assessing learners.

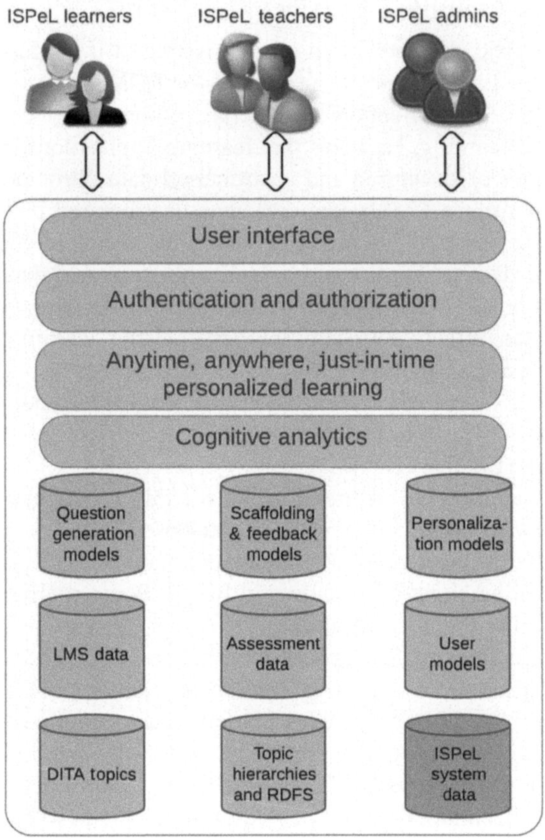

FIG. 2 ISPeL architecture for personalized learning (Gudivada, 2016).

This module also supports practice exams before the actual test. The *Scaffolding & Feedback Models* module features capability to provide guided learning and context-appropriate feedback. The *Personalization Models* component uses prerequisite dependencies among DITA topics and user models to achieve personalization. Each learner progresses at her own pace, explores learning materials in any order they choose subject to prerequisite constraints.

The cognitive analytics layer uses the data mentioned earlier together with various algorithmic and analytical techniques (e.g., data mining, machine learning, statistical analysis, natural language processing, information visualization) to provide personalized learning to students. The layer above it deals with multiple modes of personalized learning, such as anytime, anywhere, and just-in-time. The next two layers provide authentication and user interface, through which users with different credentials interact with the software application.

The ISPeL architecture shown in Fig. 2 is in tune with Siemens and Baker's call for connecting LA with EDM (Siemens and Baker, 2012). The architecture itself is holistic, but some of the individual components of the whole system are significantly challenging in their own merit. For example, for ISPeL to provide personalized learning, it is necessary to generate appropriate questions and feedback for each individual learner *automatically* and in real time. An EDM expert can help with this automation. However, this automated component must function robustly in concert with the other components of the cognitive analytics layer, which illustrates both the beauty and challenge of LA.

3 RECENT RESEARCH

In this section, we review some recent research on cognitive computing and its applications to education and learning. Many of these research results have appeared in the EDM and LAK conferences, which are the premier venues for publishing research on EDM and LA, respectively. Since EDM and LA are both very much expansive areas, it will not be possible to touch upon many of their subareas in this chapter. As a result, we will only present some recent results in a few selected problem domains. The intention is to give the reader a sampling of the progress made during the last few years. Let us first illustrate the computational challenges using the following example.

Many of our cognitive tasks involve semistructured as well as unstructured input data. Among these are tasks that we deem very much natural to us. One example is geometric word problems. Fig. 3 shows a sample problem. Problems of this type are often specified by a text description in conjunction with a diagram. In many cases, the text description alone is not sufficient to fully understand the problem, since some critical information is only given in the diagram. For example, the text description in Fig. 3 omits the important fact that AD intersects with BC at point D. Inferring such facts by combining a text description with the corresponding diagram is natural to us. An extremely challenging problem is to automate it. Very recently, Seo et al. (2015) have presented an automated solver for SAT-level word geometric problems. Their method consists of the following main components: a text parser, a diagram

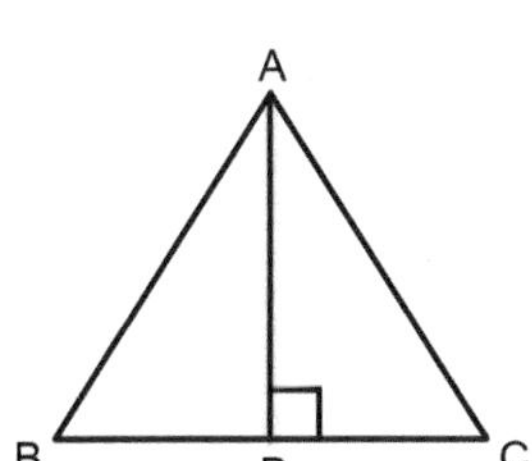

FIG. 3 A sample geometric word problem, similar to the examples given by Seo et al. (2015).

parser, and an optimizer. These components work together to produce the best possible formal representation of the problem that can be fed to a standard geometric problem solver. On unseen, official SAT geometry problems, their automated solver achieves a remarkable 49% score.

As evident from the earlier example, the full spectrum of human cognition and learning poses a grand, multiprong research challenge. This challenge is only magnified when we also consider pedagogy, technology, human judgment, social factors, and various contextual elements. We next present a survey of some recent research results in several selected categories.

3.1 Intelligent Tutoring Systems

Since the 1970s (Carbonell, 1970), ITS have been at the forefront of artificial intelligence research with a diverse array of application areas ranging from physics (Vanlehn et al., 2005) and mathematics (Matsuda and VanLehn, 2005; Melis and Siekmann, 2004) to adult education (Cheung et al., 2003) and nurse education (Hospers et al., 2003). Here, we will narrow down on some recent ITS research on learning programming. Since the publication of Anderson and Reiser's seminal work on an LISP tutoring system (Anderson and Reiser, 1985), this field has gained a lot of attention over the years. However, it is fair to say that we are still far from realizing its full potential.

Any discussion on modern-day ITS must begin with the cognitive architectures behind ITS. One of the most influential cognitive architectures is Anderson's Adaptive Character of Thought-Rational (ACT-R) theory. The central tenet of the ACT-R theory is that human cognition is the result of interactions among numerous small, indivisible units of knowledge in certain ways. First, these basic units of knowledge are of two types: declarative and procedural. For any cognitive task, the declarative units of knowledge represent the states of cognition, or in Anderson's words, "objects in the environment" (Anderson, 1996). In contrast, procedural units of knowledge, also known as *productions*, transform human cognition from one state to another.

The ACT-R theory gives the details of how these knowledge units interact, which is out of scope for this chapter. However, we should mention that ACT-R is not an abstract theory of human cognition. It is rather a concrete framework similar to a programming language. Using a programming language, we can write programs and run them. Similarly, using ACT-R, we can create a cognitive model for a specific cognitive task (e.g., computing the factorial of a number) and simulate it. The simulation details the steps in human cognitive process for that particular task under various modeling assumptions. These assumptions are usually grounded in extensive psychological experiments. We should also be noted that the ACT-R theory—in its original form—comes with a LISP interpreter where simulations can be run. Today, there are also Python and Java implementations of the ACT-R theory (Harrison, 2002; Stewart and West, 2005).

Founded on several decades of research on human cognition and intelligence, ITS is now a fast growing area in academia and industry. We now turn our attention to some cutting-edge research on ITS in a specific learning domain: programming. Several recent papers deal with the problem of helping students learn programming, in particular by giving them useful hints in real-time while they are coding. This is a vastly challenging problem, mainly because even for very simple programming tasks there are a multitude of different solution approaches, both syntactically and semantically. Even if we restrict the semantic aspect (i.e., the underlying algorithm) to a single one, the syntactic variations of implementing the algorithm present a daunting task for hint generation.

Not surprisingly, earlier work on hint generation focused on more specific problem-solving tasks, such as logic (Barnes et al., 2008; Stamper et al., 2013) and linked lists (Fossati et al., 2009). New efforts to tackle broader programming problems are at a nascent stage and use previous students' solutions to a programming task to generate hints for a new student who is working on the same task (Price and Barnes, 2015; Price et al., 2016). The basic technique in this new line of work is to first represent the previous student–tutor interactions in the form of an *interaction network* (Eagle et al., 2012). When a new student asks for a hint, that student's interaction pattern is matched with some part of the interaction network and the student is directed to an appropriate next step that ultimately leads to a solution. It is not hard to imagine the potential impact of such work on any ITS that teaches programming.

In a related work, Eagle et al. (2015) present a social network analysis of the interaction network for multiple problem domains. They found that the interaction networks exhibit power-law degree distribution. They also showed the importance of several network properties like *degree assortativity* for comparing interaction networks across different domains.

Closely connected to improving an ITS is evaluating its adaptive tutoring feature, which is very much applicable to ITS for programming. Traditionally, the literature suggests using evaluation schemes based on machine learning, such as the Performance Factors Analysis (PFA) cognitive model. This poses a challenge to course designers without a background in machine learning. A recent work addresses this issue and presents two new evaluation metrics better suited for adaptive tutoring than PFA (González-Brenes and Huang, 2015).

3.2 Clustering and Student Modeling

Clustering is a common technique in EDM for aggregating student data in order to examine student behavior. Klingler et al. (2016) have recently presented an evolutionary clustering technique for sequential student data. In comparison with previous works, their technique improves the stability of clustering for noisy data. In addition, they show that their work can be incorporated into any ITS as a black box.

Another central problem in EDM is *student modeling*. A lot of interesting research has emerged on this topic recently. One such work is on modeling learning curves. Using data from Duolingo (2016), Streeter (2015) has used probabilistic mixture models to capture the learning curves of language learners. Here, an individual's learning curve represents her error percentage over time. This is an interesting research direction, because prior to this, *knowledge tracing* has been the dominant criterion for modeling student learning (Corbett and Anderson, 1994). Streeter's work generalizes knowledge tracing and offers an elegant probabilistic model for modeling learning curves. The parameters of the model have been learned using the well-known expectation–maximization algorithm. Based on the large-scale Duolingo dataset, the mixture model outperforms many of the previous approaches, including popular cognitive models like Additive Factor Model (AFM) (Cen et al., 2006) and PFA (Pavlik et al., 2009).

Others have also worked to improve the AFM and PFA models. A recent cognitive model by MacLellan et al. (2015) has extended these in order to account for the commonly observed phenomenon that students often make mistakes on previously learned skills, which the authors have termed *slipping*. Their slipping models outperform the baseline AFM and PFA models on five datasets.

Although a lot of work has been done on modeling student learning, modeling prior knowledge has received relatively less attention. In a recent work, Nižnan et al. (2015) have presented several models to estimate prior knowledge. They started with the basic model of *Elo rating* and extended it in several ways. They evaluated the models on prior knowledge of geography, which is a subject of widely varying prior knowledge. Perhaps surprisingly, they showed that the more complex extensions of the basic Elo rating model do not outperform Elo rating by much. Although not mentioned explicitly in their paper, this could be attributed to the basic Elo rating hitting the sweet spot between overfitting and underfitting (Murphy, 2012) for that particular geography dataset.

3.3 Predicting Student Performance

Predicting student performance based on various factors has been a popular line of research in EDM. Traditionally, student performance has been measured in specific disciplines or topics, such as algebra (Stapel et al., 2016), programming in Java (Tomkins et al., 2016), or even as specific as learning fractions in mathematics (Olsen et al., 2015). Although many of the research questions on student performance are domain-specific, there are some unifying themes: How would one predict student performance? What factors are most important for a student's success? How can an ITS make use of this prediction?

Tomkins et al. (2016) answer some of these questions using their case study of a high school computer science MOOC. The MOOC is a separate course with its own evaluation, but the students taking the MOOC ultimately take the computer science Advanced Placement (AP) exam. As a result, there are two performance measures: one coming from the MOOC and the other from the AP exam. It has been empirically observed that a student's score in the AP exam is a better predictor of the student's future success than the student's performance in the MOOC. Based on their machine learning framework, Tomkins et al. identify the most important factors in MOOC for students' high and low scores in the AP exam (not MOOC). One such factor is coaching. Many of the students received coaching while taking the MOOC, while others studied independently. The ones that were coached, showed a greater level of activity in the MOOC's forum through questions, answers, and other contributions. The coached students performed better than the independent students in the MOOC. However, on the actual AP exam, the independent students scored higher. The authors also connect other features, such as textual content of the students' forum activities, to their AP exam performance.

Another new study in the context of MOOCs predicts student drop-out and proposes an intervention study (Whitehill et al., 2015). Student drop-out is an extremely common phenomenon in MOOCs (Onah et al., 2014; Rivard, 2013). Whitehill et al. note that the reasons for drop-out are not necessarily related with the quality of the course as there could be exogenous factors. More importantly, a post-course survey is ineffective in detecting the reasons for drop-outs, since the response rates for such surveys are usually very low. The authors present a classification technique for detecting student drop-outs and propose an intervention strategy in the form of early surveys in order to retain students. They used the HarvardX MOOC platform to evaluate the effectiveness of their strategy.

3.4 Affect Detection and Student Engagement

Affective states are closely related to learning and cognition. We react differently to different experiences during our learning process. Assessing affective states of students is now gaining traction within both the EDM and LA communities. As we have briefly mentioned in Section 2.1, these problems often transcend the boundary of EDM beyond traditional data mining and bring computer vision and multimodal sensory analysis into the fold of EDM research.

Based on recent research trajectories, we must start with the eyes. Episodes of mind wandering is common in various learning activities like reading (Mills and D'Mello, 2015) as well as interacting with an ITS (Mills et al., 2015). Multiple research works have shown that our eyes hold a clue to detecting whether we are paying attention or not (Bixler et al., 2016; Hutt et al., 2016). The interesting observation here is that eye-tracking has been an active research topic in HCI since the inception of HCI and is now contributing to cognitive computing in learning.

As we have mentioned earlier, students' affective states are closely tied to their learning and therefore to their performance. Based on over one thousand middle-school students' interactions with a math tutoring system, San Pedro et al. (2015) find that the fine-grained variations of affect over time ultimately impact a student's test score. For affect detection, the authors used the data of student interactions with the math tutoring system mentioned earlier. There exists a long line of prior work on such interaction-based affect detection (Pardos et al., 2013; Wang et al., 2015) and that many of the affect detectors today are video based.

Kai et al. (2015) have done a comparative study of these two major affect detection methods—video based and interaction based—for various affective states, such as boredom, confusion, delight, frustration while the students play a game called physics playground. They obtained the ground truth about the students' affective states through field observations. They showed that overall the video-based affect detector slightly outperforms the interaction-based one. This can be largely attributed to the affective state of delight for which the video-based detector is very much the superior between the two.

4 CONCLUSION

The frequent interactions between academia and industry are a remarkable feature of cognitive computing, especially when it comes to its applications in education and learning. For example, companies like IBM are embracing cognitive computing to harness the power of Big Data in multiple application areas, including education (Davis, 2016). In another success story, the ACT-R theory that we have discussed in Section 3.1 is the key driver of Carnegie Learning's *MATHia* as well as *Cognitive Tutor* software (Carnegie Learning, Inc., 2016). The Cognitive Tutor software has been adopted by the Miami-Dade county public schools in Florida since 2003, following the finding by an independent research group that the software is indeed more effective than the traditional curriculum alone. Furthermore, the new generation of LMS software like Desire2Learn (D2L) boasts predictive analytics features to identify and help at-risk students (D2L, 2016). Many online platforms for class discussion like BlikBook, which can often work in conjunction with an LMS, have built-in student behavior analytics (BlikBook, 2016). There are many other success stories stemming from the exchanges between scientific research and real-world applications.

Two premier conferences in this field—EDM and LA—are great contributors to a lot of these success stories. These two communities are driven by a grand vision and some clearly identified long-term goals. For example, we would like to build an ITS that can offer personalized learning for complex learning tasks like computer programming and adapt itself with each individual student's learning curve as well as prior knowledge. We would like to detect when a student loses engagement, both in a classroom setting and in an ITS session. Not only that, we would like to identify the causal structure behind it and design appropriate intervention strategies. We would like to design MOOCs with high student retention rates. We would also like to accurately predict student performance from behavioral data that comes in different varieties: structured, semistructured, and unstructured.

In summary, we would like to build cognitive computing applications that will have a greater understanding of how humans learn and can vastly improve the quality of learning and education. Are we there yet? No. Have we made any significant progress? Big yes. In fact, it is fair to say that the progress made so far is only the tip of the iceberg. What lie ahead are some extremely hard research problems and many untrodden research paths to solve those.

REFERENCES

Anderson, J.R., 1996. ACT: a simple theory of complex cognition. Am. Psychol. 51 (4), 355.

Anderson, J.R., Reiser, B.J., 1985. The LISP tutor. Byte 10 (4), 159–175.

Barnes, T., Stamper, J.C., Lehmann, L., Croy, M.J., 2008. A pilot study on logic proof tutoring using hints generated from historical student data. In: EDM, pp. 197–201.

Bellamy, L., Carey, M., Schlotfeldt, J., 2011. DITA Best Practices: A Roadmap for Writing, Editing, and Architecting in DITA, first ed. IBM Press, Indianapolis, IN.

Bixler, R., Mills, C., Wang, X., D'Mello, S., 2016. Automatic gaze-based detection of mind wandering during film viewing. In: Proceedings of EDM.

BlikBook, 2016. A new approach to classroom Q&A. https://www.blikbook.com.

Broekstra, J., Klein, M., Decker, S., Fensel, D., Van Harmelen, F., Horrocks, I., 2001. Enabling knowledge representation on the web by extending RDF schema. In: Proceedings of the 10th International Conference on World Wide Web, Hong Kong, WWW '01ACM, New York, NY, pp. 467–478.

Carbonell, J.R., 1970. AI in CAI: an artificial-intelligence approach to computer-assisted instruction. IEEE Trans. Man Mach. Syst. 11 (4), 190–202.

Carnegie Learning, Inc., 2016. MATHia and Cognitive Tutor personalized learning software. https://www.carnegielearning.com.

Cen, H., Koedinger, K., Junker, B., 2006. Learning factors analysis—a general method for cognitive model evaluation and improvement. In: Intelligent Tutoring Systems. Springer, New York, NY, pp. 164–175.

Cheung, B., Hui, L., Zhang, J., Yiu, S.M., 2003. SmartTutor: an intelligent tutoring system in web-based adult education. J. Syst. Softw. 68 (1), 11–25.

Corbett, A.T., Anderson, J.R., 1994. Knowledge tracing: modeling the acquisition of procedural knowledge. User Model. User Adap. Inter. 4 (4), 253–278.

D2L, 2016. Desire2Learn learning management system. www.d2l.com.

Davis, K., 2016. IBM launches industry's first consulting practice dedicated to cognitive business. http://www-03.ibm.com/press/us/en/pressrelease/47785.wss.

Duolingo, 2016. A game-based approach to learning natural languages. https://www.duolingo.com.

Eagle, M., Johnson, M., Barnes, T., 2012. Interaction networks: generating high level hints based on network community clustering. In: Proceedings of the 5th International Conference on Educational Data Mining. ERIC.

Eagle, M., Hicks, D., Peddycord III, B., Barnes, T., 2015. Exploring networks of problem-solving interactions. In: Proceedings of the Fifth International Conference on Learning Analytics and Knowledge. ACM, New York, NY, pp. 21–30.

Fossati, D., Di Eugenio, B., Ohlsson, S., Brown, C.W., Chen, L., Cosejo, D.G., 2009. I learn from you, you learn from me: how to make iList learn from students. Frontiers in Artificial Intelligence and Applications, vol. 200, pp. 491–498.

Gašević, D., Lynch, G. (Eds.), 2016. LAK '16: Proceedings of the Sixth International Conference on Learning Analytics and Knowledge, Edinburgh, United Kingdom. ACM, New York, NY. ISBN 978-1-4503-4190-5.

González-Brenes, J.P., Huang, Y., 2015. Your model is predictive—but is it useful? Theoretical and empirical considerations of a new paradigm for adaptive tutoring evaluation. In: Proceedings of EDM.

Gudivada, V., 2016. Cognitive analytics driven personalized learning. In: Educational Technology Magazine Special Issue—Big Data and Data Analytics in E-Learning, in press.

Harrison, A., 2002. JACT-R: Java ACT-R. In: Proceedings of the 8th Annual ACT-R Workshop.

Hospers, M., Kroezen, E., Nijholt, A., op den Akker, R., Heylen, D., 2003. An agent-based intelligent tutoring system for nurse education. In: Moreno, A., Nealon, J. (Eds.), Applications of Software Agent Technology in the Health Care Domain. Springer, New York, NY, pp. 143–159.

Hutt, S., Mills, C., White, S., Donnelly, P.J., D'Mello, S.K., 2016. The eyes have it: gaze-based detection of mind wandering during learning with an intelligent tutoring system. In: Proceedings of the 9th International Conference on Educational Data Mining, International Educational Data Mining Society, pp. 86–93.

Kai, S., Paquette, L., Baker, R., et al., 2015. Comparison of video-based and interaction-based affect detectors in physics playground. In: Proceedings of EDM, pp. 77–84.

Klingler, S., Käser, T., Solenthaler, B., Gross, M., 2016. Temporally coherent clustering of student data. In: Proceedings of the 9th International Conference on Educational Data Mining, International Educational Data Mining Society, pp. 102–109.

Lake, B.M., Salakhutdinov, R., Tenenbaum, J.B., 2015. Human-level concept learning through probabilistic program induction. Science 350 (6266), 1332–1338. ISSN 0036-8075. http://dx.doi.org/10.1126/science.aab3050. http://science.sciencemag.org/content/350/6266/1332.

Long, P., Siemens, G. (Eds.), 2011. LAK '11: Proceedings of the 1st International Conference on Learning Analytics and Knowledge, Banff, Alberta, Canada. ACM, New York, NY. ISBN 978-1-4503-0944-8.

MacLellan, C.J., Liu, R., Koedinger, K.R., 2015. Accounting for slipping and other false negatives in logistic models of student learning. In: Proc. EDM.

Matsuda, N., VanLehn, K., 2005. Advanced geometry tutor: an intelligent tutor that teaches proof-writing with construction. In: AIED, vol. 125. pp. 443–450.

Melis, E., Siekmann, J., 2004. Activemath: an intelligent tutoring system for mathematics. In: International Conference on Artificial Intelligence and Soft Computing. Springer, pp. 91–101.

Mills, C., D'Mello, S., 2015. Toward a real-time (day) dreamcatcher: sensor-free detection of mind wandering during online reading. In: Proceedings of the 8th International Conference on Educational Data Mining. International Educational Data Mining Society, pp. 69–76.

Mills, C., D'Mello, S., Bosch, N., Olney, A.M., 2015. Mind wandering during learning with an intelligent tutoring system. In: Artificial Intelligence in Education. Springer, New York, NY, 267–276.

Mohan, K., 2016. Visualizing K-means clustering, Stanford, CA. http://stanford.edu/class/ee103/visualizations/kmeans/kmeans.html.

Moodle, 2016. A learning management system. https://moodle.org/.

Murphy, K.P., 2012. Machine Learning: A Probabilistic Perspective. MIT Press, Cambridge, MA.

Nižnan, J., Pelánek, R., Řihák, J., 2015. Student models for prior knowledge estimation. In: Educational Data Mining.

Olsen, J.K., Aleven, V., Rummel, N., 2015. Predicting student performance in a collaborative learning environment. In: Proceedings of EDM.

Onah, D.F., Sinclair, J., Boyatt, R., 2014. Dropout rates of massive open online courses: behavioural patterns. In: Proceedings of the 8th International Conference on Educational Data Mining, International Educational Data Mining Society, pp. 211–217.

Pardos, Z.A., Baker, R.S., San Pedro, M.O., Gowda, S.M., Gowda, S.M., 2013. Affective states and state tests: investigating how affect throughout the school year predicts end of year learning outcomes. In: Proceedings of the Third International Conference on Learning Analytics and Knowledge. ACM, New York, NY, 117–124.

Pavlik, P.I., Cen, H., Koedinger, K.R., 2009. Performance factors analysis—a new alternative to knowledge tracing. In: Proceedings of the 2009 Conference on Artificial Intelligence in Education: Building Learning Systems That Care: From Knowledge Representation to Affective Modeling. IOS Press, Amsterdam, The Netherlands, pp. 531–538. http://dl.acm.org/citation.cfm?id=1659450.1659529, ISBN 978-1-60750-028-5.

Price, T.W., Barnes, T., 2015. An exploration of data-driven hint generation in an open-ended programming problem. In: Workshop on Graph-Based Data Mining held at Educational Data Mining (EDM).

Price, T., Dong, Y., Barnes, T., 2016. Generating data-driven hints for open-ended programming. In: Proceedings of the 9th International Conference on Educational Data Mining, International Educational Data Mining Society, pp. 191–198.

Rivard, R., 2013. Measuring the MOOC Dropout Rate. Inside Higher Ed, 8, https://www.insidehighered.com/news/2013/03/08/researchers-explore-who-taking-moocs-and-why-so-many-drop-out.

San Pedro, M.O.Z., Snow, E.L., Baker, R.S., McNamara, D.S., Heffernan, N.T., 2015. Exploring dynamical assessments of affect, behavior, and cognition and math state test achievement. In: Proceedings of EDM.

Seo, M., Hajishirzi, H., Farhadi, A., Etzioni, O., Malcolm, C., 2015. Solving geometry problems: combining text and diagram interpretation. In: Proceedings of the 2015 Conference on Empirical Methods in Natural Language Processing, EMNLP, pp. 17–21.

Siemens, G., Baker, R., 2012. Learning analytics and educational data mining: towards communication and collaboration. In: Proceedings of the 2nd International Conference on Learning Analytics and Knowledge. LAK '12. Vancouver, British Columbia, Canada. ACM, New York, NY, pp. 252–254.

Stamper, J., Eagle, M., Barnes, T., Croy, M., 2013. Experimental evaluation of automatic hint generation for a logic tutor. Int. J. Artif. Intell. Educ. 22 (1–2), 3–17.

Stapel, M., Zheng, Z., Pinkwart, N., 2016. An ensemble method to predict student performance in an online math learning environment. In: Proceedings of the 9th International Conference on Educational Data Mining, International Educational Data Mining Society, pp. 231–238.

Stewart, T.C., West, R.L., 2005. Python ACT-R: a new implementation and a new syntax. In: 12th Annual ACT-R Workshop.

Streeter, M., 2015. Mixture modeling of individual learning curves. In: International Conference on Educational Data Mining.

The International Educational Data Mining Society, 2016. Organizes the EDM conference series and publishes the journal of educational data mining. http://www.educationaldatamining.org/.

Tomkins, S., Ramesh, A., Getoor, L., 2016. Predicting post-test performance from student behavior: a high school MOOC case study. In: Proceedings of the 9th International Conference on Educational Data Mining, International Educational Data Mining Society, pp. 239–246.

Vanlehn, K., Lynch, C., Schulze, K., Shapiro, J.A., Shelby, R., Taylor, L., Treacy, D., Weinstein, A., Wintersgill, M., 2005. The Andes physics tutoring system: lessons learned. Int. J. Artif. Intell. Educ. 15 (3), 147–204.

Wang, Y., Heffernan, N.T., Heffernan, C., 2015. Towards better affect detectors: effect of missing skills, class features and common wrong answers. In: Proceedings of the 5th International Conference on Learning Analytics and Knowledge. ACM, New York, NY, 31–35.

White, L.W., 2013. DITA for Print: A DITA Open Toolkit Workbook. XML Press, Fort Collins, CO.

Whitehill, J., Williams, J.J., Lopez, G., Coleman, C.A., Reich, J., 2015. Beyond prediction: first steps toward automatic intervention in MOOC student stopout. Available at SSRN 2611750.

Chapter 10

Large Scale Data Enabled Evolution of Spoken Language Research and Applications

S. Jothilakshmi*[,1] and V.N. Gudivada[†]
[*]*Annamalai University, Chidambaram, Tamil Nadu, India*
[†]*East Carolina University, Greenville, NC, United States*
[1]*Corresponding author: e-mail: jothi.sekar@gmail.com*

ABSTRACT

Natural Language Processing (NLP) is an interdisciplinary field whose goal is to analyze and understand human languages. Natural languages are used in two forms: written and spoken. Text and speech are the mediums for written and spoken languages, respectively. The synergistic confluence of advances in signal processing, machine learning, cognitive computing, and big data ushered in large scale data-driven approaches to speech research and applications. This chapter provides an introductory tutorial on the core tasks in speech processing, reviews recent large scale data-driven approaches to solving problems in spoken languages, describes current trends in speech research, and indicates future research directions.

Keywords: Large scale data, Big data, Natural language processing, Speech processing

1 INTRODUCTION

Natural Language Processing (NLP) is an interdisciplinary field whose goal is to analyze and understand human languages. Natural languages are used in two forms: written and spoken. Text and speech are the mediums for written and spoken languages, respectively.

Speech signals carry information about the message to be conveyed, language identity, speaker identity, gender, and emotion. The information required to perform basic speech processing tasks is implicitly present in speech signals. Speech processing is at the intersection of digital signal processing, written form NLP, machine learning, and information retrieval.

Core tasks in speech processing include language identification, speaker recognition, speech recognition, and speech synthesis. Speech applications

Handbook of Statistics, Vol. 35. http://dx.doi.org/10.1016/bs.host.2016.07.005

typically require pipelining of core tasks. For example, Siri, a voice-activated personal assistant for iPhones, employs speech recognition and speech synthesis tasks (Apple Inc. (USA), n.d). Speech-to-speech translation (SST) and spoken document retrieval are other speech applications.

Currently, human language spoken data is generated at unprecedented levels—Big Data scale (Gudivada et al., 2015a). For example, over 300 h of video is uploaded to YouTube every minute (Smith, 2016). YouTube has over one billion users and the content spans 76 languages (YouTube, 2016). Availability of large scale data provides unique opportunities for advancing speech processing research and applications. Recently, Google open-sourced its TensorFlow AI engine (Abadi et al., 2015). It attests to the fact that algorithms and code do not provide competitive advantage compared to data.

Large datasets enable researchers to overcome problems associated with small data samples. For example, they help to relax the assumptions of theoretical models, avoid over-fitting of models to *training data*, effectively deal with noisy training data, and provide ample *test data* to validate models (Gudivada et al., 2015b).

1.1 Chapter Organization

The overarching goals of this chapter are to review the existing speech research and applications, and discuss current trends and future research in the backdrop of the availability of large scale speech datasets. It provides an introductory tutorial on the core tasks in speech processing, reviews recent large scale data-driven approaches to solving problems in spoken languages, describes current trends in speech research, and indicates future research directions.

Basic characteristics of speech signals is presented in Section 2. In Section 3, preprocessing operations that are performed on the speech signal to facilitate feature extraction are discussed. Extraction of segmental and prosodic features are described in Sections 4 and 5. Section 6 discusses mathematical models for representing and combining features. Select speech core tasks and applications are discussed in Section 7. In Section 8, publicly available resources for speech research are listed. Current trends in speech research and future research directions are discussed in Section 9. Finally, Section 10 concludes the chapter.

2 SPEECH SIGNALS

Speech signals are analog composites of sinusoidal waves of varying amplitude, frequency, and phase. Speech is produced by vibrating source of sound coupled to a resonating system (Rabiner and Schafer, 2010). For a majority of the sounds produced by humans, larynx acts as the vibrating source. Human speech results from air being pushed from the lungs into the *vocal chords* (aka *vocal folds*) and continuing passage through the *vocal tract*.

Speech is composed using different sound types, which include voiced, unvoiced, and plosive. *Voiced sounds* are produced as a result of obstruction of air flow from the lungs by vocal chords' periodic vibrations. *Unvoiced sounds* are produced when air passes through an obstruction in the vocal tract or mouth. *Plosive sounds* occur when sudden bursts of air is let out through the vocal tract through actions such as closing and rapidly releasing the vocal tract and sudden opening of the mouth.

2.1 Human Speech Frequencies

Frequencies corresponding to human speech are 50 Hz and upwards. Frequency range for voiced sounds is 50–500 Hz. Most of the speech signal energy is concentrated between 300 and 3000 Hz. However, the human ear can detect sounds in the range 20–20,000 Hz and is most sensitive in the range 300–10,000 Hz. Sounds in the frequency range 300–3400 Hz contributes most to speech intelligibility and recognition.

The periodic vibration rate of vocal folds is termed *pitch frequency*, *pitch*, or *fundamental frequency* (F_0). The typical pitch frequency for men varies from 85 to 155 Hz, and 165 to 255 Hz for women. *Pitch period* (aka *fundamental period*) is the reciprocal of the pitch. Pitch period average values are around 8 ms for men and 4 ms for women.

2.2 Formants

Through varying excitations, vocal tract can assume various shapes to produce different sounds. Each shape is characterized by a set of *formants*. The resonance frequencies of the vocal tract tube are called *formant frequencies* or simply *formants*. Formants in the sound of the human voice are particularly important due to their contribution to the intelligibility of speech. Depending on the places of articulation in the vocal tract, the following categories of formants can be produced: bilabial, labiodental, dental, interdental, retroflex, alveolar, palatoalveolar, palatal, velar, uvular, and glottal (O'Shaughnessy, 1999).

2.3 Phonemes and Syllables

The basic sounds of a human language are referred to as *phonemes*. A unit of speech is considered a phoneme if replacing it in a word results in different meaning. Phonemes are the smallest units of speech which distinguish one utterance from another in the language. *Phones* are the physical counterparts of phonemes. They refer to the instances of phonemes in actual utterances.

A *syllable* is a unit of organization for a sequence of phonemes. It is a sound that can be said without interruption. Each syllable has a vowel, and vowels can have consonants before or after them. A *word sound* is formed by a sequence of syllables.

2.4 Analog to Digital Conversion of Speech Signals

Some features of the speech signal are extracted after converting the analog signal to its digital form. This process involves *sampling* and *quantization*. Sampling refers to measuring the amplitude of the analog signal at various time points. The time interval between two successive sampling points is called the *sampling period*. *Sampling frequency* (aka sampling rate), denoted f_s, is the number of samples taken in 1 s.

Let f_m be the highest frequency present in the speech signal. According to the *sampling theorem*, the original analog signal can be reconstructed from the sequence of its samples with minimal distortion if $f_s \geq 2f_m$. The quantity $2f_m$ is referred to as the *Nyquist sampling rate*. For speech signals, the *Nyquist sampling rate* is 8 kHz, and 40 kHz for the television broadcast signals.

Quantization is the process of assigning a digital value for the amplitude of the signal at a time point. The assigned value should be closest to the original analog amplitude. The number of bits used for storing the quantized value is called the *bit resolution*. Quantized speech signal values are stored using 8 or 16 bits.

3 SIGNAL PREPROCESSING

A speech *feature* is any prominent characteristic present in the signal. Features are selected in a way to provide a compact representation of the signal for the task at hand. Also, the features must have adequate discrimination power for distinguishing between various sounds. Features are extracted in both the analog and digital domains of the speech signal. First, various *preprocessing operations* are performed on the signal to facilitate feature extraction. Shown in Fig. 1 are three common preprocessing operations: *preemphasis*, *framing*, and *windowing* (Rabiner and Juang, 1993).

3.1 Preemphasis

The energy of the high frequency components in speech signals is generally low. Therefore, they may not carry enough energy to yield useful features. *Preemphasis* is used to boost the energy of high frequency components. Assume that the analog speech signal is digitized into n discrete samples, denoted $s(1)$, $s(2)$, …, $s(n)$. Also, we use the notation $s(n)$ to denote all the samples of a digitized analog signal. The preemphasized signal, $\hat{s}(n)$, of the input signal $s(n)$ is given by:

FIG. 1 Preprocessing tasks.

$$\hat{s}(n) = s(n) - \alpha s(n-1) \tag{1}$$

where $0 < \alpha < 1$ and n is the sample number. Typical value of α is 0.95. Fig. 2A shows a speech signal and its preemphasized signal is shown in Fig. 2B.

3.2 Framing

In speech analysis, it is assumed that the signal properties change relatively slowly with respect to time. *Framing* divides the preemphasized signal $\hat{s}(n)$ into N time frames and the signal properties are assumed to remain constant within each *frame*. Also, adjacent frames are separated by M samples. The variables N and M are referred to as the *frame size* and *frame shift*. Measured on a time scale, N varies from 10 to 30 ms and $M \leq 0.5N$. A frame of the preemphasized signal is shown in Fig. 2C.

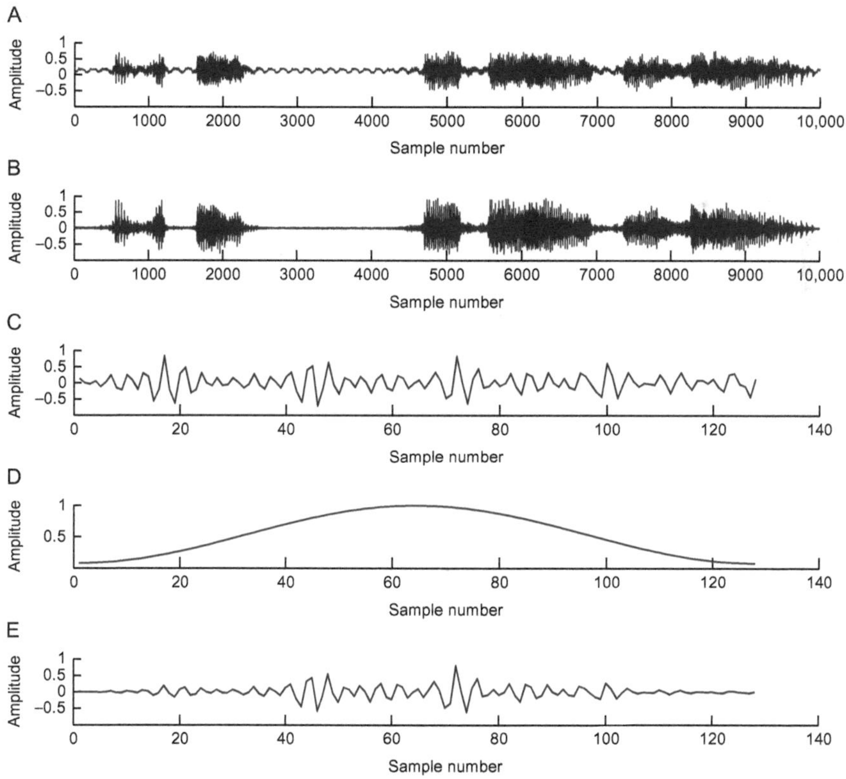

FIG. 2 Preprocessing of speech signal. (A) Speech signal. (B) Preemphasized speech signal. (C) A frame of speech signal (N = 16 ms). (D) Hamming window. (E) Preprocessed frame.

Let $\tilde{f}_l$ denote the lth frame and assume that there are L frames in the speech signal. It follows that:

$$L = 1 + \left\lfloor \frac{\text{total number of samples in the speech signal} - N}{\text{M}} \right\rfloor \tag{2}$$

$$\tilde{f}_l(n) = \hat{s}(Ml + n), \quad 0 \leq n \leq N-1, \quad 0 \leq l \leq L-1 \tag{3}$$

Framing leads to information loss due to abruptly dividing the signal at the ends. To avoid this problem, *windowing* is performed.

3.3 Windowing

Windowing minimizes the signal discontinuities at both ends of the frame. Windowing multiplies each frame with a *window function*. Assume that the window function is $w(n)$, $0 \leq n \leq N-1$. The result of windowing the signal is:

$$f_l(n) = \tilde{f}_l(n) w(n), \quad 0 \leq n \leq N-1 \tag{4}$$

The window must be selected to taper the signal to zero at the beginning and end of each frame. The *Hamming window* shown in Fig. 2D is commonly used for this purpose. It has the form:

$$w(n) = 0.54 - 0.46 \cos\left(\frac{2\pi n}{N-1}\right), \quad 0 \leq n \leq N-1 \tag{5}$$

Fig. 2E shows the hamming windowed frame. Speech signal features are computed at the frame level. Most feature extraction methods yield values that are averaged over the time points in the entire frame. Various features of the frame are represented by a *feature vector*. Segmental and prosodic are two broad classes of feature representations (O'Shaughnessy, 1999).

4 SEGMENTAL FEATURE EXTRACTION

Segmental features provide information about what the speaker has said. They are extracted at the segmental (aka phoneme) level, which corresponds to frame sizes of 10–30 ms. Table 1 summarizes most commonly used segmental features (Davis and Mermelstein, 1980; Rabiner and Schafer, 1978). The energy of a complete speech signal E is the sum of the squares of all the samples in the signal and is given by:

$$E = \sum_{n=-\infty}^{\infty} [s(n)]^2 \tag{6}$$

The energy of each analysis frame, *short time energy* (STE), is more important than the complete signal energy. STE of the lth frame is given by:

TABLE 1 Commonly Used Segmental Features

Feature	Description
Short time energy	Represents strength of the signal.
Short time average zero crossing rate	Measures the frequency content of a signal.
Short time auto correlation	Measures similarity of a signal with its delayed version.
Short time Fourier transform	Indicates different frequencies present in the signal.
Linear prediction coefficients (LPCs)	Predict a sample using weighted sum of the previous samples.
Perceptual linear prediction (PLP)	Computes LPC using psychophysics based concepts.
Cepstral coefficients	Measure rate of change of different frequency bands present in the frame.
Mel frequency cepstral coefficients (MFCC)	Result of mapping of spectral coefficients to mel-scale.

$$E_l = \sum_{n=0}^{N-1} [f_l(n)]^2 \tag{7}$$

where $f_l(n)$ is the windowed analysis frame obtained at the end of preprocessing step.

Short time average *Zero Crossing Rate* (ZCR) is defined as the weighted average of the number of times the speech signal changes sign within an analysis frame. A zero crossing is said to occur if successive samples have different algebraic signs. Short time average ZCR of the lth frame is given by:

$$Z_l = \frac{1}{2N} \sum_{n=0}^{N-1} |\text{sgn}(f_l(n)) - \text{sgn}(f_l(n+1))| \tag{8}$$

where:

$$\text{sgn}(f_l(n)) = \begin{cases} 1, & \text{if } f_l(n) \geq 0 \\ 0, & \text{if } f_l(n) < 0 \end{cases} \tag{9}$$

The rate at which zero crossings occur is a simple measure of the frequency content of a signal. High and low frequencies in the signal manifest as high and low ZCRs, respectively.

Cross correlation measures similarity of two signals as a function of the time delay between them. It is computed by summing the products of a signal sample with one from another time-shifted signal. The cross-correlation is large if the two signals have similar waveforms for some time delay. The application of cross correlation to a signal and its delayed version is called *autocorrelation*. Autocorrelation of an analysis frame is called *short time autocorrelation*.

Fourier analysis is a useful technique for analyzing signals. It decomposes the signal into constituent sinusoids of different frequencies and amplitudes. For sampled signal data, Fourier analysis is performed using the discrete Fourier transform (DFT). The fast Fourier transform (FFT) is an efficient algorithmic implementation of DFT. Short time Fourier transform (STFT) is the FFT of an analysis frame, denoted $F_l(u)$, and is given by:

$$F_l(u) = \sum_{n=0}^{N-1} f_l(n) e^{-j2\pi\left(\frac{ux}{N}\right)}, \quad 0 \leq u \leq N-1 \tag{10}$$

The STFT coefficients of a signal are known as the *spectral coefficients*. They indicate information about different frequencies present in the signal. The inverse DFT (IDFT) of $F_l(u)$, denoted $f_l(n)$, and is given by:

$$f_l(n) = \frac{1}{N} \sum_{u=0}^{N-1} F_l(u) e^{j2\pi\left(\frac{ux}{N}\right)}, \quad 0 \leq n \leq N-1 \tag{11}$$

As the name implies, the extraction of *LPCs* is based on linear prediction (LP). The latter is closely linked to the modeling of the human vocal tract system. The LP relies on the fact that a particular speech sample may be predicted by a linear weighted sum of the previous samples, and is given by:

$$f_l(n) \approx a_1 f_l(n-1) + a_2 f_l(n-2) + a_3 f_l(n-3) + \cdots + a_p f_l(n-p) \tag{12}$$

The number of previous samples p is known as the *order of prediction* and the weights a_1, a_2, ..., a_p are known as *linear prediction coefficients* (LPCs).

PLP is similar to LP analysis. PLP is based on the short-term spectrum of speech. In contrast to pure LP analysis, PLP modifies the short-term spectrum of the speech using psychophysics based concepts (Hermansky, 1990).

Cepstral coefficients are the IDFT of the log magnitude of the DFT of a given sequence of data. For the lth analysis frame, they are computed as:

$$C_l = \text{IDFT}(\log(|\text{DFT}(f_l)|)) \tag{13}$$

Each frame of the speech signal contains a specific range of frequencies known as frequency band. Cepstral coefficients provide information about rate of change of different frequency bands. The cepstral coefficients of LPC are known as *linear prediction cepstral coefficients*. Cepstral mean subtraction is performed to remove distortions in slowly varying channels. Cepstral average over a sufficiently long speech recording is subtracted from the cepstral coefficients.

Human hearing is not equally sensitive at all frequency bands. It is less sensitive at higher frequencies (above 1000 Hz). The mel-scale is based on these observations. The mapping of frequencies in mel-scale is linear below 1000 Hz and logarithmic above 1000 Hz. Mel-scale is defined as:

$$\text{Mel}(f) = 2595 \log_{10}\left(1 + \frac{f}{700}\right) \tag{14}$$

where f is the linear scale frequency and Mel(f) is the corresponding mel-scale frequency.

Mel-scale filters are used to map the spectral coefficients to mel-scale. The *mel-warped cepstral coefficients* are obtained by inserting mel-scale filters before computing the IDFT. These filters transform the frequency scale to lessen emphasis on higher frequencies. They are triangular and are equally spaced along the mel-scale (Cambridge University Engineering Department, 2016). The band edges and center frequencies of the filters are linear for low frequencies and logarithmically increase for higher frequencies.

Each short time Fourier transform magnitude coefficient is multiplied by the corresponding *filter gain* and the results are accumulated. Next, discrete cosine transform is applied to the log of the mel spectral coefficients to obtain MFCC as:

$$x(m) = \sqrt{\frac{2}{M_f}} \sum_{i=0}^{M_f - 1} E(i) \cos\left(\frac{(2i+1)m\pi}{2N}\right), \quad m = 0, \ldots, M_f - 1 \tag{15}$$

where M_f is the number of filters in the filter bank, and $E(i)$ is the log value of ith mel filter coefficient. Dynamic information can be obtained from first derivatives (aka *delta coefficients*, Δ) and second derivatives (aka *acceleration coefficients*, Δ^2) of the MFCC.

5 PROSODIC FEATURE EXTRACTION

Prosodic features, in contrast with segmental features, represent higher-level information at the syllable, word, or utterance levels. These features are extracted at the *supra-segmental* level, which corresponds to frame sizes greater than 100 ms. Table 2 shows commonly used prosodic features (O'Shaughnessy, 1999). They indicate aspects related to the naturalness of speech. Also, they carry speaker information such as emotional state and social background. Furthermore, they signify the positions of pauses based on the punctuation marks, and distinguish interrogative clauses from statements.

Extracting and representing the prosodic features is a difficult task. Prosodic feature extraction methods are differentiated based on whether or not they use an automatic speech recognizer. The methods in the first category use automatic speech recognizer to obtain syllabic and phone boundaries. However, these approaches increase the overall complexity of the system. The methods in the second category do not use automatic speech recognizer and estimate segment boundaries using speech cues.

TABLE 2 Commonly Used Prosodic Features

Feature	Description
Pitch	Measures the periodic vibration rate of vocal folds from voiced signal.
Pitch contour (aka Intonation)	Tracks pitch pattern of an utterance by detecting the direction of pitch variation with time.
Speaking rate	Counts the number of syllables spoken per second.
Duration	Measures the time taken to speak a syllable.
Stress	Indicates the relative prominence of a syllable. Stressed syllables are louder and longer in duration than unstressed.
Rhythm	Measures distance between onset points of successive vowels and the duration of voiced region within each syllable-like region.
Loudness	Measures volume of a sentence.

5.1 Non-Automatic Speech Recognizer Based Prosodic Feature Extraction

Commonly used *speech cues* are onset and offset of a syllable, and start and end of voicing. For extracting syllable-based prosodic features, the speech signal should be segmented into syllables. All spoken utterances can be considered as sequence of syllables. The syllable-like regions in continuous speech can be identified using locations of *vowel onset points* (VOPs). The latter refers to the instant at which the onset of vowel takes place in a syllable (Mary and Yegnanarayana, 2008).

There are various methods to detect *pitch* from the voiced speech signal (Babacan et al., 2013; Gerhard, 2003). Time domain methods include zero crossings and autocorrelation. Frequency domain methods include harmonic product spectrum, cepstrum, and filters. Some hybrid approaches use both time and frequency domain methods.

Intonation or pitch contour (F_0) is the pitch pattern of an utterance. The direction of F_0 changes by rising or falling with time. Pitch (F_0) contour of a speech signal is associated with VOPs locations. *Speaking rate* is the number of syllables spoken per second. It indicates how fast the user spoke. The *duration* is the time taken to speak a syllable (Mary, 2012).

The syllables in a language are categorized as strong (stressed) or weak (unstressed) syllables. They are language specific. Stress defines the rhythm of speech. It indicates the relative prominence of a syllable. Stressed syllables are louder and longer in duration. They are characterized by higher energy, larger pitch (F_0) movement, and longer duration. Stress is represented by the change in *log energy* corresponding to the voiced regions of a syllable, the F_0 contour, and duration features.

Syllabic rhythm is represented by the distance between successive VOPs and the duration of voiced region within each syllable-like region. Loudness distinguishes sentences that are spoken with different volume. Usually the average and maximum volume levels from the start to the end of the sentence are measured. Volume is generally used to show emotions such as fear or anger.

5.2 Features for Voice Activity Detection

In most speech processing systems, a voice activity detector (VAD) is used before extracting the necessary features. VAD discards silence and noise frames. Removing silence and noise frames from the training and testing signals is important to enable modeling and detecting only the relevant signal. Many algorithms are available for designing VAD (Meduri and Ananth, 2012). Energy threshold or analysis of the spectral characteristics of the audio signal is used by VAD algorithms. The VAD also classifies the speech frames into voiced or unvoiced.

6 MATHEMATICAL MODELS

All speech-processing applications operate on features, though specific features may vary from one application to another. Usually, the suitable feature set is incrementally identified by trail and error. Applications are developed using an initial set of features and the application performance is assessed. Some features in the initial set are dropped and new features are added. Sometimes, existing features are combined in some fashion to create a derived feature.

6.1 Acoustic and Language Models

Speech tasks typically need two models: acoustic and language. An *acoustic model* represents the relationship between an audio signal and linguistic units such as phonemes that make up the speech. The model is learned through supervised learning methods using labeled training data. A set of audio recordings and their corresponding transcripts comprises the training data, which is also called *speech corpus*. Creating an acoustic model essentially involves creating statistical representations such as a hidden Markov model (HMM) for each phoneme in the language.

A statistical language model (LM) constructs a probability distribution function that assigns a probability to every string in the language—strings consisting of only one word, two words, three words, and so on. Let V be a finite vocabulary for a language. Let X be the strings of words in the language over V. A language model estimates the prior probability values $P(X)$ for every X. The size of V is usually in the tens or hundreds of thousands.

The string X is broken into sentences, words, or other segments such that $X = x_1, x_2, \ldots, x_m$. In other words, $x_1, x_2, \ldots, x_m$ is the word sequence of the string X. $P(X)$ is defined using conditional probabilities as:

$$P(X) = \prod_{k=1}^{m} P(x_k | x_1, x_2, \ldots, x_{k-1}) \tag{16}$$

The parameter space of $P(x_k|x_1, x_2, \ldots, x_{k-1})$ is too large. Hence, an equivalence class determined by a function $\Psi(X_{k-1})$ is used to reduce this space. Now, Eq. (16) can be rewritten as:

$$P(X) \approx \prod_{k=1}^{m} P(x_k | \Psi(X_{k-1})) \tag{17}$$

Language modeling consists of finding appropriate equivalence classifiers Ψ. There are various methods to find the function Ψ and estimate $P(x_k|\Psi(X_{k-1}))$. The most common language modeling uses $(n - 1)$-gram equivalence classification for this purpose. It is defined as:

$$\Psi(X_{k-1}) = x_{k-n+1}, x_{k-n+2}, \ldots, x_{k-1} \tag{18}$$

Once this function is known, $P(X)$ can be computed easily from the training data. By substituting Eq. (18) in Eq. (17), we get the n-gram language model:

$$P(X) \approx \prod_{k=1}^{m} P(x_k | x_{k-n+1}, x_{k-n+2}, \ldots, x_{k-1}) \tag{19}$$

The value of n determines the type of the language model. For example, $n = 1$ gives the *unigram language model*. Similarly, by setting $n = 2$ and $n = 3$, we get the *bigram* and *trigram* language models. Under unigram model, the probability of observing a given word does not dependent on the context. In the bigram model, the probability of observing a given word depends on the immediately preceding word. In the trigram model, a word depends on the two immediately preceding words. Trigram language models are used in most practical applications.

Perplexity is a measure for evaluating language models. Conceptually, a perplexity value indicates how many choices are available to select the next word x_k given a partial sentence $X_{k-1} = x_1x_2\ldots x_{k-1}$. Therefore, a language model that has smaller perplexity is considered superior to the one that has larger perplexity. Let W denote the total number of words in the test data used for evaluation. Perplexity is defined as $2^{(-z)}$ where:

$$z = \frac{1}{W} \sum_{k=1}^{m} \log P(x_k | X_{k-1}) \tag{20}$$

Though perplexity is a useful *intrinsic evaluation* measure, it is not effective as a metric unless the test data looks just like the training data. *Extrinsic evaluation* measures compare language models based on their performance on tasks such as machine translation (MT) and speech recognition. Counts such as the number of words that are correctly translated, and the number of words in the speech that are correctly recognized are used in computing the evaluation metric.

6.2 Gaussian Mixture Models

A Gaussian mixture model (GMM), as the name suggests, is a mixture of several Gaussian distributions. Speech features are represented as vectors in an n-dimensional space. Distribution of these feature vectors is represented by a mixture of Gaussian densities. For a n-dimensional feature vector $\boldsymbol{x}$, the mixture density function for class s with model parameter λ^s is defined as:

$$p(\boldsymbol{x}|\lambda^s) = \sum_{i=1}^{\mathcal{M}} \alpha_i^s f_i^s(\boldsymbol{x})$$

The mixture density function is a weighted linear combination of $\mathcal{M}$ component unimodal Gaussian densities $f_i^s(.)$. Each Gaussian density function $f_i^s(.)$ is parameterized by the mean vector $\boldsymbol{\mu}_i^s$ and the covariance matrix $\boldsymbol{\Sigma}_i^s$ using:

$$f_i^s(\boldsymbol{x}) = \frac{1}{\sqrt{(2\pi)^{N_f}|\boldsymbol{\Sigma}_i^s|}} \exp(-\frac{1}{2}(\boldsymbol{x}-\boldsymbol{\mu}_i^s)^T(\boldsymbol{\Sigma}_i^s)^{-1}(\boldsymbol{x}-\boldsymbol{\mu}_i^s)),$$

where $(\boldsymbol{\Sigma}_i^s)^{-1}$ and $|\boldsymbol{\Sigma}_i^s|$ denote the inverse and determinant of the covariance matrix $\boldsymbol{\Sigma}_i^s$, respectively. The mixture weights $(\alpha_1^s, \alpha_2^s, \ldots, \alpha_{\mathcal{M}}^s)$ satisfy the constraint $\sum_{i=1}^{\mathcal{M}} \alpha_i^s = 1$. Collectively, the model parameters of the class s model λ^s are denoted as $\lambda^s = \{\alpha_i^s, \boldsymbol{\mu}_i^s, \boldsymbol{\Sigma}_i^s\}$, $i = 1, 2, \ldots, \mathcal{M}$. The number of mixture components is chosen empirically for a given data set. The parameters of GMM are estimated using the iterative expectation–maximization (EM) algorithm (Redner and Walker, 1984).

6.3 Hidden Markov Models

HMM is a doubly embedded stochastic process where the underlying stochastic process is not directly observable (Rabiner, 1989). A HMM not only models the underlying speech sounds but also the temporal sequencing of the sounds. It models a stochastic process by a set of states and transition probabilities between the states. Each state describes a stationary stochastic process. The transition from one state to another state describes how the process changes its characteristics over time. Thus, a HMM is used to model making a sequence of decisions on temporal basis.

There are two types of HMM: ergodic, and left to right. In the *ergodic HMM*, any state can be reached from any other state in a single step. The model provides transition probabilities from state i to state j, $a_{ij} > 0$ for all i, j. In contrast, the left to right model does not allow transitions to the states of lower indexes—$a_{ij} = 0$ when $j < i$.

Let N_x denote the number of states in the model. The individual states are denoted by $\mathbf{s} = \{s_1, s_2, \ldots, s_{N_x}\}$, and the state at time t as q_t. The state-transition probability distribution $\mathbf{A} = \{a_{ij}\}$, where:

$$a_{ij} = P[q_{t+1} = s_j | q_t = s_i], \quad 1 \le i, j \le N_x$$

a_{ij} defines the probability of transition from state s_i to s_j at time t with the constraint:

$$\sum_{j=1}^{N_x} a_{ij} = 1, \quad 1 \le i \le N_x$$

The observation probability density function, $\mathbf{B}$, in state i is given by $\mathbf{B} = \{b_i(\mathbf{O})\}$, where:

$$b_i(\mathbf{O}) = \sum_{k=1}^{\mathcal{M}} C_{ik} \ p(\mathbf{O}, \mu_{ik}, \Sigma_{ik}), \quad 1 \le i \le N_x$$

C_{ik} is the mixture coefficient for k^{th} component in state i. $\mathcal{M}$ is the number of components in a GMM, and $p(\mathbf{O}, \mu_{ik}, \Sigma_{ik})$ is a Gaussian probability density function with mean (μ_{ik}) and covariance (Σ_{ik}). The initial state distribution $\boldsymbol{\pi} = \{\pi_i\}$, where:

$$\pi_i = P[q_1 = s_i], \quad 1 \le i \le N_x$$

Given appropriate values for $N_x, \mathcal{M}, \mathbf{A}, \mathbf{B}$, and $\boldsymbol{\pi}$, the HMM can be used as a generator of observations. Also, it can be used as a model for how a given feature vector sequence $\mathbf{O} = (\mathbf{o}_1 \ \mathbf{o}_2 \ \ldots \ \mathbf{o}_T)$ was generated by an appropriate HMM, where T is the length of the feature vector sequence. The compact notation $\lambda = (\mathbf{A}, \mathbf{B}, \boldsymbol{\pi})$ is used to indicate the complete parameter set of the model.

6.4 Support Vector Machines

Support Vector Machine (SVM) is based on the principle of structural risk minimization (Burges, 1998; Campbell et al., 2006). They are used for pattern classification and nonlinear regression. For linearly separable data, SVM finds a hyperplane which separates the data with the largest margin. For linearly inseparable data, it maps the data $\boldsymbol{x}$ in the input space I into a high dimension space H—$\boldsymbol{x} \in \mathbb{R}^I \mapsto \Phi(\boldsymbol{x}) \in \mathbb{R}^H$ with kernel function $\Phi(\boldsymbol{x})$—to find the separating hyperplane.

SVM was originally developed for two-class classification problems. Subsequently, SVMs were developed for multiclass problems. Each SVM separates a single class from all the remaining classes (one-vs-rest approach).

Each SVM is trained to distinguish all instances of a single class from the instances of all other classes. During testing, the class label y of a class pattern $\boldsymbol{x}$ is determined by:

$$y = \begin{cases} n, & \text{if } d_n(\boldsymbol{x}) + t_l > 0 \\ 0, & \text{if } d_n(\boldsymbol{x}) + t_l \leq 0 \end{cases}$$

where $d_n(\boldsymbol{x}) = \max\{d_i(\boldsymbol{x})\}_{i=1}^{N_l}$, $d_i(\boldsymbol{x})$ is the distance from $\boldsymbol{x}$ to the SVM hyperplane corresponding to class i, and t_l is the classification threshold.

6.5 Artificial Neural Network Models

Artificial Neural Networks (ANNs) are a class of statistical learning algorithms used in machine learning and cognitive science domains. They are inspired by biological neural networks found in the central nervous system of animals. They are ideal for situations which require approximating a function that depends on a number of inputs, and the inputs are generally unknown.

A feed forward ANN has an input layer, output layer, and one or more hidden layers. Each layer consists of a number of processing units. Each unit represents the model of an artificial neuron. Units in one layer are connected to the units in the next layer. Weight are associated with the connections to indicate connection strength. Weight are learned using training data, which is an ordered pair consisting of input data and expected result. Both supervised and unsupervised algorithms are used to learn the connection weights. All the connection weights comprise a weight vector.

During the learning stage, ANN propagates the error backward (i.e., output units to the input units) by adjusting the weights using a back-propagation algorithm. The algorithm consists of two phases: forward- and back-propagation. In the forward-propagation phase, a pair from the training set is fed to the input layer and an output is generated based on the current weight vector. The error between the actual output and the desired output is calculated. The objective of training is to minimize this error. During the back-propagation phase, the weights are adjusted in such a way to minimize the error. These steps are repeated for all the pairs in the training data. ANNs modeled with different topologies perform different pattern recognition tasks (Haykin, 2001; Lippmann, 1989; Yegnanarayana, 1999).

6.6 Deep Neural Networks

Deep Neural Networks (DNNs), also called convolutional networks, are composed of multiple levels of nonlinear operations, such as neural nets with many hidden layers (Bengio et al., 2007; Krizhevsky et al., 2012). Deep learning methods aim at learning feature hierarchies, where features at higher levels of the hierarchy are formed using the features at lower levels (Dean et al., 2012). In 2006, Hinton et al. (2006) proved that much better results could be achieved in deeper architectures when each layer is pretrained with an unsupervised learning algorithm. Then the network is trained in a supervised mode using back-propagation algorithm to adjust weights. Current studies show that DNNs outperforms GMM and HMM on a variety of speech processing tasks by a large margin (Hinton et al., 2006).

7 SPEECH PROCESSING CORE TASKS AND APPLICATIONS

Core tasks are used as building blocks in developing speech applications. Speech recognition and speech synthesis are two such tasks. Typically, a speech application employs multiple core tasks. For example, the above core tasks are used in SST. The distinction between core tasks and applications is rather fluid.

Speech applications development requires first creating a database to hold speech data. The data comes from sources such as recording of telephone conversations, radio and television broadcast signals. It is also created by recording speech in noise-free environments. Relevant features are extracted from the data to develop and test speech algorithms. Speech data is partitioned into three nonoverlapping groups—training, development, and test. The *training data* is used to construct statistical models and estimate their parameters (see Section 6). Next, the *development data* is used to tune the model parameters. The effectiveness of the model is evaluated using the *test data*. In this section, we describe core tasks and select speech processing applications.

7.1 Speech Recognition

Automatic speech recognition (ASR) or speech-to-text system identifies spoken words in speech and converts them to written text (Juang and Furui, 2000). ASR systems extract acoustic signal characteristics from speech and determine the words in it by pattern matching (Rabiner and Juang, 1993). Acoustic and language models are used in developing ASR systems.

ASR is an attractive alternative for user interfaces to computing devices. ASR applications include call routing, automatic transcriptions, information searching, data entry, voice dialing, SST, and hands free computing for people with disabilities.

7.1.1 Functional Variations in ASR

ASR systems are classified according to the parameters and features they use—vocabulary size, speaking style, speaker mode, channel type, and transducer type. Table 3 summarizes these functional variations.

7.1.2 Mathematical Models for Speech Recognition

General-purpose speech recognition systems are based on statistical approaches for modeling both acoustics and language. A crucial issue for

TABLE 3 Functional Variations in ASR

Feature	Description	Variations
Vocabulary size	The size of the vocabulary the system can recognize.	*Small vocabulary* systems recognize less than 50 words, whereas the *large vocabulary* counterparts recognize over 1000 words. Large vocabulary systems are more versatile in that they are used in a greater number of applications.
Speaking style	Refers to whether the system recognizes only isolated words or continuous speech.	Recognizing isolated words (e.g., digits) and connected words (e.g., numbers) is relatively easier compared to continuous speech. Recognition of conversational telephone speech and broadcast news is more difficult.
Speaker mode	Refers to whether the system can recognize any speaker or only specific speakers.	*Speaker dependent* systems are trained to work with specific users, whereas the *speaker independent* systems are not tied to specific speakers.
Channel type	The range of channels the system can work with.	May range from wired and wireless telephones to broadband. Some channels may drop voice packets.
Transducer type	The range of speech input devices the system can interface with.	May encompass high quality microphones, telephones, mobile phones, or array microphones.

acoustic modeling is the selection of the speech units that represent the acoustic and linguistic information for the language. Such units can range from phones to words. Whole words are not suitable for large vocabulary systems. These systems use subword models like phones, subphones, or syllables as speech units.

Mathematical models used by speech recognition systems include HMM, GMM, dynamic time warping (DTW), neural networks (NN), and vector quantization (VQ) (O'Shaughnessy, 2008). These complex mathematical models use the information known to the system to find the hidden information. The speech recognizer selects the sequence of words that is more likely to be produced given the observed acoustic evidence. The HMM is the popular statistical tool for modeling a wide range of time series data. It has been used successfully in several speech recognition systems.

There are various approaches to statistical language modeling including Cache (Jelinek et al., 1991), class based (Brown et al., 1992), *N*-grams with modified Kneser–Ney smoothing (Chen and Goodman, 1996), normalized stupid backoff (Brants et al., 2007), structured (Chelba and Jelinek, 2000), maximum entropy language model with hierarchical softmax (Goodman, 2001b), binary maximum entropy language model (Xu et al., 2011), neural network based (Bengio et al., 2003; Morin and Bengio, 2005), recurrent neural network language model (Mikolov et al., 2010), discriminative (Roark and Saraclar, 2004), and Bayesian (Teh, 2006). Description of these models is outside the scope of this chapter.

7.1.3 Developing Speech Recognition Systems

Development of speech recognition systems is a multistep process. First, relevant features are extracted from the speech signal. Second, reference models are developed using these features. Models are needed for each sound unit. Third, feature vectors are derived from speech utterances and are presented to all the reference models. The model which gives the highest confidence measure indicates the identity of the sound unit. The sequence of the identified sound units is validated using language models. In other words, language models are used to convert sequence of sound units into text.

Conceptually, approaches to developing speech recognition systems fall into two types: template and model based. In *template-based* approach, the system is initially trained using known speech patterns. Recognition is performed first by comparing the unknown speech signals with each possible pattern learned in the training phase. Next, possible sequence of words that minimizes a distance function between the unknown patterns and the known pattern is computed. Well known template methods include the *DTW* and vector quantization (VQ). The DTW algorithm recognizes speech by measuring the similarity between two time series, which may vary in time or speed. In the VQ method, an ordered set of signal samples and parameters are coded by matching the input vector to a similar pattern or code vector (codeword) in a predefined codebook.

In the *model-based* systems, suitable features for each sound unit are extracted from the training data. Reference models need to be developed for each sound unit. Commonly used modeling techniques are GMM, HMM, NN, and SVM. Feature vectors are derived from the given test speech utterance and presented to all the reference models. The model which gives the highest confidence indicates the identity of the sound unit.

To achieve good results in speech recognition and SST, scalable language and acoustic modeling techniques are needed. Goodman (2001a) investigated the performance of various language modeling techniques on large data sets. As noted in Chelba et al. (2010), large training datasets improve the performance of statistical language and acoustic models (Chelba et al., 2013b). Mikolov et al. (2011) demonstrated that recurrent neural network models scale well to datasets that are hundreds of millions of words in size. However, it takes weeks to train such systems.

7.1.4 Speech Recognition Algorithms and Systems

A distributed framework for large scale language modeling is discussed in Jyothi et al. (2012). The framework utilizes 87,000 h of speech data to train the model. This data amounts to approximately 350 million words. A distributed perceptron training algorithm based on scalable MapReduce framework (Dean and Ghemawat, 2008; White, 2015) is used to deal with large training dataset. A benchmark corpus comprising one billion words of training data to measure progress in statistical language modeling is discussed in Chelba et al. (2013a). This dataset is useful for evaluating various language models.

In Chelba et al. (2013b), it is shown that the accuracy of a state-of-the-art automatic speech recognition system can be increased by training the model on large datasets. The back-off acoustic model is used in this work. The model is estimated using the MapReduce distributed computing infrastructure (Chelba et al., 2012; Dean and Ghemawat, 2008). The model comprises 1.1 million states and 40 million Gaussians.

DNNs have achieved great success in acoustic modeling. High-quality acoustic models depend on the availability of large and reliably transcribed training datasets recorded in different acoustic environments. In Kapralova et al. (2014), an approach to developing high-quality, unsupervised datasets to train DNN models for large vocabulary speech recognition is presented. They construct a large corpora of about 20 million utterances with automatic transcription close to human transcriber quality using Google's speech recognition tools. The transcription quality is improved by using a slower and more accurate off-line speech recognition system. DNN based acoustic models are trained with 15,000 h of speech. The system achieved 10% relative improvement in large vocabulary dictation and voice-search for Brazilian, Portuguese, French, Italian, and Russian languages.

7.1.5 Evaluating Speech Recognition Systems

One evaluation metric is based on the number of words in the speech that are correctly recognized. Speech recognition systems experience three types of word recognition errors: substitution, insertion, and deletion. A *substitution error* occurs when a word in the utterance is transcribed as a different word. An *insertion error* occurs when a word that appears in the transcription is not in the utterance. A *deletion error* occurs when a word in the utterance is completely missed in the transcription. Word error rate (WER) is defined as the ratio of sum of all three types of errors and the total number of words in the reference transcript. Systems with smaller WER are preferred over the ones with larger WER.

7.2 Speech Synthesis

Speech synthesis, aka Text-to-Speech (TTS), is the process of converting given input text to synthetic speech (Dutoit, 1997). Applications of TTS include SST, voice-based dialog systems, and telephone inquiry systems. Limited domain TTS are successfully deployed in several commercial systems such as kiosks for inquiring train and flight schedules (Black and Lenzo, 2000; Raghavendra and Prahallad, 2010). Speech synthesis involves three phases—text analysis, prosodic analysis, and concatenating speech units.

7.2.1 Text Analysis

Text analysis transforms the input text into speakable forms using text normalization, linguistic and phonetic analysis. *Text normalization* converts numbers and symbols into words, replaces abbreviations by their corresponding whole phrases, among others. The goal of *Linguistic analysis* is to understand the content of the text. Statistical methods are used to find the most probable meaning of the text. A *grapheme* is a letter or a sequence of letters that represent a sound (i.e., phoneme) in a word. *Phonetic analysis* converts graphemes into phonemes. Generation of sequence of phonetic units for a given standard word is referred to as *letter to phoneme* or *text to phoneme* rule. The complexity of these rules and their derivation depends on the nature of the language.

7.2.2 Prosodic Analysis and Concatenating Speech Units

Prosodic analysis adds naturalness and intelligibility to the speech by incorporating prosodic information. Prosody contains the rhythm of speech, stress patterns, and intonation.

In the last phase, speech is synthesized by concatenating the speech units. Methods for concatenation include formant, concatenative, and articulatory. In *formant synthesis*, periodic and nonperiodic source signals are generated and passed through a resonator circuit or a filter that models the human vocal tract. Formant synthesis is flexible and relatively easy to implement. However, it often generates unnatural sounds.

Concatenative synthesis is based on integrating segments of recorded *speech units* which are stored in a database (Kishore and Black, 2003; Kominek and Black, 2003). It uses phones, diphones, triphones, syllables, words, and sentences as basic units (Clark et al., 2007; Hunt and Black, 1996). Diphones extend from the central point of one phone to the central point of the following one. They contain the transitions between adjacent phones. A triphone is a phoneme with a specific left and right context. It has the structure: half phoneme–phoneme–half phoneme.

Articulatory synthesis is the most sophisticated method in terms of model structure and computation. It aims to model the complicated human speech production mechanisms as closely as possible (Qinsheng et al., 2011). The articulatory model produces speech by adjusting the position of lips and tongue, lung pressure, tension of vocal cords, and other parameters. This requires obtaining a large number of control parameters in various ways by analyzing natural speech.

7.2.3 Building Speech Synthesis Systems

Building speech synthesis systems require a speech units corpus. Natural speech must be recorded for all units—for example, all phonemes—in all possible contexts. Next, the units in the spoken speech data are segmented and labeled. Finally, the most appropriate speech units are chosen (Black and Cambpbell, 1995).

Generally, concatenative synthesis yields high quality speech (Thomas et al., 2006). With a large speech units corpus, high quality speech waveforms can be generated. Such synthesized speech preserves naturalness and intelligibility. Separate *prosody modeling* is not necessary for speech unit selection due to the availability of many units corresponding to varied contexts.

When phones are used as basic units, the size of the database will be less than 50 units for most languages. However, the phones provide very limited coarticulation information across adjacent units. Hence it fails to model the dynamics of speech sounds. Diphones and triphones will minimize the discontinuities at the concatenation points and capture the coarticulation effects. But a single example of each of them is not sufficient to produce good quality speech.

The advantage of using syllables as basic units is that they have fairly long duration when compared to phonemes or diphones (Nageshwara Rao et al., 2005). Also, segmentation becomes easier. Furthermore, since the boundaries of most of the syllables are low energy regions due to consonants, the distortion is significantly smaller (Venugopalakrishna et al., 2008).

In addition to speech unit methods, statistical parametric speech synthesis methods are also used (Tokuda and Zen, 2015; Zen et al., 2009, 2013). These methods do not require recorded speech in any phonetic and prosodic contexts. They generate the average of some sets of similarly sounding speech segments. First, they extract features of speech from a dataset. Second, the features and training data are used in building hidden Markov, neural network, or DNN model. Third, speech parameters are generated for a given word sequence to be synthesized from the set of estimated models to maximize their output probabilities. Finally, a speech waveform is reconstructed from the estimated parameters of speech.

Hybrid approaches to speech synthesis are developed by combining unit selection and statistical parametric methods. It is possible to retain the advantages of both and eliminate the drawbacks by combining them suitably. Multiple synthesized speech variations in multiple languages can be produced if large datasets are available. Very large datasets require substantial computing resources which limits unit selection techniques. However, this limitation can be overcome with distributed high performance computing frameworks such as Hadoop (White, 2015).

7.3 Speaker Recognition

The goal of automatic speaker recognition systems is to identify or confirm the identity of a person based on her voice. There are two aspects to this task—speaker identification and verification (Bimbot et al., 2004). *Speaker identification* involves recognizing a speaker from a set of known speakers. *Speaker verification* is an authentication task, which determines whether a person is who she claims to be.

Speaker recognition systems can be text dependent or text independent. In *text-dependent* systems, the speaker has to utter a fixed phrase for enrollment and verification (Kinnunen and Li, 2010). A feature vector is extracted from the utterance. Voice password based access control is an example for text-dependent system. In contrast, a text-independent system recognizes a person from whatever words she utters. *Text-independent* recognition is more difficult but flexible. Another variation in speaker recognition systems is based on whether the set of people that the system can recognize is fixed *a priori*. A system that can only recognize a speaker from a predetermined list is called *closed-set* identification system; otherwise, it is an *open-set* system.

An *i-vector* is a compact representation of a speaker's utterance—typically a vector of 400–600 dimensions after projection into a low-dimensional subspace (Dehak et al., 2011a). i-vector based speaker recognition systems were introduced recently (McLaren and Van Leeuwen, 2011). The model is trained using the factor analysis method.

7.3.1 Speaker Search

Identifying a speaker from a large population of speakers is referred to as the *speaker search* problem. GMM based systems work well (Reynolds and Rose, 1995) for this problem. The speaker search method proposed in Apsingekar and Leon (2009) is based on GMMs. During training, GMM-based speaker models are constructed. Next, they are clustered using the *k*-means algorithm to reduce the search space. When searching for a speaker, only the speaker models in the selected clusters are examined. Such clusters are identified using a scoring metric.

A speaker search algorithm based on kernelized locality-sensitive hashing (LSH) is proposed in Jeon and Cheng (2012). MFCC and delta coefficients are used as acoustic features. A total of 9422 speakers from the SPEECON database (Iskra et al., 2002) is used for development and testing. Of these, 1500 speakers are used for training and 7922 for testing. Nearest neighbor search using Hamming distance is used to find the candidate matches for speaker search. They are rank ordered by using the utterance comparison model.

Another speaker search system is proposed in Schmidt et al. (2014). The algorithm is tested on a large dataset from YouTube, which comprises 998 speakers from 1111 videos. Each video is at least 30 min in duration. Speaker identification is based on combining LSH (Har-Peled et al., 2012) and i-vectors. The LSH enables fast searching whereas the i-vectors improve the identification accuracy.

7.3.2 Implementing Speaker Recognition Systems

Speaker recognition systems are comprised of two functional components—enrollment and recognition. The *enrollment component* creates reference models for each speaker from the corresponding speaker's utterance. Reference models are represented by a sequence of feature vectors. The *identification component* matches the sequence of feature vectors of the speaker to be identified with those reference models in the system. Verification decisions are made by comparing the distance between the feature vectors of the speaker and the claimed entity. System performance degrades as the number of speakers increases.

Speaker recognition systems are implemented by using a wide range of techniques (Reynolds, 2002). We discuss two of them: template matching, and nearest neighbor. In *template matching*, there is a template in the system database for each speaker to be recognized. The template of a speaker is constructed from a sequence of feature vectors, which are extracted from the utterance of a fixed phrase by the speaker. Speakers are identified and verified by computing the distance between templates. These distance scores are computed using DTW technique. The DTW finds an optimal match between two sequences of feature vectors (Sakoe, 1978).

The *nearest neighbor* method recognizes a speaker by measuring the cumulative distance between each feature vector from the sequence of feature vectors to their k nearest neighbors in the speaker's reference sequence vectors. The speaker reference model can also be created by using mathematical models like neural networks, HMM, GMM, and SVM. These models are explicitly trained for each speaker. During testing, the model which gives the highest confidence is used to identify the speaker. According to the published results, HMM based systems perform well. Variations of these two basic techniques are widely used. A detailed survey of speaker identification systems is given in Kinnunen and Li (2010).

7.3.3 Evaluating Speaker Recognition Systems

Speaker identification systems are evaluated using an identification accuracy metric. Speaker verification systems are evaluated using two types of errors—false rejection rate (FRR) and false acceptance rate (FAR). *False rejection* occurs when the system rejects a valid speaker, and *false acceptance* when the system accepts an imposter speaker. Both types of error are a function of the *decision threshold* used in the template matching and nearest neighbor methods. Choosing a high threshold of acceptance will result in high FRR (low FAR). Likewise, choosing a low threshold will result in high FAR (low FRR).

The threshold value is determined from a *decision-error trade-off* (DET) curve, which is a plot between FAR and FRR. The FAR and FRR values of a verification system define different points on the DET curve. The threshold values for false acceptance and FRRs are selected when these rates are equal. This value is referred to as the *equal error rate* (EER). It indicates that the proportion of false acceptances is equal to the proportion of false rejections. Lower EER values imply higher accuracy of the system. Another important metric is the *detection cost* function. It is defined as a weighted sum of the miss and false alarm error probabilities, and is given by:

$$C_{\text{Det}} = C_M P_T P_{M|T} + C_{\text{FA}}(1 - P_T)(P_{\text{FA|KT}} P_K + P_{\text{FA|UT}}(1 - P_K)) \quad (21)$$

where T, original speaker of the speech segment; C_M, cost of a false rejection; C_{FA}, cost of a false acceptance; P_T, a priori probability that the speaker is the target speaker; P_K, a priori probability that the nontarget speaker is one of the evaluation target speakers; and P_U, a priori probability that the nontarget speaker is not from the evaluation target speakers.

7.4 Spoken Language Identification

A language identification (LID) system labels the language from speech utterances (Muthusamy et al., 1994). Humans have the ability to identify the language just by hearing 1 or 2 s of speech of a familiar language. LID systems are developed to replicate this human capability. The need for reliable LID systems is ever growing due to increased human interaction with computing devices through natural language interfaces. LID applications include spoken language translation, multilingual speech recognition, spoken document retrieval, and public information centers. LID systems should feature multilingual capability to effectively function in a globalized world.

LID techniques are similar to speaker recognition in terms of model formulation, methodologies, and performance metrics (Li et al., 2013). Current LID systems use certain amount of language specific information either implicitly or explicitly, but they differ in the amount of the information used. Performance and complexity of LID systems are proportional to the amount of linguistic information used.

The systems which do not require phone recognizers—segmented and labeled speech data—are called *implicit LID systems*. In other words, they require only the raw speech training data along with the true identity of the language spoken. Language models and language-specific information are derived from the training data. Current implicit LID systems differ mainly in the type of features selected for discriminating languages. In contrast, *explicit LID systems* require segmented and labeled speech signals as training data for the languages to be recognized.

7.4.1 Developing LID Systems

Linguistic and perceptual cues differentiate languages (Zissman and Berkling, 2001). LID systems are developed by considering these cues. Based on the cues used, approaches to developing LIDs are labeled as acoustic–phonetic, phonotactic, prosodic, and lexical (Ambikairajah et al., 2011).

Like other speech applications, LID systems are developed in two phases—training and testing. First, suitable features for each language are extracted from the training data. Second, using these features reference models are developed. Such models include GMM, HMM, VQ, SVM, and NN. During testing phase, the feature vectors that are extracted from the speech utterance in the test data are presented to all the reference models. The language that corresponds to the model which gives the highest confidence is considered to be the language of the speech utterance.

7.4.2 Acoustic–Phonetic Approach

The *phoneme sets* used in different languages differ, even though many languages share a common subset (Kirchhoff, 2006). Acoustic–phonetic approaches make use of these variations. They use spectral based features like MFCC to capture the phonetic differences between languages. Especially shifted delta cepstrum features work well for LID (Torres-carrasquillo et al., 2002).

7.4.3 Phonotactic Approach

The rules that govern the sequencing of phonemes vary from one language to another. *Phonotactic* approaches are based on these sequencing rules. These approaches build phone *n*-gram models for each language (Jurafsky and Martin, 2009). They are similar to the *language models* used for written text, but segments the speech waveform into a sequence of phones. The latter describes the probability of a subsequence of *n* phones.

The LIDs which are based on phonotactic approaches are called phone recognition followed by language modeling (PRLM) systems (Zissman, 1996). This is because a phone *n*-gram model employs a phone recognizer followed by an *n*-gram language models for the target languages. The phone recognizer converts speech units into a sequence of sound tokens using a common phoneme inventory. A sound token describes a distinct acoustic–phonetic attribute. A token may be a speech frame, phone, syllable, or lexical word. The phone *n*-gram model describes each target language in terms of phone *n*-gram statistics. A PRLM is called a *parallel phone recognition language modeling* when multiple phone recognizers are employed (Zissman, 1993). Each one is trained for a specific language.

7.4.4 Lexical Approach

Lexical approaches make use of the syntactic variations in languages. Words, roots, and other lexicons differ from one language to another. Also, each language has its own vocabulary. Furthermore, the word order in sentences varies across languages. Lexical approaches use large vocabulary continuous speech recognition systems in parallel for different target languages

(Schultz et al., 1996). The system that gives the highest confidence is used to identify the language. Since the lexical approaches are not cost effective, acoustic–phonetic and phonotactic approaches are commonly used in practice.

7.4.5 Evaluating LID Systems

Metrics for evaluating LID systems include accuracy, DET curve, and EER. Note that these metrics are same as the ones used for evaluating the speaker recognition systems. *Average detection cost* is another metric for LID systems and is defined as:

$$C_{\text{avg}} = C_M P_T \frac{1}{L}\sum_{l=1}^{L} P_M(L_l) + C_{\text{FA}}(1 - P_T)\frac{1}{L}\sum_{l=1}^{L}\left[\frac{1}{L-1}\sum_{m\neq l} P_{\text{FA}}(L_l, L_m)\right] \quad (22)$$

where L, number of languages; T, original speaker of the speech segment; C_M, cost of a false rejection; C_{FA}, cost of a false acceptance; $P_M(L_l)$, FRR (the error when a test segment of language L_l is rejected when it is spoken in that language); $P_{FA}(L_l, L_m)$, FAR (the error when a test segment of language L_m is accepted when it is spoken in language L_l); and P_T, a priori probability that the language is the target language.

7.4.6 Recent LID Systems

A LID system that uses DNNs is investigated in Lopez-Moreno et al. (2014). This study uses a very large corpus named Google 5M LID dataset. As demonstrated in this work, DNNs trained with more data outperform the i-vector systems. This improvement is attributed to DNN and the Google 5M LID dataset. Another study which used a frame-by-frame approach demonstrated the suitability of DNNs for real-time applications (Gonzalez-Dominguez et al., 2015).

Since LID task is similar to the speaker identification, recently the popular *i-vector* based speaker recognition technique is applied to LID (Dehak et al., 2011b; Martínez et al., 2011). Recent success of DNNs for acoustic modeling warranted their application to LID systems as well (Rahman Mohamed et al., 2012).

7.5 Speech-to-Speech Translation

An SST system converts speech from a source language to a target language (Zhou et al., 2013). Applications of SST include speech information retrieval, interactive navigation, dictation, and summarization and archiving.

SST systems integrate three distinct technologies: ASR, MT, and TTS (Sakti et al., 2009). First, the source language speech is recognized and then transcribed to source language text using ASR. Second, the source language text is translated into target language text using MT techniques. Finally, a TTS system is used to convert target language text to speech in the same language. ASR and TTS synthesis are already described in Sections 7.1 and 7.2.

7.5.1 Machine Translation

An MT system translates text from one natural language to another (Weaver, 1955). Lexical and syntactic ambiguities are the two primary challenges in MT. A word in the *source language*, depending on its part-of-speech tag or word sense, will translate to different words in the *target language*. Another challenge comes from differing word orders in the source and target languages. For example, English word order is subject–verb–object, whereas subject–object–verb is the order in Indian languages such as Telugu.

Classical approaches to MT are rule based. Language knowledge is endorsed by linguistic experts and is encoded in the form of rules. But formulating these rules is expensive in terms of time and labor. Furthermore, these approaches fail to attain satisfactory performance for large-scale applications. Finally, these systems are difficult to adapt to new contexts. In contrast, current approaches to MT are data-driven and employ machine learning techniques.

Statistical MT approaches automatically acquire knowledge from large training datasets. They view MT as a supervised learning problem and use a parallel corpus as a training set of translation examples. For instance, IBM's early work in this area used a parallel corpus of French–English translation of the Canadian parliament proceedings. The parallel corpus comprised 1.7 million sentences of length 30 words or less.

7.5.2 Approaches to SST

Most of the current approaches to SST employ statistical models (Nakamura et al., 2006; Zhou et al., 2002). Translating spoken language is much more difficult than text translation. Spoken language contains ungrammatical and colloquial expressions. Furthermore, it does not include punctuation, which is vital to disambiguating the meaning of sentences. SST challenges include speech recognition and MT in adverse environments, and lack of training data and linguistic resources for under-studied languages (Ettelaie et al., 2013). Another challenge is designing scalable algorithms and building models that perform well.

Given the right amount of training data, developing speech translation components is rather straightforward. As indicated above, SST systems integrate three modules. The output of one module is used as input to the next. Therefore, errors of a module strongly affect the performance of the next module (Hashimoto et al., 2011). For example, the MT component which performs well with correct source language text degrades considerably due to speech recognition errors. SST performance can be enhanced by improving the interface between ASR and MT systems.

MASTOR is an IBM SST system. It provides real-time translation of free-form speech (Gao et al., 2002, 2006). Asian Speech Translation Advanced Research (A-STAR) consortium developed the first SST system for Asian languages. Currently, the system handles eight Asian languages and English.

7.5.3 Evaluating SST Systems

WER is a commonly used metric for evaluating SST systems. Bilingual Evaluation Understudy (BLEU) is another metric for SST evaluation developed by IBM. BLEU uses a modified form of *precision* to compare a candidate translation against multiple reference translations. National Institute of Standards and Technology uses a modified form of the BLEU metric. Metric for Evaluation of Translation with Explicit ORdering is another metric, which is designed to alleviate some of the deficiencies of BLEU. Finally, LEPOR is an automatic, language-independent, evaluation metric with tunable parameters.

7.6 Speaker Diarization

A speaker diarization system answers the question "who spoke when" in an audio recording (Tranter and Reynolds, 2006). It identifies the number of speakers and creates a list of speech time intervals for each speaker. Information such as the number of speakers and their identities is not provided to the system. Speaker diarization consists of two tasks: speaker segmentation and speaker clustering (Moattar and Homayounpour, 2012). *Speaker segmentation* automatically partitions a conversation involving multiple speakers into homogeneous segments. In the latter, each segment corresponds to only one speaker. *Speaker clustering* groups together all the segments that correspond to the same speaker.

Diarization applications include speech recognition, searching and indexing of audio archives, automatic transcriptions, summarization, parsing, and MT. There are three primary domains which have been used for speaker diarization research and development: broadcast news audio, recorded meetings, and telephone conversations (Miró et al., 2012). The data from these domains differ in many ways: quality of the recordings, amount and types of nonspeech sources, number of speakers, duration and sequencing of speaker turns, and style/spontaneity of the speech. Each domain presents unique challenges (Mirghafori and Wooters, 2006).

In telephone conversations, typically only two people are involved and there are few overlapping segments. In contrast, broadcast news comprises more speakers. This is mostly prepared speech with a large part of studio quality voice. The hardest domain is recorded meetings. Since the speech is spontaneous, overlapping voices, disfluencies, distant speakers (in case of table microphones), and background noise pose serious challenges.

7.6.1 Speaker Diarization Algorithms

These algorithms fall into three classes based on how they perform speaker segmentation and clustering tasks. The classes are step-by-step, integrated, and mixed (Meignier et al., 2006; Reynolds et al., 2009). *Step-by-step algorithms* perform the following tasks in sequential order: (1) find the speaker change points, *segmentation*; (2) grow similar audio segments using clustering and determine the number of speakers from the clusters, *clustering*. Integrated algorithms perform both segmentation and clustering tasks simultaneously. Mixed category ones first apply the step-by-step algorithm. Next, a resegmentation process using an integrated algorithm refines the segment boundaries, clusters, and the number of speakers.

7.6.2 Speaker Segmentation

Algorithms for speaker segmentation are grouped into three classes: model based, metric based, and hybrid (Kotti et al., 2008). In the *model-based* approach, for each class, a set of models is derived from and trained using a training corpus. The classes include female and male speakers, silence, music, laughter, breathing, and noise. The incoming speech stream is classified using these classes. It is assumed that a speaker change is likely to occur at the time when the model's identification decision changes from one class to another. Prior knowledge is a prerequisite to initialize the speaker models.

In the *metric-based* approach, two adjacent windows are selected from the speech stream and their dissimilarities are evaluated. Speaker change is detected if two windows are dissimilar. The dissimilarity is quantified by measuring the acoustic distance. The latter is measured using metrics like the Kullback–Leibler Distance (KLD), Generalized Likelihood Ratio (GLR), Mahalanobis distance, Bhattacharya distance, and Bayesian Information Criterion (BIC) (Rabiner and Juang, 1993). The metric-based approaches are unsupervised and do not require knowledge of speakers or training data.

Hybrid algorithms combine both metric and model-based techniques. Metric-based segmentation is used initially to presegment the input audio signal. The resulting segments are then used to create a set of speaker models. Next, model-based resegmentation is used to obtain a more refined segmentation.

7.6.3 Clustering

Speaker clustering refers to unsupervised classification of speech segments based on speaker voice characteristics. It identifies all speech segments uttered by the same speaker in an audio recording and assigns a unique label to them. Both deterministic and probabilistic approaches are used for speaker clustering. Deterministic approaches use a metric like KLD, GLR, or BIC to cluster similar audio segments, whereas nondeterministic ones employ GMMs, or HMMs.

7.6.4 Speaker Diarization Systems

They are broadly grouped into two classes: bottom-up and top-down. In the *bottom-up* approach, the number of initial clusters created is more than the number of expected speakers (Nguyen et al., 2009; Wooters and Huijbregts,

2008). The number of clusters are then successively merged until there is only one cluster for each speaker. In the *top-down* approach, the number of initial clusters is a small number and often just one (Fredouille et al., 2009). Initially, the entire audio stream is modeled with a single speaker, and new speaker models are added incrementally.

A speaker diarization system development involves multiple steps. First, the data is preprocessed depending on the domain. For example, noise reduction for a recorded meeting data is performed. Second, the speech segments are demarcated using Voice Activity Detection (VAD) tool. Third, necessary features are extracted for the detected speech segments. Fourth, the number of clusters are initialized based on the type of the system—many for the bottom-up approach and one for the top-down approach. Fifth, speaker segmentation and clustering tasks are performed either in sequential order or parallelly. Sixth, the clusters are iteratively merged or new ones are introduced using distance metric. The split or merge process is controlled using a termination criterion.

In Huijbregts and Van Leeuwen (2012), a diarization method is proposed for long recordings. They divide the recordings into smaller chunks. Then diarization method described in Ajmera et al. (2002) is performed on each chunk. A speaker model is created for each speaker cluster. Each speaker model is then scored against other speaker models. Using bottom-up clustering, similar segments from different clusters are linked together. A multi modal diarization method for speech and video recordings of meetings is discussed in Ferras et al. (2014). This approach builds on the one presented in Ferras and Boudard (2012). The dataset consists of over 100 meetings.

7.6.5 Evaluating Speaker Diarization Systems

Diarization error rate (DER) and equal error rate (EER) are the two metrics used for evaluating the performance of diarization. A reference of speaker information for the data is required for analyzing DER. The reference is generated from manual transcription of the data. It is used to compare the result given by the algorithm. The DER is defined as $DER = MS + FA + SE$, where Missed speech (MS), speaker of the speech segment is present in the reference but missing in hypothesis; false alarm (FA), speaker present in the speech segment is in the hypothesis but missing in the reference; speaker error (SE), the reference speaker is not the hypothesized one.

8 RESOURCES FOR SPEECH RESEARCH

As discussed earlier, many machine learning based approaches to speech processing critically depend on training and test data. Datasets come in different forms and sizes. Datasets range from raw speech corpora, to speech with manual or automatic annotations, words and their relationships (morphological, semantic, translational, evolutionary), grammars, and parameters of statistical models (e.g., grammar weights).

The Linguistic Data Consortium (LDC, 2016) is the major source of data for speech processing research. It is an open consortium of universities, libraries, corporations, and government research laboratories. It creates and distributes a wide array of language resources to promote and enable speech research and applications. In October 2014, LDC released 8500 h of the United Nations Proceedings Speech dataset in six languages—Arabic, Chinese, English, French, Russian, and Spanish.

In 2006, Google released a *trillion-word corpus* with frequency counts for all sequences up to five words long. Since the data came from unfiltered Web pages, it has various kinds of errors including grammatical and spelling as well as incomplete sentences. Furthermore, the data is not annotated carefully with hand-corrected part-of-speech tags. Given its sheer size, this corpus could be used for building statistical models for language modeling, MT, and speech recognition. Google 5M LID dataset contains 5 million utterances collected from different Google Services for training and testing systems for automatic LID (Gonzalez-Dominguez et al., 2015).

In 2013, Google released a large dataset of *syntactic n-grams* derived from English Google Books (Goldberg and Orwant, 2013). Syntactic n-grams are structures in which word contexts are based on their positions in a syntactic parse tree, and not their sequential order in the sentence. Therefore, words that are far apart from each other in the sentence can be close to each other syntactically. The syntactic n-grams dataset contains over 10 billion unique items covering a wide range of syntactic structures (Google, 2016b). It also includes a temporal dimension. A large benchmark dataset containing *one billion words* for evaluating language modeling techniques is described in Chelba et al. (2013a). An approach to large-scale speaker recognition using a dataset collected from the Google Tech Talk channel on YouTube is discussed in Schmidt et al. (2014). The dataset contains 1111 videos with 998 distinct speakers. Each video contains at least 30 min of speech.

Frameworks and tools such as hidden Markov model Toolkit (HTK) (Cambridge University Engineering Department, 2016), VOICEBOX (Brookes, 2014), Praat (Praat, 2016), and openSMILE (OpenSMILE, 2016) are used to develop speech applications. The HTK is used for speech recognition and synthesis. The VOICEBOX is a speech processing toolkit based on MATLAB. The Praat is freeware for speech signal analysis and reconstruction. The openSMILE is a toolkit for extracting audio feature in real time. More speech processing tools and resources are available at Carnegie Mellon University (2016), Idiap Research Institute (2016), and VoiceBox (2016).

Google recently open-sourced a machine learning library called *TensorFlow*. It is well suited for developing machine learning-centric applications (TensorFlow, 2016). SyntaxNet (SyntaxNet, 2016) is a neural network framework embedded in TensorFlow. SyntaxNet comes with a pretrained parser, called Parsey McParseface, for analyzing English language text. Both TensorFlow and SyntaxNet can be leveraged to advance speech processing research.

The latest findings of speech processing research are reported in various scientific forums such as the IEEE International Conference on Acoustics, Speech and Signal Processing (ICASSP), International Speech Communication Association (ISCA) online archive, and Annual Conference of the International Speech Communication Association (INTERSPEECH). These forums also provide latest information about speech processing tools and resources.

9 TRENDS AND RESEARCH DIRECTIONS

The rapid rise of Big Data presents both challenges and opportunities to speech researchers. Extracting useful information from spoken language data is challenging. Both traditional machine learning and cognitive computing approaches require scalable parallel computing infrastructure to cope with massive data volumes.

The current research in speech processing is driven by statistical machine learning approaches than linguistic theory (Johnson, 2009). Statistical approaches use large training data and validate the view that simpler models often outperform more complex ones. For example, simple *n*-gram models or linear classifiers based on millions of specific features perform better than elaborate models that try to discover general rules (Halevy et al., 2009).

Based on the experiences with statistical methods in speech recognition and MT, memorization is a good practice if lots of training data is available. Earlier approaches relied on elaborate rules for the relationships between syntactic and semantic patterns in the source and target languages. In contrast, current statistical MT models employ large memorized phrase tables that give candidate mappings between specific source- and target-language phrases.

Performance of many speech processing applications depends on the quality of the language models. The amount of training data is one of the factors that determines the effectiveness of the language models. One billion words of training data (Chelba et al., 2013a; Google, 2016a) and syntactic *n*-grams data (Goldberg and Orwant, 2013; Google, 2016b) can be used to improve accuracy. Until now, most successful approaches to language modeling depended on sequential data (e.g., *n*-grams) as large-scale datasets of syntactic counts were not readily available.

Typically, speech processing systems require large volumes of high quality training data. Acoustic models are constructed using manually transcribed data. However, manual transcription of large datasets is expensive. Without large datasets, systems that support a broad range of languages scale poorly. This is another area where Big Data plays a significant role. Availability of large scale training data for speech processing research is ushering in a paradigm of experimentation with hybrid models and ensemble approaches.

Turning to practical applications, functional features of Siri, Google Now, and Cortana showcase the state-of-the-art in natural language interfaces. Siri is a virtual assistant available for Apple's iOS devices. It is an embodiment of several decades of research of the NLP community. It uses sequential inference and contextual awareness to understand and respond to user voice commands. Siri understands and speaks over 15 languages including various dialects of English and Spanish. Google Now and Cortana are similar applications for Android and Windows Phone devices. These practical applications provide the impetus for industry-driven speech research and innovative applications.

10 CONCLUSIONS

Speech based interfaces are revolutionizing the way humans interact with computers. Google conducted a mobile voice study in October 2014 to understand search habits of American smart phone users across different age groups. According to this study, 55% of the teenagers use hands-free search every day, and 76% of the respondents think voice search is great for multitasking.

Voice search is used for tasks such as seeking driving directions, dictating text messages, making phone calls, and checking weather. This trend will increase in the future as voice-based interfaces become mainstream. Moreover, the need for fast, efficient, accurate, and robust means of speech based applications is of growing importance for commercial, forensic, and government applications. Speech based interaction with various household appliances such as refrigerators and washing machines in the near future will be a reality. Even more exciting will be speech based interfaces to subsystems in a car such as cruise control, audio systems, navigational systems, climate control, and windshield wiper blades.

Speech research into scalability, multilingual interfaces, personalization, portability, and provenance can greatly benefit from the availability of extremely large datasets. Large scale data provides unprecedented opportunities to advance the speech processing field as well as the way we interact with computing devices, home appliances, and other machines.

REFERENCES

Abadi, M., Agarwal, A., Barham, P., Brevdo, E., Chen, Z., Citro, C., Corrado, G.S., Davis, A., Dean, J., Devin, M., Ghemawat, S., Goodfellow, I., Harp, A., Irving, G., Isard, M., Jia, Y., Jozefowicz, R., Kaiser, L., Kudlur, M., Levenberg, J., Mané, D., Monga, R., Moore, S., Murray, D., Olah, C., Schuster, M., Shlens, J., Steiner, B., Sutskever, I., Talwar, K., Tucker, P., Vanhoucke, V., Vasudevan, V., Viégas, F., Vinyals, O., Warden, P., Wattenberg, M., Wicke, M., Yu, Y., Zheng, X., 2015. TensorFlow: large-scale machine learning on heterogeneous systems. Software available from http://tensorflow.org/.

Ajmera, J., Bourlard, H., Lapidot, I., McCowan, I., 2002. Unknown-multiple speaker clustering using HMM. In: Proceedings of International Conference on Spoken Language Processing (ICSLP) 2002, pp. 573–576.

Ambikairajah, E., Li, H., Wang, L., Yin, B., Sethu, V., 2011. Language identification: a tutorial. IEEE Circuits Syst. Mag. 11 (2), 82–108.

Apple Inc. (USA), n.d. A voice based personal assistant and knowledge navigator in iPhones. http://www.apple.com/ios/siri/. Retrieved July 4, 2016.

Apsingekar, V.R., Leon, P.L.D., 2009. Speaker model clustering for efficient speaker identification in large population applications. IEEE Trans. Audio Speech Lang. Process. 17 (4), 848–853.

Babacan, O., Drugman, T., D'Alessandro, N., Henrich, N., Dutoit, T., 2013. A comparative study of pitch extraction algorithms on a large variety of singing sounds. In: 2013 IEEE International Conference on Acoustics, Speech and Signal Processing (ICASSP). IEEE, pp. 7815–7819.

Bengio, Y., Ducharme, R., Vincent, P., Janvin, C., 2003. A neural probabilistic language model. J. Mach. Learn. Res. 3, 1137–1155.

Bengio, Y., Lamblin, P., Popovici, D., Larochelle, H., 2007. Greedy layer-wise training of deep networks. In: Schölkopf, B., Platt, J., Hoffman, T. (Eds.), Advances in Neural Information Processing Systems 19. MIT Press, Cambridge, MA, pp. 153–160.

Bimbot, F., Bonastre, J.F., Fredouille, C., Gravier, G., Magrin-Chagnolleau, I., Meignier, S., Merlin, T., Ortega-García, J., Petrovska-Delacrétaz, D., Reynolds, D.A., 2004. A tutorial on text-independent speaker verification. EURASIP J. Appl. Sign. Process. 2004, 430–451.

Black, A.W., Cambpbell, N., 1995. Optimising selection of units from speech database for concatenative synthesis. In: Proceedings of EUROSPEECH, pp. 581–584.

Black, A.W., Lenzo, K.A., 2000. Limited domain synthesis. In: Proceedings of International Conference on Spoken Language Processing (ICSLP), Beijing, China.

Brants, T., Popat, A.C., Xu, P., Och, F.J., Dean, J., 2007. Large language models in machine translation. In: Proceedings of the Joint Conference on Empirical Methods in Natural Language Processing and Computational Natural Language Learning, pp. 858–867.

Brookes, M., 2014. VOICEBOX: speech processing toolbox for MATLAB. http://www.ee.ic.ac.uk/hp/staff/dmb/voicebox/voicebox.html.

Brown, P.F., DeSouza, P.V., Mercer, R.L., Pietra, V.J.D., Lai, J.C., 1992. Class-based n-gram models of natural language. Comput. Linguist. 18 (4), 467–479.

Burges, C.J.C., 1998. A tutorial on support vector machines for pattern recognition. Data Min. Knowl. Disc. 1384-5810. 2 (2), 121–167.

Cambridge University Engineering Department, 2016. Hidden markov model toolkit (HTK). http://htk.eng.cam.ac.uk/.

Campbell, W.M., Campbell, J.P., Reynolds, D.A., Singer, E., Torres-Carrasquillo, P.A., 2006. Support vector machines for speaker and language recognition. Comput. Speech Lang. 20 (5), 210–229.

Carnegie Mellon University, 2016. Open source speech software. http://www.speech.cs.cmu.edu/.

Chelba, C., Jelinek, F., 2000. Structured language modeling. Comput. Speech Lang. 14 (4), 283–332.

Chelba, C., Schalkwyk, J., Brants, T., Ha, V., Harb, B., Neveitt, W., Parada, C., Xu, P., 2010. Query language modeling for voice search. In: Proceedings of the 2010 IEEE Workshop on Spoken Language Technology, pp. 127–132.

Chelba, C., Xu, P., Pereira, F., Richardson, T., 2012. Distributed acoustic modeling with back-off n-grams. In: 2012 IEEE International Conference on Acoustics, Speech and Signal Processing (ICASSP), pp. 4129–4132.

Chelba, C., Mikolov, T., Schuster, M., Ge, Q., Brants, T., Koehn, P., Robinson, T., 2013a. One billion word benchmark for measuring progress in statistical language modeling. Comput. Res. Reposit. 1–6. abs/1312.3005.

Chelba, C., Xu, P., Pereira, F., Richardson, T., 2013b. Large scale distributed acoustic modeling with back-off n-grams. IEEE Trans. Audio Speech Lang. Process. 21 (6), 1158–1169.

Chen, S.F., Goodman, J., 1996. An empirical study of smoothing techniques for language modeling. In: Proceedings of the 34th Annual Meeting on Association for Computational Linguistics, pp. 310–318.

Clark, R.A., Richmond, K., King, S., 2007. Multisyn: open-domain unit selection for the festival speech synthesis system. Speech Comm. 49 (4), 317–330.

Davis, S.B., Mermelstein, P., 1980. Comparison of parametric representations for monosyllabic word recognition in continuously spoken sentences. IEEE Trans. Acoust. Speech Signal Process. 28, 357–366.

Dean, J., Ghemawat, S., 2008. Mapreduce: simplified data processing on large clusters. Commun. ACM 51 (1), 107–113.

Dean, J., Corrado, G., Monga, R., Chen, K., Devin, M., Le, Q.V., Mao, M.Z., Ranzato, M., Senior, A.W., Tucker, P.A., Yang, K., Ng, A.Y., 2012. Large scale distributed deep networks. In: Advances in Neural Information Processing Systems 25, pp. 1232–1240.

Dehak, N., Kenny, P., Dehak, R., Dumouchel, P., Ouellet, P., 2011. Front-end factor analysis for speaker verification. IEEE Trans. Audio Speech Lang. Process. 13 (4), 788–798.

Dehak, N., Torres-Carrasquillo, P.A., Reynolds, D.A., Dehak, R.A., 2011. Language recognition via i-vectors and dimensionality reduction. In: Conference of the International Speech Communication Association (Interspeech). ISCA, pp. 857–860.

Dutoit, T., 1997. An introduction to text-to-speech synthesis. Kluwer Academic Publishers, Norwell, MA.

Ettelaie, E., Georgiou, P.G., Narayanan, S.S., 2013. Unsupervised data processing for classifier-based speech translator. Comput. Speech Lang. 27 (2), 438–454.

Ferras, M., Boudard, H., 2012. Speaker diarization and linking of large corpora. In: Spoken Language Technology (SLT) Workshop. IEEE, pp. 280–285.

Ferras, M., Masneri, S., Schreer, O., Bourlard, H., 2014. Diarizing large corpora using multi-modal speaker linking. In: Conference of the International Speech Communication Association (Interspeech). ISCA.

Fredouille, C., Bozonnet, S., Evans, N.W.D., 2009. The LIA-EURECOM RT09 speaker diarization system. In: RT09, NIST Rich Transcription Workshop, Melbourne, FL.

Gao, Y., Zhou, B., Diao, Z., Sorensen, J., Erdogan, H., Sarikaya, R., 2002. A trainable approach for multilingual speech-to-speech translation system. In: Proceedings of Human Language Technology (HLT), San Diego, CA.

Gao, Y., Gu, L., Zhou, B., Sarikaya, R., Afity, M., Juo, H.K., Zhong, W., Zhu, Y.D., Prosser, C., Zhang, W., Besacier, L., 2006. IBM mastor system: multilingual automatic speech to speech translator. In: Proceedings of the Workshop on Medical Speech Translation at HLT-NAACL, pp. 57–60.

Gerhard, D., 2003. Pitch extraction and fundamental frequency: history and current techniques. Department of Computer Science, University of Regina.

Goldberg, Y., Orwant, J., 2013. A dataset of syntactic-Ngrams over time from a very large corpus of English books. In: Second Joint Conference on Lexical and Computational Semantics (*SEM), Volume 1: Proceedings of the Main Conference and the Shared Task: Semantic Textual Similarity. ACL, Atlanta, GA, pp. 241–247.

Gonzalez-Dominguez, J., Lopez-Moreno, I., Moreno, P.J., Gonzalez-Rodriguez, J., 2015. Frame by frame language identification in short utterances using deep neural networks. Neural Netw. 64, 49–58.

Goodman, J.T., 2001a. A bit of progress in language modeling. Comput. Speech Lang. 15 (4), 403–434.

Goodman, J.T., 2001b. Classes for fast maximum entropy training. Comput. Res. Reposit. 561–564. cs.CL/0108006.

Google, 2016. 1-billion-word-language-modeling-benchmark. https://code.google.com/p/1-billion-word-language-modeling-benchmark/.

Google, 2016. Syntactic N-grams. http://storage.googleapis.com/books/syntactic-ngrams/index.html.

Gudivada, V., Baeza-Yates, R., Raghavan, V., 2015. Big data: promises and problems. IEEE Comput. 48 (3), 20–23.

Gudivada, V., Rao, D., Raghavan, V., 2015. Big data driven natural language processing research and applications. In: Govindaraju, V., Raghavan, V., Rao, C.R. (Eds.), Big Data Analytics. Elsevier, New York, NY, pp. 203–238.

Halevy, A., Norvig, P., Pereira, F., 2009. The unreasonable effectiveness of data. IEEE Intell. Syst. 24 (2), 8–12.

Har-Peled, S., Indyk, P., Motwani, R., 2012. Approximate nearest neighbor: towards removing the curse of dimensionality. Theory Comput. 8 (14), 321–350.

Hashimoto, K., Yamagishi, J., Byrne, W., King, S., Tokuda, K., 2011. An analysis of machine translation and speech synthesis in speech-to-speech translation system. In: Proceedings of IEEE International Conference Acoustics, Speech and Signal Processing (ICASSP), pp. 5108–5111.

Haykin, S., 2001. Neural Networks: A Comprehensive Foundation. Pearson Education, Singapore.

Hermansky, H., 1990. Perceptual linear predictive (PLP) analysis of speech. J. Acoust. Soc. Am. 57 (4), 1738–1752.

Hinton, G.E., Osindero, S., Teh, Y.W., 2006. A fast learning algorithm for deep belief nets. Neural Comput. 18 (7), 1527–1554.

Huijbregts, M., Van Leeuwen, D.A., 2012. Large-scale speaker diarization for long recordings and small collections. IEEE Trans. Audio Speech Lang. Process. 20 (2), 404–413.

Hunt, A., Black, A.W., 1996. Unit selection in a concatenative speech synthesis system using a large speech database. In: 1996 IEEE International Conference on Acoustics, Speech, and Signal Processing, 1996. ICASSP-96. Conference Proceedings, pp. 373–376.

Idiap Research Institute, 2016. Resources for language research and applications development. http://www.idiap.ch/scientific-research/resources.

Iskra, D.J., Grosskopf, B., Marasek, K., Van den heuvel, H., Diehl, F., Kießling, A., 2002. SPEECON–speech databases for consumer devices: database specification and validation. In: Proceedings of the Third International Conference on Language Resources and Evaluation. http://www.lrec-conf.org/proceedings/lrec2002/pdf/177.pdf.

Jelinek, F., Merialdo, B., Roukos, S., Strauss, M., 1991. A dynamic language model for speech recognition. In: Proceedings of the DARPA Workshop on Speech and Natural Language, pp. 293–295.

Jeon, W., Cheng, Y.M., 2012. Efficient speaker search over large populations using kernelized locality-sensitive hashing. In: 2012 IEEE International Conference on Acoustics, Speech and Signal Processing (ICASSP), pp. 4261–4264.

Johnson, M., 2009. How the statistical revolution changes computational linguistics. In: Proceedings of the EACL 2009 Workshop on the Interaction Between Linguistics and Computational Linguistics: Virtuous, Vicious or Vacuous? Athens, Greece. ACL, Stroudsburg, PA, pp. 3–11.

Juang, B.H., Furui, S., 2000. Automatic recognition and understanding of spoken language–a first step toward natural human-machine communication. Proc. IEEE 0018-9219. 88 (8), 1142–1165.

Jurafsky, D., Martin, J.H., 2009. Speech and Language Processing, second Prentice Hall, Englewood Cliffs, NJ.

Jyothi, P., Johnson, L., Chelba, C., Strope, B., 2012. Large-scale discriminative language model reranking for voice-search. In: Proceedings of the NAACL-HLT 2012 Workshop: Will We Ever Really Replace the N-gram Model? On the Future of Language Modeling for HLT, pp. 41–49.

Kapralova, O., Alex, J., Weinstein, E., Moreno, P., Siohan, O., 2014. A big data approach to acoustic model training corpus selection. In: Conference of the International Speech Communication Association (Interspeech). ISCA.

Kinnunen, T., Li, H., 2010. An overview of text-independent speaker recognition: from features to supervectors. Speech Comm. 52 (1), 12–40.

Kirchhoff, K., 2006. Chapter 2–language characteristics. In: Schultz, T., Kirchhoff, K. (Eds.), Multilingual Speech Processing. Academic Press, Burlington, MA, pp. 5–31.

Kishore, S.P., Black, A.W., 2003. Unit size in unit selection speech synthesis. In: Proceedings of EUROSPEECH, pp. 1317–1320.

Kominek, J., Black, A., 2003. CMU ARCTIC databases for speech synthesis. Language Technologies Institute, School of Computer Science.

Kotti, M., Moschou, V., Kotropoulos, C., 2008. Review: speaker segmentation and clustering. Signal Process. 88 (5), 1091–1124.

Krizhevsky, A., Sutskever, I., Hinton, G.E., 2012. Imagenet classification with deep convolutional neural networks. In: Pereira, F., Burges, C.J.C., Bottou, L., Weinberger, K.Q. (Eds.), Advances in Neural Information Processing Systems 25. Curran Associates, Inc, pp. 1097–1105.

Li, H., Ma, B., Lee, K.A., 2013. Spoken language recognition: from fundamentals to practice. Proc. IEEE 101 (5), 1136–1159.

Linguistic Data Consortium, 2016. Language resources. https://www.ldc.upenn.edu/language-resources.

Lippmann, R.P., 1989. An introduction to computing with neural nets. IEEE ASSP Mag. 4, 4–22.

Lopez-Moreno, I., Gonzalez-Dominguez, J., Plchot, O., Martinez, D., Gonzalez-Rodriguez, J., Moreno, P., 2014. Automatic language identification using deep neural networks. In: 2014 IEEE International Conference on Acoustics, Speech and Signal Processing (ICASSP), pp. 5337–5341.

Martínez, D., Plchot, O., Burget, L., Gleombek, O., Matejka, P., 2011. Language recognition in ivectors space. In: Conference of the International Speech Communication Association (Interspeech). ISCA, pp. 861–864.

Mary, L., 2012. Automatic extraction of prosody for speaker, language and speech recognition. In: Extraction and Representation of Prosody for Speaker, Speech and Language Recognition, Springer Briefs in Electrical and Computer Engineering, Springer, New York, NY, pp. 19–33.

Mary, L., Yegnanarayana, B., 2008. Extraction and representation of prosodic features for language and speaker recognition. Speech Comm. 50 (10), 782–796.

McLaren, M., Van Leeuwen, D.A., 2011. Improved speaker recognition when using i-vectors from multiple speech sources. In: 2011 IEEE International Conference on Acoustics, Speech and Signal Processing (ICASSP). IEEE, pp. 5460–5463.

Meduri, S., Ananth, R., 2012. A Survey and Evaluation of Voice Activity Detection Algorithms: Speech Processing Module. LAP Lambert Academic Publishing, Germany.

Meignier, S., Moraru, D., Fredouille, C., Bonastre, J.F., Besacier, L., 2006. Step by step and integrated approaches in broadcast news speaker diarization. Comput. Speech Lang. 20, 303–330.

Mikolov, T., Karafiát, M., Burget, L., Cernocký, J., Khudanpur, S., 2010. Recurrent neural network based language model. In: 11th Annual Conference of the International Speech Communication Association. ISCA, pp. 1045–1048.

Mikolov, T., Deoras, A., Povey, D., Burget, L., Cernocký, J., 2011. Strategies for training large scale neural network language models. In: IEEE Workshop on Automatic Speech Recognition & Understanding (ASRU 2001), pp. 196–201.

Mirghafori, N., Wooters, C., 2006. Nuts and flakes: a study of data characteristics in speaker diarization. In: Proceedings of International Conference on Acoustics, Speech and Signal Processing (ICASSP).

Miró, X.A., Bozonnet, S., Evans, N.W.D., Fredouille, C., Friedland, G., Vinyals, O., 2012. Speaker diarization: a review of recent research. IEEE Trans. Audio Speech Lang. Process. 20 (2), 356–370.

Moattar, M.H., Homayounpour, M.M., 2012. A review on speaker diarization systems and approaches. Speech Comm. 54 (10), 1065–1103.

Morin, F., Bengio, Y., 2005. Hierarchical probabilistic neural network language model. In: AISTATS'05, pp. 246–252.

Muthusamy, Y.K., Barnard, E., Cole, R.A., 1994. Reviewing automatic language identification. IEEE Signal Process. Mag., 33–41.

Nageshwara Rao, M., Thomas, S., Nagarajan, T., Murthy, H.A., 2005. Text-to-speech synthesis using syllable like units. In: Proceedings of National Conference on Communication (NCC), IIT Kharagpur, India, pp. 227–280.

Nakamura, S., Markov, K., Nakaiwa, H., Kikui, G., Kawai, H., Jitsuhiro, T., Zhang, J.S., Yamamoto, H., Sumita, E., Yamamoto, S., 2006. The ATR multilingual speech-to-speech translation system. IEEE Trans. Audio Speech Lang. Process. 14 (2), 365–376.

Nguyen, T.H., et al., 2009. The IIR-NTU speaker diarization systems for RT 2009. In: RT09, NIST Rich Transcription Workshop, Melbourne, FL.

OpenSMILE, 2016. The munich versatile and fast open-source audio feature extractor. http://www.audeering.com/research/opensmile.

O'Shaughnessy, D., 1999. Speech Communications: Human and Machine, second. Wiley-IEEE Press, New Jersey.

O'Shaughnessy, D., 2008. Invited paper: automatic speech recognition: history, methods and challenges. Pattern Recogn. 41 (10), 2965–2979.

Praat, 2016. Doing phonetics by computer. www.praat.org.

Qinsheng, D., Jian, Z., Lirong, W., Lijuan, S., 2011. Articulatory speech synthesis: a survey. In: 2013 IEEE 16th International Conference on Computational Science and Engineering, pp. 539–542.

Rabiner, L.R., 1989. A tutorial on hidden Markov models and selected applications in speech recognition. Proc. IEEE 77 (2), 257–287.

Rabiner, L., Juang, B.H., 1993. Fundamentals of Speech Recognition. Prentice-Hall, Upper Saddle River, NJ.

Rabiner, L., Schafer, R., 1978. Digital Processing of Speech Signals. Prentice-Hall, Englewood Cliffs, NJ.

Rabiner, L., Schafer, R., 2010. Theory and Applications of Digital Speech Processing. Prentice-Hall, Upper Saddle River, NJ.

Raghavendra, E., Prahallad, K., 2010. A multilingual screen reader in Indian languages. In: National Conference on Communications (NCC), Chennai, India.

Rahman Mohamed, A., Dahl, G.E., Hinton, G.E., 2012. Acoustic modeling using deep belief networks. IEEE Trans. Audio Speech Lang. Process. 20 (1), 14–22.

Redner, R.A., Walker, H.F., 1984. Mixture densities, maximum likelihood and the EM algorithm. SIAM Rev. 26, 195–239.

Reynolds, D., 2002. An overview of automatic speaker recognition technology. In: 2002 IEEE International Conference on Acoustics, Speech, and Signal Processing (ICASSP). IV, 4072–4075.

Reynolds, D.A., Rose, R.C., 1995. Robust text-independent speaker identification using Gaussian mixture speaker models. IEEE Trans. Speech Audio Process. 3 (1), 72–83.

Reynolds, D.A., Kenny, P., Castaldo, F., 2009. A study of new approaches to speaker diarization. In: Conference of the International Speech Communication Association (Interspeech). ISCA, pp. 1047–1050.

Roark, B., Saraclar, M., 2004. Discriminative language modeling with conditional random fields and the perceptron algorithm. In: Proceedings of Association for Computational Linguistics (ACL), pp. 47–54.

Sakoe, H., 1978. Dynamic programming algorithm optimization for spoken word recognition. IEEE Trans. Acoust. Speech Signal Process. 26, 43–49.

Sakti, S., Kimura, N., Paul, M., Hori, C., Sumita, E., Nakamura, S., Park, J., Wutiwiwatchai, C., Xu, B., Riza, H., Arora, K., Luong, C.M., Li, H., 2009. The asian network-based speech-to-speech translation system. In: Automatic Speech Recognition & Understanding (ASRU). IEEE, pp. 507–512.

Schmidt, L., Sharifi, M., Lopez Moreno, I., 2014. Large-scale speaker identification. In: 2014 IEEE International Conference on Acoustics, Speech and Signal Processing (ICASSP), Florence, pp. 1650–1654.

Schultz, T., Rogina, I., Waibel, A., 1996. LVCSR-based language identification. IEEE Trans. Acoust. Speech Signal Process. 2, 781–784.

Smith, C., 2016. By the numbers: 130+ amazing youtube statistics. http://expandedramblings.com/index.php/youtube-statistics/.

SyntaxNet, 2016. An open source neural network framework for tensorflow for developing natural language understanding (NLU) systems. https://github.com/tensorflow/models/tree/master/syntaxnet.

Teh, Y.W., 2006. A hierarchical bayesian language model based on Pitman-Yor processes. In: Coling/ACL, 2006. 9. Association for Computational Linguistics, Stroudsburg, PA, pp. 985–992.

TensorFlow, 2016. An open source software library for numerical computation using data flow graphs. https://www.tensorflow.org/.

Thomas, S., Nageshwara Rao, M., Murthy, H.A., Ramalingam, C.S., 2006. Natural sounding TTS based on syllable-like units. In: Proceedings of 14th European Signal Processing Conference, Florence, Italy.

Tokuda, K., Zen, H., 2015. Directly modeling speech waveforms by neural networks for statistical parametric speech synthesis. In: 2015 IEEE International Conference on Acoustics, Speech and Signal Processing (ICASSP), pp. 4215–4219.

Torres-carrasquillo, P.A., Singer, E., Kohler, M.A., Deller, J.R., 2002. Approaches to language identification using gaussian mixture models and shifted delta cepstral features. In: Proceedings of International Conference on Spoken Language Processing (ICSLP), pp. 89–92.

Tranter, S.E., Reynolds, D.A., 2006. An overview of automatic speaker diarization systems. IEEE Trans. Audio Speech Lang. Process. 14 (5), 1557–1565.

Venugopalakrishna, Y.R., Vinodh, M.V., Murthy, H.A., Ramalingam, C.S., 2008. Methods for improving the quality of syllable based speech synthesis. In: Proceedings of Spoken Language Technology (SLT) Workshop, Goa, pp. 29–32.

VoiceBox, 2016. Conversational voice technology. http://www.voicebox.com/.

Weaver, W., 1955. Translation. In: Locke, W.N., Booth, D.A. (Eds.), Machine Translation of Languages. The MIT Press, pp. 15–23.

White, T., 2015. Hadoop: The Definitive Guide, fourth. O'Reilly.

Wooters, C., Huijbregts, M., 2008. The ICSI RT07s speaker diarization system. In: Multimodal Technologies for Perception of Humans, Springer-Verlag, Berlin, Heidelberg, pp. 509–519.

Xu, P., Gunawardana, A., Khudanpur, S., 2011. Efficient subsampling for training complex language models. In: Proceedings of EMNLP.

Yegnanarayana, B., 1999. Artificial Neural Networks. Prentice-Hall of India, New Delhi.

YouTube, 2016. Viewership statistics. http://www.youtube.com/yt/press/statistics.html.

Zen, H., Tokuda, K., Black, A.W., 2009. Review: statistical parametric speech synthesis. Speech Comm. 51 (11), 1039–1064.

Zen, H., Senior, A., Schuster, M., 2013. Statistical parametric speech synthesis using deep neural networks. In: 2013 IEEE International Conference on Acoustics, Speech and Signal Processing (ICASSP), pp. 7962–7966.

Zhou, B., Gao, Y., Sorensen, J., Diao, Z., Picheny, M., 2002. Statistical natural language generation for speech-to-speech machine translation systems. In: Proceedings of International Conference on Spoken Language Processing (ICSLP). 3, Denver, CO.

Zhou, B., Cui, X., Huang, S., Cmejrek, M., Zhang, W., Xue, J., Cui, J., Xiang, B., Daggett, G., Chaudhari, U., Maskey, S., Marcheret, E., 2013. The IBM speech-to-speech translation system for smartphone: improvements for resource-constrained tasks. Comput. Speech Lang. 27 (2), 397–418.

Zissman, M.A., 1993. Automatic language identification of telephone speech messages using phoneme recognition and n-gram modeling. In: Proceedings of International Conference on Acoustics, Speech and Signal Processing (ICASSP), pp. 309–402.

Zissman, M.A., 1996. Comparison of four approaches to automatic language identification of telephone speech. IEEE Trans. Speech Audio Process. 4 (1), 31–44.

Zissman, M.A., Berkling, K.M., 2001. Automatic language identification. Speech Comm. 35 (1-2), 115–124.

Chapter 11

The Internet of Things and Cognitive Computing

F.D. Hudson and E.W. Nichols
University Corporation for Advanced Internet Development, White Plains, NY, United States

ABSTRACT
The Internet of Things (IoT) and Cognitive Computing are concepts that have been developing since the 1950s with varying nomenclature. Enabled by advances in technology such as the development of lower cost, lower power, smaller microprocessors and microcontrollers, cellular/wireless chips, and the enablement of more widespread machine-to-machine communications, IoT technologies are now more widely deployed. Advances in Artificial Intelligence (AI) and question-answering systems (Ferrucci et al., 2010; Simmons, 1970) have enabled development of cognitive computing systems which can be applied to numerous use cases. There is an opportunity to leverage IoT and cognitive computing technologies together with data analytics (http://social-innovation.hitachi.com) to develop deeper insights from the vast amount of data being created by IoT to create value for people, cities, and industry. In this chapter, we explore the state of the art and future opportunities to apply IoT and cognitive computing together in IoT uses cases for smart cities and connected healthcare, to improve efficiencies, outcomes, and the human experience.

Keywords: IoT, Internet of Things, Cognitive computing, Insights, Big data, Smart city/cities, Connected healthcare

1 INTRODUCTION

The Internet of Things (IoT) and Cognitive Computing are concepts that have been developing since the 1950s with varying nomenclature. Through advances in technology such as the development of lower cost, lower power integrated circuits and sensors, and the enablement of more widespread machine-to-machine (M2M) communications, IoT technologies are now more widely deployed. Similarly, advancements in Artificial Intelligence since the 1950s, natural language question-answering (QA) systems since the 1960s (Simmons, 1970), and Deep Question Answering (DeepQA (Ferrucci et al., 2010)) systems more recently have enabled development of cognitive computing systems which are being applied to IoT and healthcare. DeepQA is an

Handbook of Statistics, Vol. 35. http://dx.doi.org/10.1016/bs.host.2016.07.009

extensible architecture that can be used as a foundation for combining, deploying, evaluating, and advancing a wide range of algorithmic techniques to rapidly advance the field of QA, such as is leveraged in the IBM Watson system. There is an opportunity to leverage IoT and cognitive computing technologies more broadly and deeply together with data analytics (http://social-innovation.hitachi.com) to create deeper insights from the vast amount of data being created by IoT to bring value to people, cities, and industries (https://www.oreilly.com/ideas/how-the-internet-of-things-impacts-big-data-strategies) of the world, today and in the future.

The key messages in this chapter are as follows:

- IoT and cognitive computing are ideas that began in the 1960s that were implemented in steps and evolved, and now can be applied to use cases such as smart cities and healthcare (http://www.forbes.com/sites/85broads/2014/03/26/4-ways-the-internet-of-things-is-transforming-healthcare/#27563b243fbe).
- IoT creates large volumes of data which can be analyzed by cognitive computing systems to provide insights and recommendations to inform machine and human actions and potential outcomes.
- Cognitive computing provides the opportunity for perpetual optimization of systems and outcomes leveraging data from the IoT.

When information technology and computing was in its infancy in the early 1900s, tabulating systems were the innovations of that time. Over the past 100 years or so, we have seen an evolution from centralized tabulating systems, to programmable computers, which were then distributed creating challenges in managing a dispersed set of computing entities. The world has since evolved rather rapidly, with the advent of the world wide web and the Internet, enabling computing, people, and content to be networked around the world. At the same time, the engineering world was developing electronically enabled physical systems such as electronic systems in vehicles, appliances, and building systems. The ability to connect physical systems with information technology has enabled the engineering and information technology domains to come together to enable the IoT. At the same time, artificial intelligence and QA systems (Ferrucci et al., 2010; Simmons, 1970) have evolved into cognitive computing systems (Kelly and Hamm, 2013), with the ability to gain insight from a diverse set of data resources (Hurwitz et al., 2015). Together, IoT and cognitive computing can access, integrate, analyze, and leverage data to improve outcomes across many industries including smart cities, and healthcare (see Fig. 1).

2 THE IOT—DEFINITION AND HISTORY

The IoT is a term representing the billions of physical devices across the world that have digital sensors, and how they are interconnected leveraging the Internet or other network technology (https://newsroom.cisco.com/feature-content?type=webcontent&articleId=1208342).

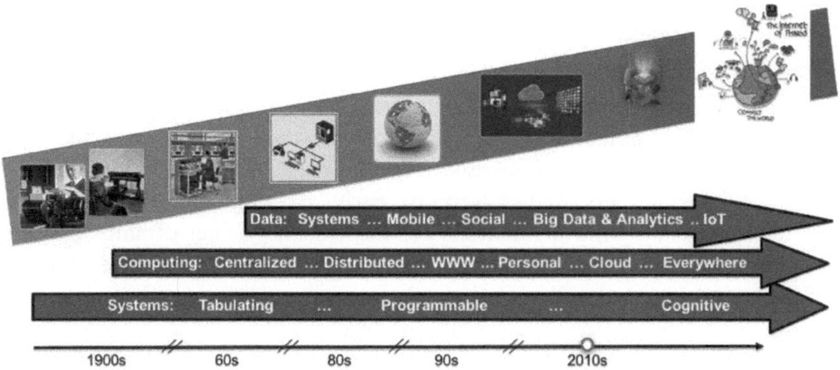

FIG. 1 Advances in technology and cultural evolution are ushering in a new era of computing enabled by the Internet of Things and Cognitive Computing. *Internet2 Chief Innovation Office.*

There were an estimated 13B connected devices in 2015, which is projected to grow to more than 38B devices by 2020, a rise of over 285% (Juniperresearch, 2015). These connected devices include vehicles, appliances, lighting, health and wellness devices, manufacturing systems, buildings, bridges, tunnels, food containers, electric meters, security systems, cameras, wearable devices, drones, and many more.

These many devices have digital sensors that collect data that the device then transmits to other devices or to a centralized management system. The public Internet or private networks connecting these devices provide the communications between these devices or "things."

The IoT represents the confluence of Information Technology (IT) and Operational Technology (OT), in a simple equation form as IT+OT=IoT. Information Technology is often referred to as computer systems, which capture and store information such as information from retail systems that report what is purchased in a store, to information about the amount of money in your bank account, to the information about the number of people in a country determined by a census. In fact, the first information technology systems were tabulating systems which counted the people in the United States census around the year 1900. Operational technology refers to the physical systems in our world such as airplanes, trains, automobiles, buildings, and medical devices. The sensors and microprocessors of the IT domain that have been integrated into the physical systems of the OT domain enable the IoT.

From a historical perspective, the confluence of informational and operational technology has been a key element of certain engineering fields since the mid-20th century. US military aircraft were updated with Electronic Counter Measure (ECM) capabilities beginning in the 1960s. ECM was combined with IT and OT elements of aircraft operations. It provided new levels of insight regarding the location of friend or foe aircraft and ground vehicles to enable improved risk management and defense.

The US Navy was ahead of its time—and enabled what was essentially an IoT type of airborne command and control system (Navair, 2016) in the 1960s—when they deployed the E-2C Hawkeye aircraft. This was long before the Internet and IoT was a recognized term. The E-2C aircraft manufactured by Northrop Grumman, with a radar dome attached to the upper fuselage, was deployed in January 1964 and is still operational in 2016. The E-2C Hawkeye provides all-weather airborne early warning, battle management, and command and control functions for the US Navy fleet of ships and aircraft. Its capabilities include surface surveillance coordination, offensive and defensive air-based decisions, close-air support coordination, time critical coordination, and search and rescue airborne coordination and communications relay. The E-2C Hawkeye utilizes computerized radar, friend or foe identification, and electronic surveillance sensors to provide early warning and threat analysis. This system of systems for airborne command and control was a very early deployment of what we now call the IoT. "Things" were other aircraft and vehicles. The "network" was the airborne peer-to-peer and ground communications system. Especially as we consider opportunities for improved management of various IoT systems in this day and age, and the need to mitigate the threat of cybersecurity and foes, the way in which the US Navy developed and enabled these technologies was certainly extraordinary.

Today's Internet of things deployments include "greenfield," a new system opportunity, and "brownfield," or retrofitting of existing systems and devices. This approach was also taken in the airborne command and control systems. Established aircraft platforms were updated to include ECM technology, for autonomous management in the context of the field of battle, and to be part of the larger system of systems (Meilich, 2006) which was being measured, monitored, and managed by the E-2C Hawkeye. One example is the F-111A Raven aircraft that the Air Force contracted with Grumman Aerospace to modify in 1975 to include an electronic platform. The evolution from the F-111A to the EF-111A included the addition of the electronics, to serve as IT to complement the OT. When retrofitting the aircraft with electronics, they added an electronic jamming system, a self-protection system, and a threat warning system. To accommodate the 6000 pounds of new electronics, Grumman added a narrow, 16-foot long canoe-shaped radome under the fuselage and a pod mounted on top of the vertical stabilizer on the tail of the aircraft.

The example of the F-111A aircraft being retrofitted with electronics to become the EF-111A enabled "brownfield" or existing operational systems to realize the benefits of IT and OT together, or IoT. The "brownfield" or established aircraft platform was updated to communicate in the electronic command and control system in the sky. This meant updating existing aircraft with new electronics and IT to improve information capture, along with analysis and leverage of the information to improve insights and potential outcomes. The new information captured through the electronic systems was analyzed by humans and systems, providing insight to enable operational systems and the humans operating them to improve outcomes. The aircraft or "high-velocity moving platforms" such as the EF-111A had standalone systems such as jamming and self-protection to provide autonomous, actionable insight. The E-2C Hawkeye served as a centralized command and control system, gathering data from multiple friend and foe aircraft through radar, analyzed and shared the data with other friendly aircraft, enabling the development of information-rich recommendations for operational actions to enable coordinated and improved risk management and mission outcomes.

The F-14 Tomcat aircraft was a "greenfield" system created after the F-111A which also participates actively in the E-2C Hawkeye electronically enabled airborne command and control management system.

This is similar to how we add sensors and cybersecurity warning systems to existing operational systems and devices for IoT system and risk management. For instance, when outfitting a smarter building, sensors and cameras can be added to current buildings to measure occupancy and temperature to improve energy use, and to monitor the people and actions taking place to improve security and safety.

Cognitive computing systems can analyze IoT data on an ongoing basis, leveraging analytical algorithms which are refined from continuous feedback loops assessing recommendations and actual outcomes, providing the opportunity for perpetual optimization. This can include natural language processing and visualization tools which can be customized and optimized for human interaction, for instance via speech, audio, video, words, images, and graphics (Hurwitz et al., 2015; Kelly and Hamm, 2013) (see Fig. 2).

The human ability to ingest, digest, and make sense of all the data coming from applications such as the internet of things in a smart city or healthcare use case is limited and becoming increasingly challenging with the increase

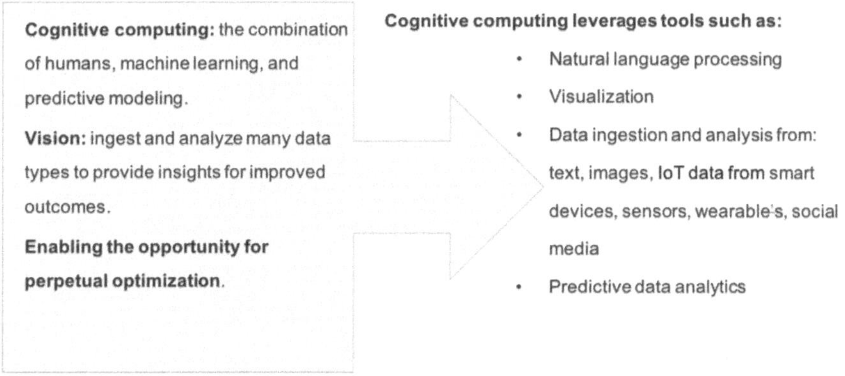

FIG. 2 Cognitive computing combined with IoT increases insights and the potential for perpetual optimization. *Internet2 Chief Innovation Office.*

in data being created. A medical doctor in the United States may not have access to or the time to read research journals and articles regarding new discoveries, insights, treatments, and outcomes occurring around the world in many languages on a daily basis (Kelly and Hamm, 2013). A cognitive computing system could be fed all the data available in a certain domain and leverage that to enable a medical doctor to access all the research material around a certain subject, ascertain the sources of the data, combine that data with context regarding the patient symptoms and history, family history, and environmental information to provide contextual insights and recommendations to the medical doctor to provide more informed medical care.

Cognitive computing provides the ability to gather, ingest, digest, and analyze vast amounts of data, both the constant influx of new and real-time data combined with leveraging historical curated data. Cognitive systems have the capability to learn, provoke, analyze, resolve, and present alternative solutions with rational and evidence to support each recommendation in a manner that is relevant to the question and user (Hurwitz et al., 2015).

3 THE ROLE OF BIG DATA

Big data created by IoT and other mechanisms, and analyzed by cognitive computing systems, can be characterized from five perspectives, known as the 5Vs of big data (Demchenko et al., 2014; see Fig. 3):

- Volume, as standalone data elements or in aggregate
- Variety, including alphabetic, numeric, audio, video, image
- Velocity, how quickly it is being created and captured
- Veracity, dealing with the uncertainty that the data created is a true reflection of the facts
- Value, analyzing the data, determining the value of a data element and the data in aggregate

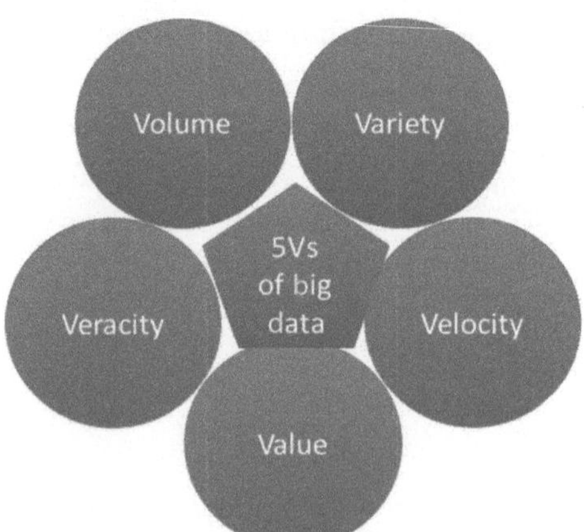

FIG. 3 The 5Vs of big data.

As we consider use cases of IoT-enabled smart cities and IoT-enabled healthcare, we will posit an aspirational future dealing with the volume, variety, velocity, and veracity of big data, to increase the value and impact which IoT and cognitive computing could enable together.

4 BIG DATA CHALLENGES AND OPPORTUNITIES FOR IOT AND COGNITIVE COMPUTING

Cognitive computing is especially powerful when combined with the Internet of things and people where the volumes of data are so large, and the veracity or uncertainty of the data needs to be vetted. The confluence of IoT and cognitive computing provides the opportunity to analyze the data, in order to develop insights and recommendations in a timely manner on large volumes of data that need to be vetted for accuracy, and then communicated to systems and humans to improve outcomes in a timely manner.

There are many challenges and opportunities regarding the creation, analysis, and leverage of all this data. The big data challenges and opportunities we will consider regarding potential leverage of IoT and cognitive computing together can be classified in the 5Vs of big data (Demchenko et al., 2014):

- Volume
  - How can we leverage cognitive computing to analyze the vast amounts of IoT data in a timely manner to enable systems and humans to leverage the resulting insights to improve outcomes?
- Variety
  - How do we synthesize various data types and formats such as numerical, textual, video, images, audio, and social media to develop deeper insights?
- Velocity
  - How do we combine the speed of real-time data with the insight from historical data to develop real-time recommendations?
- Veracity
  - How do we know if the IoT data is a true reflection of the facts vs. an opinion, incorrect translation, or if it was altered at any point in the lifecycle of the use of the data?
  - How do we know if the user identity, device, or data has been compromised?
- Value
  - How do we create true value from the data?
  - How do humans and systems interact to best leverage the data?

Now that we know the questions related to the 5Vs of big data, let's look at how we deal with these challenges and opportunities, and the discovery and invention yet to be done.

4.1 Volume

The total amount of data created in the world annually is doubling in size every two years, and is expected to reach 44 zettabytes (10^{21} bytes) or 44 trillion gigabytes by 2020. IDC (Turner, 2014a) predicts that in 2020, 10% of that data will be from IoT. IoT is creating a growing percentage of the world's new data, and that will continue into the future.

Big data can refer to large amounts of little data, like the large amounts of data generated by the sensors in the billions of devices in the IoT (Turner, 2014b). Big data can also refer to big data elements on their own such as medical images and streaming video images.

Traditionally, some of the largest data sets have been related to high energy physics and nuclear fusion which is near the exascale range (Bunn and Newman, 2003) (an exabyte is 10^{18} bytes), such as the data generated by the Large Hadron Collider (LHC) at CERN in Switzerland, or the National Synchrotron Light sources operated by US Department of Energy laboratories. These data sets are shared around the world for the advancement of science through high-speed networks such as ESnet (the Energy Science Network operated by the US Department of Energy). ESnet brings high energy physics data from the LHC through multiple 100 Gigabit per second (Gbps) and 10 Gbps network links for total capacity of 340 Gbps that traverses the Atlantic Ocean to the United States (ESnet, 2015), where ESnet links to and partners with Internet2, a national research and education network that connects over 300 universities along with national labs, regional networks, and industry partners, so the data can be distributed to scientists and students to be analyzed, enabling scientific discovery. There are National Research & Education Networks (NRENs) around the planet by region and country that connect together to enable sharing of big data in pursuit of scientific discovery. These NRENs are connected to high-performance computing (HPC) and high-throughput computing (HTC) facilities to send and analyze large amounts of data to glean insights from the data. NRENs are providing high-speed data transport for scientific research and education around the world and will increasingly be used for climate science, smart cities, and genomics (ESnet, 2015). Cognitive computing techniques are being applied to the IoT (Forbes, 2016) to correlate various data types, analyze the data against historical patterns, do predictive analytics, and create a series of recommendations that are then presented to humans to choose the optimal recommendation, or to machines to act upon the recommendations. Cognitive computers are already being applied to medicine (IBM, 2016). Sixteen cancer institutes are working with IBM Watson to help doctors translate DNA insights into personalized treatment options for patients. Researchers from Baylor College of Medicine are using IBM Watson to develop solutions for automated hypothesis generation, and at Mayo Clinic to help doctors match patients to relevant clinical trials. Dr. Larry Norton, a world-renowned oncologist at Memorial Sloan-Kettering Cancer Center, believes that the IBM Watson cognitive computing system will be able to synthesize volumes of medical and patient information to help physicians more quickly and easily identify treatment options for complex health conditions. In this way, there is an ecosystem of computing and data, from HPC, to HTC, to distributed and mobile devices, to the cognitive computers that make sense of all the data to provide insight and wisdom for healthcare and life sciences. For instance, when a patient presents with a lump that could be cancerous, a Doctor might suggest a mammogram which takes a week or two to schedule, a few days to get the results, so delaying treatment if it is cancer by a number of weeks. If a cognitive computer has access to the Doctor's findings, plus the patient's DNA information, family history, and recent clinical research that shows high odds of cancer based on DNA, family history, or other parameters associated with the patient, the Doctor could go right to a recommended biopsy to reduce testing costs and improve mortality rates. This is wisdom gained from insight beyond the information available from the initial exam which is presented by a cognitive computing system.

The increasing use of digital technology for healthcare and life sciences research including genomics has created data sets that are already in the petascale range (a petabyte is 10^{15} bytes). It is expected that healthcare and life sciences data will have high growth, and could equal or eventually surpass the size of the traditional big data sets of scientific domains such as high energy physics and cosmology. Fueled by the quest for precision medicine (National Institute of Health, 2016), the healthcare and life sciences researchers and clinicians will increasingly leverage many data types together, including genomics, clinical research, pure research, clinical care, environmental, demographic, IoT wearable, and patient experiential data (Collins and Varmus, 2015). The data will be in textual, image, numerical, sensor based, and video and audio forms. The Internet of medical things is already creating data constantly, through wearable consumer devices that track your exercise, sleep, motion, and vital signs including blood pressure, temperature, heart rate, and respiration. The Internet of prescribed and regulated medical things are also creating data, including mobile wearables such as glucose meters, insulin pumps, and defibrillators. There are also wired devices in medical settings such as the infusion pumps, pulse/oxygen meters, and many devices in a hospital or clinic, or in an ambulance. All of these devices create data, rather constantly.

To provide the most insightful recommendations from the widest range of pertinent data sources, a cognitive computing system needs to be able to ingest, digest, and correlate volumes of data quickly, especially in a real-time IoT environment such as clinical healthcare and smart city systems such as smart transportation systems and electric vehicles. The state of the art of current real-time traffic applications leverages real-time Global Positioning System (GPS) data, ingestion, and digestion of new data from vehicles and users to provide updated real-time traffic route recommendations (Lee et al., 2010). Cognitive computers are focused on more broad and data-rich environments, such as the management of an entire city (Batty et al., 2012), so need to correlate a wide variety of data. A cognitive computer, for instance, could be a smart city enabler, ingesting and digesting historical traffic information based on time of day, type of day (weekday, weekend, holiday), and traffic volumes, along with weather data, public safety information, social media data, and a citizen's plans for the day, to optimize transportation suggestions.

The fact that the IoT devices in any of these use cases are all digitally enabled leads to the fact that they are all hackable. The IoT introduces a wide range of new security risks and challenges to the IoT devices themselves, their platforms and operating systems, their communications, and systems to which they are connected (http://www.gartner.com/newsroom/id/3221818). Security is key and needs to be designed in from the beginning of the IoT device and system design (Xu et al., 2014) to protect the data, the humans, and the physical infrastructure from both information attacks and physical tampering. IoT devices are not just creating data, but they can also be hacked to cause physical harm such as in connected healthcare devices such as insulin pumps (Radcliffe, n.d.) and embedded heart devices, and in connected vehicles which have been remotely hacked causing their brakes to fail and engine to turn off on a highway (Hackers Remotely Kill a Jeep on the Highway, 2015). Increased focus on Trust, Identity, Privacy, Protection, Safety, and Security (TIPPSS) is being developed through partnerships between IEEE, Internet2, and the US National Science Foundation (NSF) working with government, laboratory, industry, research, and education leaders around the world. The TIPPSS focus needs to be on the hardware, software, and services related to life critical IoT devices. A "defense in depth" strategy as suggested by the US Food and Drug Administration (U.S. Food and Drug Administration, 2013) is required for new devices and devise already in the market, including ensuring trusted users through identification and authentication. We need to protect the patients, through device- and system-level security, for data at rest, as in a device, and data in motion, as it is transported across a network or accessed wirelessly such as with a Bluetooth or a Wi-Fi connection.

The hardware capability to store the data and the software to manage it is all part of a complete cognitive computing system. While we have discussed data sets that are in the petascale (10^{15} bytes) and exascale (10^{18} bytes) ranges, by 2025 we will be dealing with zettabytes (10^{21}) and by 2050 could be in the range of yottabytes (10^{24} bytes), and brontobytes (10^{27} bytes) of data in aggregate around the world. A lot of this data growth is due to the IoT. In order to be able to analyze and leverage these vast amounts of data, cognitive computing systems need to be scalable in their ability to gather, store, and analyze the data to provide insights from the vast amounts of information.

4.2 Variety

In many IoT use cases including smart cities and healthcare, there is a broad variety of data types and formats, and a variety of protocols for data storage and transmission. All this diverse data needs to be analyzed for their particular data points and insights, and then correlated to develop a comprehensive view. Furthermore, the variety of data could potentially increase in the future as new devices and use cases are developed.

The data types we consider today could include:

- Alphabetical, from books, reports, journals, online and hardcopy, social media, email, enterprise systems, physical and digital archives, in many languages
- Numerical data, from science experiments, enterprise systems such as banks and financial systems, healthcare devices
- Video, from cameras, personal devices, cloud services, television and cable feeds
- Images, from stationary and mobile cameras, news outlets, television and cable services, institutional or personal cameras, medical images such as MRIs or X-rays, images from space
- Audio, from radio, cloud services, social media, public safety personnel, cameras, emergency management, sensors on light posts/buildings/people/vehicles
- Location/GPS data, from vehicles, people, animals

The need to synthesize these various data types and formats will be challenging yet doable. For instance, in a smart city, cameras can identify suspicious data points and "things" in the Internet of things based on certain parameters created by public safety and law enforcement. It could be a vehicle that is parked illegally in a crowded area without a driver for an extended period of time that is determined to be a potential safety and security threat. Cameras on light posts or buildings or people can transmit images that can be correlated to determine what the driver looked like, potentially do facial recognition to identify them, and transmit that picture to law enforcement. The vehicle image and tag information can be correlated with motor vehicle records, drivers' license affiliations, passport connections, law enforcement records, and more to develop a risk profile for a given insight from pattern recognition from connected cameras. The cognitive computing system could potentially correlate all this data to provide insights that could save lives. Of course these surveillance systems open the debate about security vs privacy. One such case is the City of Chicago which has an extensive and integrated network of government video surveillance cameras, according to former US Homeland Security Secretary Michael Chertoff. The City probably has access to somewhere between 10,000 and 20,000 publicly and privately owned surveillance cameras. In downtown Chicago, virtually every segment of the public way is under constant video surveillance. These cameras have powers that greatly exceed ordinary powers of human observation, including automatic tracking of cars, and magnification of small objects at great distances. In February 2011, the American Civil Liberties Union of Illinois (ACLU) published the first large-scale, independent analysis regarding this growing threat to our privacy (Schwartz, 2013).

4.3 Velocity

Early forms of cognitive computing such as in IBM Watson had a database that was developed after ingesting vast amounts of data, for instance all of Wikipedia, and architected to support certain types of questions, to begin with for the *Jeopardy!* televised game show in 2011. That type of system required days to ingest, digest, and format the data in the database.

Cognitive computing can ingest questions that it can then bring data to bear to answer. Cognitive computers will continue to mature to the level of being able to actually learn from new data in real time, correlating historical and real-time data to provide the most insightful input for time critical applications, including healthcare, public safety, and city operations. An aspirational view of a smart city leveraging IoT with cognitive computing will be explored later on in this chapter.

4.4 Veracity

Veracity relates to ascertaining if the data received is factual and accurate. There are multiple aspects when considering the veracity of the data: first, determining the data provenance, meaning having facts with certainty that show the data was created by the device we expected to create it, that it has not been compromised in any fashion and has been protected end to end from its creation to transmission to the IoT data collection system and complementary cognitive computing system; second, determining how factual the data is if it was created by humans and might be an opinion or translation of factual information or circumstances.

The veracity of the data is critical, especially for life affecting use cases such as in smart cities use cases, whether water monitoring, or peer-to-peer smart vehicle systems in a smart city, or in healthcare systems. Validating that the data received is the true data created, ensuring it is factual, and determining if it was altered or hacked at the device, data or network layer is a challenge that is still to be fully addressed.

For instance, there could be sensors in citizen or municipal water sources such as a reservoir that transmit data about the water, let's say pH or detection of any dangerous chemicals. A bad actor could potentially hack the sensor, data transmission link, network, router, or cloud service to disable the transmission of data from the sensor identifying a poisonous substance in the water or could transmit signals that suggest everything is fine when in fact there could be a dangerous substance in the water. We need to provide an end-to-end trust and security solution that provides two-way communications to ensure a trusted connection and device, identify the device and sender, ensure privacy and protection of the data and resources, and ensure safety and security of our resources and citizens. This TIPPSS requirement is ubiquitous in IoT to secure our critical infrastructure and resources to protect the citizens. Cognitive computing could provide pattern recognition capabilities that learn from critical resource-based sensors in a system of systems (Delicato et al., 2013) and network of municipalities to ensure any anomalistic or dangerous situations identified are communicated to ensure we are all protecting our infrastructure and citizens.

We must consider the veracity of data if it is user-generated data such as tweet, email, or other social media input, if it is factual or the creator's interpretation or opinion. An opinion is valuable when it adds further context, or if you are dealing with sentiment and potential risks arising from human behavior or other potentially escalating issues. But if you want to know about a public safety situation, you need the facts to manage it appropriately.

Dealing with the veracity of data is a very complex subject in IoT, and cognitive computing overall. Wherever the data comes from, knowing the provenance, ensuring it is factual is critical when it is the base for recommendations, especially in life-altering situations like smart transportation, water monitoring, and healthcare. The challenge of veracity should be addressed by a combination of technology and process improvements. We need to solve the end-to-end trust and security challenges of IoT to better ensure that the data is factual, that it is from the expected device, and that it has not been altered or compromised. If the data is cultivated from social media sites, it must be determined if it is indeed factual information or rather, subjective interpretations of other data. These are just a few examples of the questions that people and machines must investigate on an ongoing basis in order to channel and leverage information in a useful and effective way.

To solve these data veracity challenges, there is a need for increased research and development in ensuring TIPPSS in IoT, which can be assessed by cognitive computing. TIPPSS, namely Trust, Identify, Privacy, Protection, Safety, and Security, are critical elements to ensure the certainty of IoT data (Condra, 2015). This is a key element of the IEEE, Internet2, and NSF collaborations as identified in the End-to-End Trust and Security for IoT Workshop in February 2016 (IEEE Standards Association, 2016) cosponsored by these organizations, and the following day in the IEEE Experts in Technology and Policy Workshop on Internet Governance, Cybersecurity, and Privacy (IEEE, 2016).

Cognitive computers can look for examples of anomalistic behavior, or patterns in the data that confirm or deny an insight. This ability can provide additional information to ensure the Trust, Identity, Privacy, Protection, Safety and Security for IoT data and related applications.

Cognitive computing can analyze data to determine if there are patterns or anomalies in the data, and then request additional data to better determine illogical or anomalistic data. This can then be assessed by the human expert using the data. In healthcare, this would be the healthcare provider. This is a prime example of why cognitive computing systems will often provide recommendations to humans and expert systems, but the humans make the final decision. This is a concern in areas like healthcare where there is regulation regarding healthcare. In other areas, such as traffic and transportation recommendations, the cognitive computing system does not need to be certified to provide recommendations.

One of the interesting considerations as cognitive computing develops is regarding when the computer is making recommendations dealing with a human in time critical systems, is the computer or the human thinking fast or slow. In Nobel Laureate Daniel Kahneman's book "Thinking Fast and Slow," (Kahneman, 2011) the author refers to experts thinking fast such as a chessmaster walking past eight chess games and quickly suggesting the best next move for each game, in a matter of seconds. This level of expertise comes from experience and success. As the cognitive computers ingest and digest volumes of data much more vast than humans can ingest and digest, and as the computers quickly integrate real-time data, identify patterns, compare the data to historical data, and make recommendations, when do we cross the line regarding when the computer is more prepared to make the best decision rather than the human, where the computer is thinking fast and the human is thinking slow. This is one of the concerns with cognitive computing surpassing human thought and action. In the aerospace industry, some aircraft navigation systems think faster than the pilots can react, so the aircraft system makes decisions. We need to think about how this affects the leverage of cognitive computing with the IoT for various use cases. For instance, in a regulated industry such as healthcare, if the cognitive computing connection can assess family history, patient history, the newest clinical research, and a reaction the patient is having from a medical event or perhaps a drug interaction, when does the computer make the decision on care? These questions get to the ethics and regulated standards regarding cognitive computing as we progress into the future.

4.5 Value

The projected value from the data and the analysis of the data needs to be assessed prior to development of a cognitive computing and IoT system, to ensure that the system is architected and utilized to provide that value. The user and human interface needs to be assessed as well, and developed in a sociotechnical model to ensure the human interface, natural language, and visualization, enabling the user to leverage the cognitive computing insights to inform their decisions to improve outcomes.

For instance, in a smart city scenario, if there is insight from an IoT system that there is a major accident or a water main break closing roads in a certain area of the city so you want to alert drivers to avoid that area, the alert generated and human interface needs to connect to an application the drivers and commuters are using and can easily and quickly understand. Then, we would want the cognitive computing system to assess the volumes of drivers/vehicles, compare that to historical data, predict potential traffic challenges, and then provide redirect information to autonomous vehicles and drivers to reduce traffic congestion and improve public safety. We would also want the computing system to assess the TIPPSS elements of the data. Is the data from a known source? Has it been tampered with? How do we determine this? Much of this is science, research, and discovery that the research and education community needs to address.

Big data can be accessed by cognitive computing, for "sense making," analyzing the data to identify patterns and anomalies to "make sense" of the data. This leads to "insight computing," where sufficient data is gathered to identify patterns, trends, and anomalies, to develop new insight that informs humans and systems that can make decisions to improve outcomes. These outcomes could be healthcare outcomes, operational smart city outcomes, industrial system outcomes, or personal outcomes. They could lead to improved healthcare-based actions to improve health and longevity, improved city operations, improved safety, and industrial system output. As cognitive computing is an intriguing and expanding era in information technology, the potential is vast.

5 IOT USE CASES AND OPPORTUNITY TO LEVERAGE COGNITIVE COMPUTING

5.1 Smart Cities—Personal Transportation Planning

The IoT is a network of billions of devices, connected to each other in a system of systems. A city today is a system of systems (Delicato et al., 2013), perhaps not all well coordinated yet. For example, let's take a closer look at the modern transportation system.

An IoT-enabled smarter transportation system is interconnected in a system of systems for a smarter city. Cities around the world are on a journey to a smarter city. The city of Copenhagen in Denmark has invested in an efficient, reliable, and highly integrated public transport network, making the levels of congestion and pollution one of the lowest of the major international cities. The number of car trips in the central part of Copenhagen fell from 351,000 in 1970 to 284,900 in 2010 (http://www.smartcityexpo.com/en/copenhagen-smart-city). Such a smarter city could be a system of systems that include:

- City transportation systems such as trains, subways, and buses
- Personal transportation systems, such as cars, trucks, and bicycles
- Tolling systems
- Bridges
- Water pipes
- Buildings

- Public safety systems including fire, police, and ambulance
- Cameras for public safety and traffic
- Electric grid, including the meters, energy, and pricing incentives
- Personal devices, such as smartphones, helmets, and wearables
- Social media data
- Overall city operations

There is data created by all these systems (http://www.theatlantic.com/sponsored/hitachi-building-tomorrow/big-data-and-the-internet-of-things/678/). There is also a plethora of historical data, and when combined with real-time data, it is possible to analyze this data by a cognitive, learning system. It can even predict and communicate how to improve outcomes for the city operations and the citizens.

The data types could include:

- Text, from social media, email, city archives, and systems
- Video, from cameras and personal devices
- Audio, from social media, public safety personnel, cameras on light posts, fire trucks, police vehicles, bridges, and buildings
- Location/GPS data, from city and personal vehicles and people

5.1.1 A Scenario

Imagine a suburban commuter living in Westchester County, New York trying to get to a meeting in Brooklyn, New York. There is an opportunity here, even though the task seems rather mundane: to optimize the commute time for the individual, improve the efficiency of city operations thereby reducing cost, improve city and citizen safety, and enhance the user experience resulting in the increased satisfaction of city workers and citizens.

Without the benefit of the Internet of Things and Cognitive Computing working together in a complementary and integrated system of systems, this is how the commuter may proceed (Fig. 4):

- The commuter looks on an app/website/paper schedule to determine total commute time either driving, using various forms of mass transportation, walking, or a mixture, to get to her desired destination from home.
- The commuter obtains input from radio, TV, and real-time travel apps to determine total expected commute time in context of real-time data, and the potential uncertainties (traffic congestions, transit delays, etc.).
- The commuter plans the details of the trip choosing multiple forms of transport, driving from home to the suburban train station, utilizing the commuter rail line from the suburbs to Manhattan, taking a subway from Manhattan to Brooklyn, and then walking to the meeting from the Brooklyn subway station.
- The commuter deals with travel uncertainties (U) with today's actions (A):

FIG. 4 Multiple unintegrated systems inform commuter needs.

- (U) Is there traffic between home and the suburban train station? How does that change the departure and arrival timing?
  - Check an app, not integrated with the train app. Reroute as needed.
- (U) Is there a parking spot at the train station? If not will this commuter have to drive home and take a cab to the train station? How much time might that take?
  - No app to check parking spot availability, they may drive and hope for the best.
- (U) Is the train delayed? If so, what are my other options? Should the commuter wait or go to another train line and station?
  - Checks the train app; it provides real-time data. If it is delayed, there are no recommendations regarding waiting, taking an earlier or later train, or going to a different station.
- (U) Once arrived at Grand Central Terminal (GCT) in NYC, how long will it take for the next leg of the trip from GCT to Brooklyn?

- (U) How long will it take to find an Uber or a cab? How long will the drive be from GCT to Brooklyn?
  - The commuter looks for a cab outside, checks Uber, and then decides which is faster, though they do not know the approximate cost for the cab to enable her decision, the commuter does have a fare estimate from Uber and assumes the taxi would be about the same amount. Will it be?
- (U) When will the next subway come to GCT? Will that subway run all the way to her destination, or will it require passengers to disembark due to water or congestion in the subway tunnel? Then what?
  - Goes to the subway platform and waits for her subway
  - If the subway forces her off at an earlier station, she must either find another subway or take a bus or a cab or Uber from there, reactively
- (U) When will the next bus come by GCT? Does she have to transfer? When will that next bus come? When will they actually get to Brooklyn?
  - Waits for the bus and hopes for the best

○ (U) Once in Brooklyn, how long will it take to get to the office?
- Is there construction requiring rerouting of the walk?
- Are there public safety measures requiring rerouting?
  - Emerges from her subway/bus/taxi and deals with the potential construction and public safety issues reactively

○ (U) Overall uncertainties for the entire commute:
- Are there water main breaks closing streets, subways, or buildings?
- Will rain or snow or other storms delay transportation?
- Do the police or fire department or emergency management have certain streets blocked or closed?

This is a complicated scenario, and the reality in 2016.

5.1.2 Aspirational View

If we could design a system using IoT and Cognitive Computing to create a smarter city, it could improve efficiency and effectiveness of the transportation system, as well as improve human experience, productivity, and safety.

Consider this same scenario: a suburban commuter living in Westchester County, New York has a meeting in Brooklyn, New York. This is a 38-mi. commute, with a minimum of 90 min for any commute option during rush hour.

Tomorrow, with the benefit of Cognitive Computing and the Internet of Things (CCIoT) working together in a complementary and integrated system of systems, this is how the commuter would proceed (Fig. 5):

- The commuter looks on the app to determine total commute time either driving, using various forms of mass transportation, walking, or a mixture, to get to her desired destination from home.

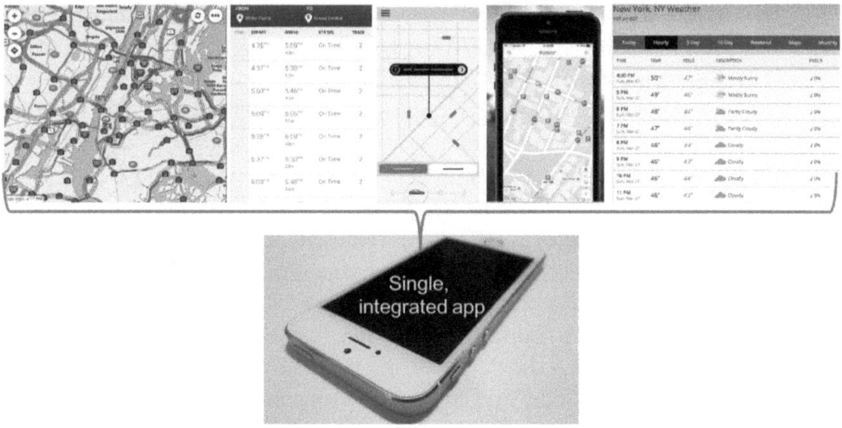

FIG. 5 Multiple integrated systems inform commuter transportation.

- ○ (CCIoT) The commuter chooses the optimal route from a Personal Transportation Planning (PTP) app, which provides a recommended integrated travel experience. The app provides total expected commute time in context of real-time data including driving, parking, walking, trains, subways, buses, Ubers, taxis, including risk factors such as expected increase in traffic based on real-time GPS and predictive analytics based on similar situational analysis from prior days in the city. It integrates real-time sensor data from trains, automobiles, and planned and potential predictive closures from city operations and public safety based on prior day realities, expected construction or public safety issues, and potential weather issues.

- The commuter considers uncertainties (U) and takes actions (A), or the PTP app provides recommendations:
  - ○ (U) Is there traffic between home and the suburban train station? How does that change her departure and arrival timing?
    - ▪ (PTP) The integrated PTP app integrates traffic and weather information to predict the best route and provides an updated home departure time and route.
  - ○ (U) Is there a parking spot at the train station? If not will she have to drive home and take a cab to the train station? How much time might that take?
    - ▪ (PTP) The integrated PTP app includes where the parking spots are at the train station, where they are projected to be based on historical and predictive analytics, and a backup plan if things change once the commuter gets close to the train station.
  - ○ (U) Is the train delayed? If so, what are my other options? Should she wait or go to another train line and station?

- (PTP) The PTP tool proactively pushes real-time updates to the commuter via text/voice/ADA based on changing commute options, suggesting the fastest route by car, train, and other options, suggesting departure times, or recommending an earlier or later train, or going to a different station.

- (U) Once she arrives at GCT in NYC. How long will it take for the next leg of the trip from GCT to Brooklyn?
  - (PTP) The progressive PTP app recommends the best and alternative options from there, leveraging analytics from IoT data including sensors, crowd-sourced data, the Department of Transportation, City Operations, Fire Dept., Police Dept., camera feeds from the streets and other areas, subway, bus, taxi, Uber, walking estimated times, weather delays now and expected during the commute, and any construction, water main break, street closing, building closing or congestion issues, recommending the optimal route every step of the way.

As you can see, the combination of the Internet of Things and Cognitive Computing together leverages historical and real-time information from many devices, both moving and fixed, personal and infrastructure connected, to predict and recommend actions for improved outcomes. This can improve city efficiency, and the experience and productivity of the people. This can also improve safety for the citizens and city and emergency management workers, for instance if the PTP recommendations steer the citizens away from potentially dangerous street closings, building closings, weather-related shutdowns, and emergency management situations.

Cities such as Singapore and Rio de Janeiro in Brazil (YouTube, 2012a) have been developing smart cities for years, leveraging the IoT and getting involved in the beginnings of smart city analytics. As cognitive computing is applied to these smart cities applications, improved analytics could provide perpetual optimization as new devices, sensors, users, context, and applications are developed and deployed. This can result in improvements in traffic congestion, public safety, and city operations.

5.2 Smart Cities—City Transportation Management

Let's consider a smart city from the perspective of the city operations using the state of the art and the science in 2016. As communicated in the author's TEDx (YouTube, 2012b) talk in 2012, when we look at IoT and cities, there is an opportunity to leverage sensor data, pricing incentives, historical and predictive analytics, and integration of home, building, grid, tolling, public transportation, and automotive systems to improve city transportation management. This requires the use of IoT, and cognitive computing, to be accomplished. Potential improved outcomes could include reduced traffic congestion, reduced energy use, reduced vehicle emissions, improved air quality, and improved public safety.

To date there have been incentives-based systems in intelligent transportation, smart grid, and building management. The opportunity is to leverage IoT and cognitive computing technologies together to enable a real-time data and insight-driven smart city to improve city operations, the citizen experience, reduce traffic congestion, energy use and vehicle emissions, improving air quality and public safety.

Let's look at the separate smart city systems, then the opportunity to leverage them together in new ways with the help of IoT and cognitive computing.

An Intelligent transportation system was designed and deployed in Stockholm in 2007 when the Swedish Road Administration began a pilot road charging system with the goals to reduce traffic congestion in the inner city and improve the environment through the reduction of ambient pollution and CO_2 emissions. They knew they needed to do this in partnership with the people of Stockholm to ensure the system met their needs and would in fact be utilized so the desired outcomes would be achieved. They analyzed the situation and decided on a congestion pricing and tolling system where during certain high-traffic times of the day there would be increased tolls for vehicles driving along certain routes. This would create an incentive for people not to drive on those roads then. The city still wanted people to have a way to get to work in the city, and to travel to the city to purchase entertainment, goods, and services that enable the city to thrive. Hence, they planned increased mass transportation. Research-level technologies of high-fidelity Optical Character Recognition were deployed to improve the ability to read the license plates to ensure the vehicles were charged and toll collection would be successful. The comprehensive plan was presented to the City of Stockholm, so the citizens knew that there would be a pilot stage for this congestion pricing/tolling system. After the pilot program, with congestion pricing/tolling and increased mass transportation, the pilot program ended, and the City published a report on the results. The pilot was so successful; Stockholm residents overwhelmingly supported full adoption of the system. The city of Stockholm is well known for its traffic congestion tax, which drove a 25% reduction in car use and a 14% reduction in emissions from road traffic. (IBM, 2009) In keeping with best practices for integrating service delivery across transport modes, Stockholm implemented the traffic congestion tax as part of a holistic transport plan that also increased bus services and park-and-ride facilities, and uses an integrated ticketing system that links the major modes of transportation. Due to the system of systems (Delicato et al., 2013; Meilich, 2006) approach, providing mass transportation to enable travelers with alternative means for transportation, ridership in mass transportation increased by 60,000 people per day (Asmartplanet, 2009).

Smart grid systems can also be a player in smarter transportation management. Smart grid systems offer the potential to reduce energy use, reduce the need for capital intensive investments as we plan for the potentially increasing peak load on the grid especially in regulated markets such as the United

States, and improve citizen and household efficiencies and the environment. One example is applied research led by the US Department of Energy Pacific Northwest National Laboratory (PNNL) (Pratt, 2007) Pacific Northwest GridWise™ Demonstration project on the Olympic Peninsula in the state of Washington and elsewhere in the region, with support from the US Department of Energy and in partnership with the Bonneville Power Administration, PacifiCorp, Portland, General Electric, IBM, Invensys Controls, Clallam County PUD #1, the City of Port Angeles, and Whirlpool Corporation. This pilot project demonstrated how smart grid technology could integrate a wide variety of smart grid resources—deployed among residential, commercial, and institutional consumers—with the moment-by-moment operation of the power grid. Residential customers were shown to respond positively to incentives to shift from a fixed-price contract to a time-differentiated contract and adjust their usage accordingly to lower their power bills. When given a choice between a time-of-use/critical peak price and a 5-min real-time price, a substantial fraction pursued the greater savings opportunities presented by the real-time contract. The home energy management automation technology provided by the project allowed customers to fully automate their response to price signals, adjusting their consumption to capitalize on savings opportunities, while providing the flexibility to change or override selected preferences for relative comfort vs economy at any time. The system consisted of a smart meter, a gateway between the broadband service and home computer, a smart thermostat, and a water heater load control module, all communicating wirelessly within the home. The customers used their home computer and Internet connection to program the responsiveness of their thermostats and water heaters.

At the core of the project was a local, substation-level retail market. The price signals superimposed the costs of wholesale electricity, transmission congestion, and a local retail market that closed every 5 min to strictly manage a distribution constraint (albeit an artificial one). The market functionality developed was implemented within a middleware system that used Internet-based communications. The same market signaling mechanism was used to aggregate the response of a commercial building, a backup generator, a microturbine, and a municipal water pumping system. PNNL observed some remarkable capabilities of this two-way communication-based demand management network. It was shown to cap net demand at an arbitrary level, 16% less than the normal peak demand, for days on end. This represents real capital cost savings when a $10M substation can be deferred or downsized.

They also were able to easily synchronize thermostatically controlled loads to follow the grid's need for regulation services. Demand resources easily responded over the short term because the excursions from normal set points were so small that there was minimal if any discomfort. Market closing costs to buy this kind of response were very low as a result. The implication is that demand can provide an ancillary service very analogous to regulation, and is likely to do so at costs far lower than what power plants charge to ramp up and down. Mass-produced residential appliances alone generally make up about 20% of demand, so they can provide a vast and inexpensive "safety net" for the power grid. A large influx of plug-in hybrid and electric vehicles could represent an even higher percentage of total demand.

Now let's consider managing a smart city as a system of systems leveraging intelligent transportation systems, smart grid, city operations including water systems, public safety, and incentive pricing. This is a complex, real-time enabled system of systems, leveraging the IoT and increasingly the opportunity for cognitive computing (Fig. 6).

Let's start with the premise of the Stockholm goals—reduce traffic congestion, reduce carbon emissions, and improve the environment. Necessary components of the plan and rollout included thinking holistically about the city and its citizens.

Envision a smart city operations cognitive computing system with data from all pertinent and potentially available devices, IoT elements, and systems. Then address the opportunities to create value, dealing with the uncertainties.

The systems that would be leveraged in an integrated way could include:

- Transportation Systems
  - Tolling systems—with flexible pricing capabilities
  - Buses—with GPS sensors
  - Cars/trucks—with GPS sensors
  - Trains
- Smart grid
  - Power grid including traditional and renewable energy sources
  - Electric meters, with incentive pricing and demand response capabilities
  - Hybrid/electric vehicle charging systems with demand response and incentive pricing capabilities
  - Residential and commercial

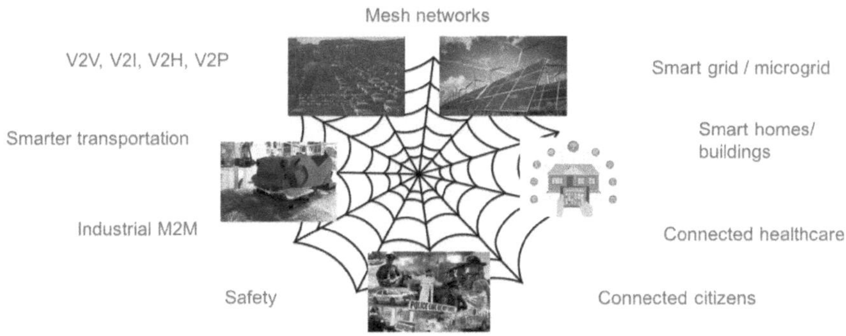

FIG. 6 System of Systems approach to a smart city.

- City services
  - Water systems—including sensors on the pipes to detect leaks
  - Public safety systems—fire, police, emergency management
  - Surveillance systems—cameras, police

It is possible to integrate this IoT-enabled environment with cognitive computing to develop predictive and real-time analytics to engage the citizens, city operations, homes, commercial buildings, and vehicles, with a natural language interface for the citizens, explaining the options they have to have an improved citizen experience based on temporal context of city operations and hazards (Batty et al., 2012). The value could be improved citizen experience, improved efficiency in the city, improved safety and emergency management, reduced energy costs to residents, reduced capital investments by energy sources, and reduced emissions due to less time sitting in traffic.

The cognitive computing systems would have predictive analytics based on history and real-time data of traffic patterns for that type of day:

- Weekday which is a working day, with different patterns Monday through Friday
- Government designated holiday, and percent of residents affected
- Religious holiday, perhaps affecting fewer residents, with geographic considerations based on communities
- Weekend day
- Time of year traffic patterns, for instance Fridays in summer vs winter, might have different traffic patterns

The ecosystem of smart city services including the city department of transportation, tolling systems, smart grid management, and incentive pricing team for residential and commercial properties could perhaps work together to change traffic patterns and reduce traffic congestion and vehicle emissions by:

- Developing a progressive incentive pricing scheme for plug-in hybrid/electric vehicle charging to distribute the lower cost charging times, thereby incenting vehicle owners to plug in at certain times of day, and therefore drive at another time
- For instance, one in three people in each square mile of the city could be given 50% cost reduction in their EV charging between 7 a.m. and 8 a.m., thereby incenting them to drive before or after that time; another third of the population given the 50% cost reduction between 8 a.m. and 9 a.m., and so on
- The complementary pricing signals could be available at commercial buildings as well, to manage the traffic patterns, reduce traffic congestion, and improve the environment
- This would require partnerships and sharing of data across the ecosystem described earlier

Then real-time data would be added to the system, altering expectations for traffic patterns and issues:

- Vehicle locations, buses, cars, trucks, trains
- Weather such as rain, and potential flooding
- Snow and ice, and potential closed businesses and schools
- Plowing of snow and salting of ice, where by when
- Water main breaks, and flooding issues for streets, buildings, subways, tunnels
- Road closings, due to accidents or construction, planned or emergency
- Emergency management, such as fires or police activity
- Crowd-sourced information, regarding numbers of people in various locations
- Social media, relaying status, and new events or situations

The cognitive computing systems would gather all this information, curate it to determine which data has a high degree of certainty vs uncertainty, consider the implications of the factual data and uncertain data, take actions to further vet the uncertain data, and in real time update the predictive models, and send updates and alerts:

- To emergency management, to confirm with system and in situ resources, before rerouting suggestions are sent broadly to citizens, systems, and vehicles
- To citizens to reroute their trip, or suggest they work at home or another location in an unaffected area
- To vehicles, to reroute their trip
- To businesses, to have them alert their occupants

There could be sharing of data across the ecosystem of smart city systems across multiple cities in a region, in a country, or around the world, to enable these cities and systems to learn from each other, develop benchmarks and incentives to improve conditions, and continuously optimize traffic patterns learning from actual behavior of the drivers and systems.

There are intelligent transportation systems already in use(Hudson, 2016) such as the OneBusAway (OneBusAway.com) application developed by students at the University of Washington, which was first deployed at the University and now provides real-time information for metro-area bus systems not only in Puget Sound but also in other cities across the United States. These are the early building blocks for the intelligent transportation systems and integrated smart city systems of the future.

5.3 Healthcare—The Internet of Medical Things and Precision Medicine

The opportunity to leverage the IoT with cognitive computing together provides great possibilities in healthcare (Valera et al., 2010), especially in the journey to precision medicine. Precision medicine is an emerging approach for disease treatment and prevention that takes into account individual variability in genes, environment, and lifestyle for each person (Pratt, 2007). Precision medicine will enable a new era of medicine in which researchers, providers, and patients work together to develop individualized care.

While some advances in precision medicine have been made, the practice is not currently in use for most diseases or patient treatment. Some diseases are difficult to diagnose and even more difficult to treat safely. We must gain better insights into the biological, environmental, and behavioral influences on these diseases to make a difference for the millions of people around the world who suffer from them. We must leverage information on different diseases around the world to better care for the "Citizens of the World" who travel widely. Global medicine needs to increase insight related to the humans who might have different antibodies due to their environment and origin, and to improve care while dealing with the reality of the availability of different treatments in their home country due to local regulations after contracting a disease in another country. Perhaps care is different for natives contracting diseases in their home lands; perhaps they have antibodies to more easily combat a locally contracted disease than people not indigenous to that part of the world where they contracted an illness. The goal of precision medicine is to harness the data from genomics, clinical and pure research, patient history, demographics, connected health and patient lifestyle, environmental data, life sciences research data, and clinical care data, including the vast amount of data distributed around the world, to improve healthcare and medical outcomes (see Fig. 7).

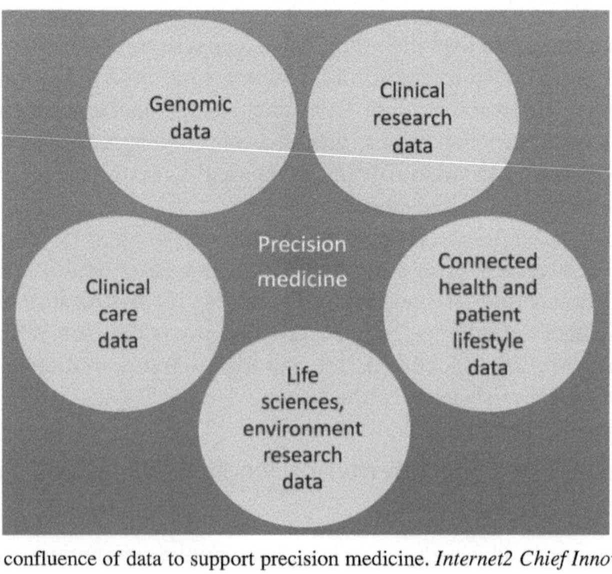

FIG. 7 The confluence of data to support precision medicine. *Internet2 Chief Innovation Office.*

We can bring together many data points and sources from historical and recent research in healthcare, life sciences, pharmacology, cancer, environment, and medicine to create a more complete array of data to inform more precise diagnosis and treatment. The IoT captures information regarding a patient and the environment. This can include data from wearable consumer technologies that monitor and report vital signs such as heart rate, blood pressure, respiration, and exercise, as well as medical devices including infusion pumps, glucose meters, implantable heart devices, pulse/oxygen monitors, and environmental sensors identifying and reporting particles in the air, elements, and substances in the water or earth where people live or visit, or where the food they eat is produced. Combine the IoT data with genomics data, family history, information from basic research and clinical research, and the clinical care data from an Electronic Medical Record and patient history including text, images, video, audio, and the doctor's physical assessment, and we have a lot of data that can provide deep insight regarding a patient, the potential outcomes from various medical actions, and recommend potential treatments to affect patient outcomes. Cognitive computers could ingest and digest vast amounts of data from the patient, their environment, and pertinent data from around the world, in varying formats, in multiple languages, to identify patterns in a patient's personal medical history, family history, genomic cohorts, regional cohorts, as well as treatments from around the world, to provide insight regarding the patient's ailment, potential treatments, and potential outcomes. A cognitive computer can leverage all this data to provide doctors and patients with more insight regarding current clinical and pure research from around the world than they can ingest and digest in a timely manner. The potential value from IoT and cognitive computing could be to improve healthcare, improve quality of life, lengthen life spans, avoid accidental deaths from allergic reactions, and enable people to take more control of the life they live and the effects on their health and well-being.

We can consider a potential use case and aspirational view of precision medicine leveraging the IoT and cognitive computing to improve healthcare.

Many people are becoming "Citizens of the World," traveling extensively to study, vacation, work, and live. The opportunity and incidence are increasing for a traveler to contract an illness not indigenous to a traveler's native country, which cannot be easily diagnosed or treated in their native land upon returning home. Add to that the lack of treatments for new illnesses, for illnesses contracted in another country where local remedies are not available around the world, and we have a growing challenge in global healthcare.

The use case today for a Citizen of the World is a world traveler who was born and lived in the United States and traveled to 20 countries over their lifetime to date. This is a real use case based on the author's experience, but without the benefit of IoT devices to avoid health risks and potentially improve outcomes. They travel in this use case to another country with high levels of pollution, contracted an ailment misdiagnosed many times, then diagnosed, treated with drugs which created an anaphylactic allergic reaction, and nearly died. Let's consider the use case without the benefit of IoT and cognitive computing, and then an aspirational view with both.

The Citizen of the World visits a country with severe pollution for a week of business, and there develops trouble breathing while exercising. The ailment does not prohibit the person from traveling home. Once home in the United States, the symptoms become worse, until they are experiencing severe fatigue, coughing, and crackling when breathing. The first doctor diagnoses allergies in week 1, which the patient has never had. The next doctor diagnoses asthma in week 2, which the patient has never had. The patient then has an X-ray and is diagnosed with left upper lung pneumonia, 3 weeks after the initial symptoms. This particular form of pneumonia is most common in geriatric patients who are lying prone to allow the infection to be in the upper lung. This patient, however, has been traveling, exercising, and upright most of the time. This anomaly intrigues the doctors, but does not cause them to consider other potential diagnoses. The patient presents as having viral pneumonia, with no treatment to combat it. The patient becomes sicker and then presents as having bacterial pneumonia. Once again the doctors are not concerned with this changed diagnosis. Antibiotics are provided, but the patient's condition does not show improvement. The patient then goes to a complex disease doctor, who asks for any allergies, but does not ask for family history of allergies to sulfa-based drugs. The patient states that they are allergic to thimerosal, which is a mercury-based substance. The doctor prescribes an antibiotic and a sulfa-based drug, the patient experiences anaphylaxis and goes to the hospital for emergency care, and the patient survives.

Let's consider this Citizen of the World use case leveraging the IoT and cognitive computing. The citizen lands in a pollution heavy environment in a country on the other side of the world. Her cell phone or mobile device provides a pollution alert regarding parts per million of pollutants; the alert is sent to the patient. The patient then dons a mask to reduce exposure. The gym where the citizen is exercising has a similar pollution alert from the local environmental protection agency or pollution sensors connected to the building management system, a cognitive computer determines if the potential for negative breathing effects has increased beyond a certain predefined threshold. If the cognitive system believes the unsafe, threshold has been reached or will be reached in the next 4 h, then the building manager is alerted, or the building can automatically turn on an additional stronger air filtration system to reduce the particles in the exercise room. The trainers and patrons in the gym are alerted of the increased pollution risk and then the trainers bring masks to the gym patrons. This alone could reduce the potential negative impact on the pulmonary function of the citizen.

The citizen then meets with someone who is feeling sick but is not outwardly symptomatic, and they shake hands numerous times. An alert is sent to the citizen's personal fitness device that a new germ indigenous to the country she is visiting has been transferred to her skin, which is detected when the citizen puts her hand on her personal medical device on her wrist after shaking hands with a native of the country she is visiting. The citizen is provided with treatment recommendations via her mobile device to prophylactically defend against this germ, which is known to create pneumonia-like symptoms. The citizen is told where to go to get this treatment; she does, is treated, and does not get sick.

Alternatively, the patient goes back to the United States and then becomes sicker. Her personal medical device data are uploaded to her electronic medical record, and the cognitive computer system sends an alert to her doctor to prescribe medicine to combat the illness infiltrating her body. The cognitive computer knows that American patients exposed to the similar germs cause pneumonia-like systems and severe, life-threatening coughs. The United States and visited country have created an integrated data base that meets global healthcare privacy standards, including illnesses, symptoms, and treatments based on the DNA of individuals, their indigenous environmental situation, the germs from each country, treatments, risk, and previous outcomes from actual cases. The cognitive computer system identifies the pattern or anomaly from this data which is affecting the patient, and alerts the special care provisions for a "Citizens of the World Global Healthcare Consortium" to allow treatments that are only approved for such cases, based on the locally available and special circumstance medicines in the United States. The patient had been feeling run down, thought she was just tired from the travel and 18-h work days. Without the combined IoT and cognitive computing analysis and alerts, she would just continue to get sicker. The alert sends her to the doctor, where they review her symptoms, prescribe the medicine to treat the oncoming illness, and deliver it to her based on confirmed GPS coordinates from her personal medical device. A drone delivers the medicine and validates the patient's identity through an audio and biometric scan confirmed by the cognitive system through the drone. She takes the medicine as prescribed and begins to feel better.

There is already wearable technology with biosensors that can transmit patient information regarding body temperature, heart rate, blood pressure, and blood/oxygen level, which can enable identification of a possible infection or worse (Ajami and Teimouri, 2015). This technology can enable the identification of increases in the body temperature or other signs of a possible infection in a patient. Wearable biosensors capable of continuous vital signs monitoring and feedback to the user including the patient and doctor will be significantly effective in timely prevention, diagnosis, treatment, and control of diseases. As the Internet of Medical Things progresses, researchers, scientists, engineers, and medical professionals can work together to solve these challenges to provide early diagnostics to prevent and then to treat diseases.

Now let's consider how cognitive computing could reduce the odds of anaphylaxis as described in the treatment case earlier. If a doctor is considering prescribing a sulfa-based drug, patient history and family history can be analyzed by a cognitive computing system, and if there is a match on either the patient or family, then the sulfa-based drug can be denied. The cognitive computer could also analyze sulfa-based negative reactions that have been documented in clinical experience from medical practitioners, emergency rooms, patient input, and electronic medical records, correlating these insights with any other patterns such as thimerosal or mercury-based allergies since thimerosal is a preservative used in vaccinations and other situations. While medical practices might suggest that there is no correlation between thimerosal and sulfa-based drug allergies, it could be that clinical experience suggests otherwise. Cognitive computers could find these patterns to potentially reduce the risk of accidental allergic reactions to medicine. It is possible that the cognitive computing system working with the Internet of Medical Things could inform a personal medical device which might sound or speak an alert based on data sent to it by a cognitive computing system regarding the risk in sulfa-based medicine, so the doctor or patient would check the personal medical device before prescribing or accepting a sulfa-based drug.

The leverage of the Internet of Medical Things in patient care is a key focus of the medical community, US FDA (U.S. Food and Drug Administration, 2013), and now IEEE. The author keynoted a workshop and spoke on a panel at the first IEEE (The first IEEE Conference, 2016) Conference on Connected Health, Applications, Systems and Engineering in June 2016 regarding the need for increased TIPPSS in connected healthcare. A keen focus on security to ensure these devices are used safely, with trust and identity authentication for any use case, is critical.

The use of cognitive computing for pattern recognition, anomaly recognition, development, and communication of insights by analyzing vast amounts of data that is beyond the capabilities of humans to do at all or in a timely manner can provide recommendations that could improve healthcare, life expectancy, and quality of life for people around the world.

6 FUTURE OPPORTUNITIES FOR IOT AND COGNITIVE COMPUTING

The opportunity to leverage the Internet of things and cognitive computing together will increase over time as the experience increases and state of the art improves in both areas. The IoT will become more ubiquitous, there will be many more new devices we have not even thought of yet, that will help us in new ways. As these new devices create data, the cognitive computers will need to determine how to ingest and digest the data, in different formats, with different protocols, in different programming and natural languages, and must deliver their insights and recommendations in consumable ways to humans, animals, and systems to improve outcomes. There will be some systems that require human intervention, like healthcare where a certified care provider must be in the loop to use their expertise to make the best decision, supported by deeper and wider insights due to cognitive computing and its leverage of the IoT. Some systems will provide a M2M interface, so the insights from a cognitive system can be used to communicate to a machine to take an action, for instance, the example in healthcare earlier where a high pollution alert can turn on an additional air filtration system. This is all part of the process updates that have to be designed as we leverage IoT and cognitive computing data and insights together.

Additional research and development is needed to improve the leverage and real-time learning from all data types, including image, sense, smell, audio, video, numerical, textual, and holographic in the future, for both historical and real-time data together, in various languages, and provide a user interface that allows the insights to be leveraged appropriately in a timely fashion (Rowland et al., 2015). The ability to leverage the data coming from various IoT devices, the holistic analysis, and integration of the data, and the ability to send signals back to devices to leverage insight from the data require interoperability between the devices, data, and applications, which is a significant area of research (Blackstock and Lea, 2013; Lea and Blackstock, 2014; Rajaraman et al., 2015). The end-to-end architecture of the Internet of Things at the sensor, security, cloud, and terrestrial level is also being researched from numerous perspectives (daCosta and Henderson, 2014). Since IoT and cognitive computing will evolve and can be improved for a long time to come, the research and development opportunities will exist for many years into the future.

REFERENCES

Ajami, S., Teimouri, F., 2015. Features and application of wearable biosensors in medical care. J. Res. Med. Sci. 20 (12), 1208–1215. http://dx.doi.org/10.4103/1735-1995.172991. http://www.ncbi.nlm.nih.gov/pmc/articles/PMC4766830/.

Batty, M., Axhausen, K., Giannotti, F., Pozdnoukhov, A., Bazzani, A., Wachowicz, M., Ouzounis, G., Portugali, Y., 2012. Smart cities of the future. Eur. Phys. J. Spec. Top. 214 (1), 481–518. Springer-Verlag.

Blackstock, M., Lea, R., 2013. Toward Interoperability in a Web of Things. In: Proceedings of the 2013 ACM Conference on Pervasive and Ubiquitous Computing Adjunct Publication.

Condra, G., 2015. A plea for incremental work in IoT security. In: Proceedings of the 5th International Workshop on Trustworthy Embedded Devices. ISBN: 978-1-4503-3828-8.

daCosta, F., Henderson, B., 2014. Rethinking the Internet of Things: A Scalable Approach to Connecting Everything. Apress Media LLC, New York.

Delicato, F., Pires, P., Batista, T., Cavalcante, E., Costa, B., Barros, T., 2013. Towards an IoT ecosystem. In: Proceedings of the First International Workshop on Software Engineering for Systems-of-Systems. ACM, New York.

Demchenko, Y., de Laat, C., Membrey, P., 2014. Defining architecture components of the Big Data Ecosystem. In: Collaboration Technologies and Systems (CTS), 2014 International Conference. http://dx.doi.org/10.1109/CTS.2014.6867550. http://ieeexplore.ieee.org/stamp/stamp.jsp?tp=&arnumber=6867550&isnumber=6867522.

Ferrucci, D., Brown, E., Chu-Carroll, J., Fan, J., Gondek, D., Kalyanpur, A., Lally, A., Murdock, J., Nyberg, E., Prager, J., Schlaefer, N., Welty, C., 2010. Building Watson: an overview of the DeepQA project, AI Mag. 31 (3), Fall, 59–79. http://www.aaai.org/Magazine/Watson/watson.php.

Hurwitz, J., Kaufman, M., Bowles, A., 2015. Cognitive Computing and Big Data Analytics. Wiley, Hoboken, NJ.

Kahneman, D., 2011. Thinking, Fast and Slow. Farrar, Straus and Giroux, New York.

Kelly III, J., Hamm, S., 2013. Smart Machines: IBM's Watson and the Era of Cognitive Computing. Columbia University Press, New York.

Lea, R., Blackstock, M., 2014. Smart cities: an IoT-centric approach. In: Proceedings of the 2014 International Workshop on Web Intelligence and Smart Sensing. ACM, New York.

Lee, W.-H., Tseng, S.-S., Shieh, W.-Y., 2010. Collaborative real-time traffic information generation and sharing framework for the intelligent transportation system. Inform. Sci. 0020-0255. 180 (1), 62–70. http://dx.doi.org/10.1016/j.ins.2009.09.004. http://www.sciencedirect.com/science/article/pii/S0020025509003879.

Meilich, A., 2006. System of systems (SoS) engineering & architecture challenges in a net centric environment. In: 2006 IEEE/SMC International Conference on System of Systems Engineering. http://dx.doi.org/10.1109/SYSOSE.2006.1652264. http://ieeexplore.ieee.org/stamp/stamp.jsp?tp=&arnumber=1652264&isnumber=34635.

Navair.navy.mil, 2016. E-2C Hawkeye. http://www.navair.navy.mil/index.cfm?fuseaction=home.displayPlatform&key=FAD3E30B-AD5C-441A-8925-624BD0BA45F1.

Pratt, R., 2007. In: Summary of Comments: FERC Technical Conference on Demand Response. ferc.gov. Available at: http://www.ferc.gov/EventCalendar/Files/20070501124211-Pratt,%20PNNL.pdf.

Rajaraman, V., Misra, P., Dhotrad, K., Warrior, J., 2015. Enabling plug-n-play for the Internet of things with self describing devices. In: Proceedings of the 14th International Conference on Information. Processing in Sensor Networks. ACM, New York.

Rowland, C., Goodman, E., Charlier, M., Light, A., Lui, A., 2015. Designing Connected Products: UX for the Consumer Internet of Things. O'Reilly Media, Sebastopol, CA.

Schwartz, A., 2013. Chicago's video surveillance cameras: a pervasive and poorly regulated threat to our privacy. Northwest. J. Technol. Intellect. Prop. 11(2). Article 4.

Simmons, R., 1970. Communications of the ACM: Natural Language Question-Answering Systems, 13th ed. ebook, http://brenocon.com/simmons70overview_qa.pdf.

Turner, V., 2014a. The Digital Universe of Opportunities. emc.com. Available at: http://www.emc.com/leadership/digital-universe/2014iview/digital-universe-of-opportunities-vernon-turner.htm.

Turner, V., 2014b. The Digital Universe of Opportunities: Rich Data and the Increasing Value of the Internet of Things. http://www.emc.com/leadership/digital-universe/2014iview/digital-universe-of-opportunities-vernon-turner.htm.

U.S. Food and Drug Administration, 2013. Cybersecurity for Medical Devices and Hospital Networks: FDA Safety Communication. http://www.fda.gov/MedicalDevices/Safety/AlertsandNotices/ucm356423.htm.

Valera, A.J.J., Zamora, M.A., Skarmeta, A.F.G., 2010. An architecture based on Internet of things to support mobility and security in medical environments. In: 7th IEEE Consumer Communications and Networking Conference. http://dx.doi.org/10.1109/CCNC.2010.5421661. http://ieeexplore.ieee.org/stamp/stamp.jsp?tp=&arnumber=5421661&isnumber=5421566.

Xu, T., Wendt, J.B., Potkonjak, M., 2014. Security of IoT systems: design challenges and opportunities. In: 2014 IEEE/ACM International Conference on Computer-Aided Design (ICCAD), San Jose, CA, pp. 417–423. http://dx.doi.org/10.1109/ICCAD.2014.7001385. http://ieeexplore.ieee.org/stamp/stamp.jsp?tp=&arnumber=7001385&isnumber=7001313.

ADDITIONAL READING

Asmartplanet.com/blog, 2009. Lessons from Stockholm: Congestion Charging and (ITS) Conference. Available at: http://asmarterplanet.com/blog/2009/09/lessons-from-stockholm-congestion-charging-and-the-intelligent-transportation-systems-conference.html.

Bunn, J., Newman, H., 2003. Data Intensive Grids for High Energy Physics. California Institute of Technology. http://pcbunn.cacr.caltech.edu/grids/gridbook.htm.

Collins, F., Varmus, H., 2015. A new initiative on precision medicine. N. Engl. J. Med. 372, 793–795. DOI: http://dx.doi.org/10.1056/NEJMp1500523. http://www.nejm.org/doi/full/10.1056/nejmp1500523#t=article.

ESnet, Energy Sciences Network, 2015. ESnet Extends 100G Connectivity Across Atlantic—New Capabilities Improve Network Speeds Between U.S. Research Sites, European Facilities. https://www.es.net/news-and-publications/esnet-news/2014/esnet-extends-100g-connectivity-across-atlantic/.

Forbes.com, 2016. Five Ways Cognitive Computing Will Power the Internet of Things. http://www.forbes.com/sites/ibm/2015/12/15/five-ways-cognitive-computing-will-power-the-internet-of-things/#73b1cd7566c7.

Hackers Remotely Kill a Jeep on the Highway, 2015. Two hackers have developed a tool that can hijack a Jeep over the internet. WIRED senior writer Andy Greenberg takes the SUV for a spin on the highway while the hackers attack it from miles away. https://www.youtube.com/watch?v=MK0SrxBC1xs.

http://iihtofficialblog.blogspot.com/2014/07/5-vs-of-hadoop-big-data.html?m=1.

Hudson, F.D., 2016. The Internet of things is here. EDUCAUSE Rev.

IBM.com, 2009. New IBM Research Reveals Critical Gap for Managing Congestion Problems in Cities Worldwide. Available at: http://www-03.ibm.com/press/us/en/pressrelease/28468.wss.

IBM Watson Teams up with American Cancer Society to Pit Cognitive Computing Against Cancer, 2016. http://www.healthcareitnews.com/news/ibm-watson-teams-american-cancer-society-pit-cognitive-computing-against-cancer.

IEEE, 2016. IEEE Experts in Technology and Policy (ETAP) Forum on Internet Governance, Cybersecurity, and Privacy. Washington, DC. http://internetinitiative.ieee.org/images/files/resources/reports/etap-washingtondc-report.pdf.

IEEE Standards Association, 2016. IEEE Trust and Security Workshop for the Internet of Things (IoT), Washington, DC, Meeting Recap. http://standards.ieee.org/events/iot/presentations/ieee_end_to_end_trust_meeting_recap.pdf.

Juniperresearch.com, 2015. 'Internet of Things' Connected Devices to Almost Triple to over 38 Billion Units by 2020. Available at: http://www.juniperresearch.com/press/press-releases/iot-connected-devices-to-triple-to-38-bn-by-2020.

National Institute of Health, 2016. About the Precision Medicine Initiative Cohort Program. nih.gov. Available at: https://www.nih.gov/precision-medicine-initiative-cohort-program.

OneBusAway.org.

Radcliffe, J. Hacking Medical Devices for Fun and Insulin: Breaking the Human SCADA System. http://www.aicas.com/cms/sites/default/files/BH_US_11_Radcliffe_Hacking_Medical_Devices_WP.pdf.

The first IEEE Conference on Connected Health: Applications, Systems and Engineering Technologies, June 2016. http://conferences.computer.org/chase/.

YouTube.com, 2012a. Eduardo Paes: The 4 Commandments of Cities. https://www.youtube.com/watch?v=B8Z2G7d2kzs.

YouTube.com, 2012b. TEDxNJIT—Florence Hudson—Sustainability on a Smarter Planet. https://www.youtube.com/watch?v=P44VaTEYBF0.

Index

Note: Page numbers followed by "*f*" indicate figures, and "*t*" indicate tables.

D

E

F

G

H

I

L

M

N

FSC
www.fsc.org
MIX
Papier | Fördert gute Waldnutzung
FSC® C083411

Kolibri 360 GmbH
Ferdinand-Jühlke-Straße 7
99095 Erfurt, Deutschland
produktsicherheit@kolibri360.de